Hybridoma Technology in Agricultural and Veterinary Research

Hybridoma Technology in Agricultural and Veterinary Research

Norman J. Stern
and
H. Ray Gamble

Editors

Proceedings from a symposium held October 30-November 1, 1983 at the University of Maryland, University College, Center of Adult Education, College Park, Maryland.

Sponsored by

The Beltsville Agricultural Research Center
Northeastern Region
Agricultural Research Service
United States Department of Agriculture

Rowman & Allanheld
PUBLISHERS

ROWMAN & ALLANHELD

Published in the United States of America in 1984
by Rowman & Allanheld, Publishers
(A division of Littlefield, Adams & Company)
81 Adams Drive, Totowa, New Jersey 07512

Copyright © 1984 by

Library of Congress Cataloging in Publication Data
Main entry under title:

Hybridoma technology in agricultural and veterinary
  research.

  "Proceedings from a symposium held October 30–
November 1, 1983, at the University of Maryland, Uni-
versity College, Center of Adult Education, College
Park, Maryland, sponsored by the Beltsville Agri-
cultural Research Center, Northeastern Region, Agri-
cultural Research Service, United States Department of
Agriculture."
  Includes index.
  1. Hybridomas—Congresses. 2. Antibodies, Monoclonal
—Congresses. 3. Plants—Disease and pest resistance—
Congresses. 4. Veterinary immunology—Congresses.
I. Stern, Norman J. II. Gamble, H. Ray. III. Beltsville
Agricultural Research Center.
QR185.8.H93H92 1984   636.089'537   84-17795
ISBN 0-8476-7362-6

84 85 86 / 10 9 8 7 6 5 4 3 2 1

Printed in the United States of America.

Contributors and their Affiliations

Douglas F. Antczak
J. A. Baker Institute for Animal
    Health
New York State College of
    Veterinary Medicine
Cornell University
Ithaca, New York 14853

David L. Brandon
USDA, Western Regional
    Research Center
800 Buchanan Street
Berkeley, California 94710

Thomas M. Buchanan
Immunology Research Laboratory
Seattle Public Health Hospital
1131 14th Avenue South, 11th
    Floor
Seattle, Washington 98114

George D. Cain
Department of Zoology
University of Iowa
Iowa City, Iowa 52242

Harry D. Danforth
USDA, Animal Parasitology
    Institute
Building 1040, BARC-East
Beltsville, Maryland 20705

William C. Davis
Department of Veterinary
    Microbiology and Pathology
College of Veterinary Medicine
Washington State University
Pullman, Washington 99164

Solke DeBoer
Agriculture Canada
6660 Northwest Marine Drive
Vancouver, BC, CANADA V6T1X
    1X2

Suzanne L. Epstein
Transplantation Biology Section
NIH, National Cancer Institute
Building 10, Room 4B17
Bethesda, Maryland 20205

William C. Fleenor
Department of Animal Science
Agricultural Sciences Building 38
University of Arizona
Tuscon, Arizona 85721

H. Ray Gamble
USDA, Animal Parasitology
    Institute
Building 1040, BARC-East
Beltsville, Maryland 20705

Richard A. Goldsby
Biology Department
Amherst College
Amherst, Massachusetts 11040

Edward L. Halk
Agrigenetics Corporation
5649 East Buckeye Road
Madison, Wisconsin 53716

Hei-Ti Hsu
American Type Culture Collection
12301 Parklawn Drive
Rockville, Maryland 20852

Mark Jacobs
Biology Department
Swarthmore College
Swarthmore, Pennsylvania 19081

Ramon Jordan
USDA, Horticultural Science
   Institute
Building 004, Room 101,
   BARC-West
Beltsville, Maryland 20705

Roger H. Lawson
USDA, Horticultural Science
   Institute
Building 004, Room 101,
   BARC-West
Beltsville, Maryland 20705

Peter T. Lomedico
Department of Molecular Genetics
Research Division, Hoffman
   LaRoche Inc.
Nutley, New Jersey 07110

Joan K. Lunney
USDA, Animal Parasitology
   Institute
Building 1040, BARC-East
Beltsville, Maryland 20705

Yvonne E. McHugh
Hana Biologics, Inc.
2323 5th Street
Berkeley, California 94710

Ernest D. Marquez
Department of Microbiology
Pennsylvania State University
College of Medicine
Hershey, Pennsylvania 17033

Keiko Ozato
Laboratory of Developmental and
   Molecular Immunity
National Institute of Child Health
   and Human Development
National Institutes of Health
Bethesda, Maryland 20205

Terry W. Pearson
Department of Microbiology and
   Biochemistry
University of Victoria
Victoria, BC, CANADA V8W 2Y2

Lee H. Pratt
Department of Botany
University of Georgia
Athens, Georgia 30602

Matthew D. Scharff
Albert Einstein College of Medicine
   of Yeshiva University
1300 Morris Park Avenue
Bronx, New York 10461

Judith V. Schollmeyer
USDA, Meat Animal Research
  Center
Clay Center, Nebraska 68933

Alan F. Sher
Immunology and Cell Biology
  Section
Laboratory of Parasitic Diseases
National Institutes of Health
Building 5, Room 114
Bethesda, Maryland 20205

S. Srikumaran
Biology Department
Amherst College
Amherst, Massachusetts 01002

Richard A. Van Deusen
USDA, National Veterinary
  Services Laboratory
Box 844
Ames, Iowa 50010

M. H. V. Van Regenmortel
Institut de Biologie Moléculaire et
  Cellulaire du C.N.R.S.
15, rue Descartes
Strasbourg, France

U. Zimmerman
Arbeitsgruppe Membranforschung
  am Institut für Medizin
Kernforschungsanlage Jülich
  GmbH
Postfach 1913, D-5170 Jülich
Federal Republic of Germany

Organizers

Stern, Dr. Norman
USDA, ARS, BARC, ASI, MSRL
Bldg. 201, Rm. 10, BARC-East
Beltsville, MD 20705

Gamble, Dr. Ray
USDA, ARS, BARC, API
Bldg. 1044, BARC-East
Beltsville, MD 20705

Greisbach, Dr. Rob
USDA, ARS, BARC
Bldg. 004, Rm. 103, BARC-West
Beltsville, MD 20705

Guidry, Dr. Al
USDA, ARS, BARC
Bldg. 173, Rm. 100, BARC-East
Beltsville, MD 20705

Administrators

Klassen, Dr. Waldemar
Acting Area Director, BARC
USDA, ARS, BARC
Bldg. 003, Rm. 227, BARC-West
Beltsville, MD 20705

Putnam, Dr. Paul
Director, BARC
USDA, ARS, BARC
Bldg. 003, Rm. 227, BARC-West
Beltsville, MD 20705

Program Assistants

Baker, Ms. Jacki
USDA, ARS, BARC
Bldg. 1044, BARC-East
Beltsville, MD 20705

Kazmi, Ms. Shahanah
University of Maryland
College Park, MD 20742

Woolery, Ms. Sara
University of Maryland
College Park, MD 20742

Barrett, Charles J.
USDA, ARS, BARC
Bldg. 004, Rm. 103, BARC-West
Beltsville, MD 20705

Contents

PREFACE

NORMAN J. STERN

Meat Science Research Laboratory
Animal Science Institute, Agricultural Research Service
U.S. Department of Agriculture, Beltsville, MD 20705

In 1975, the scientific community became aware of
monoclonal antibodies. The paper by Milstein and Kohler
describing the fusion of immune splenocytes with myeloma cells,
and the subsequent production of monoclonal antibodies, has
greatly affected immunology and widely varied areas of
biochemical research. The techniques, approaches, and uses of
hybridoma technology have provided scientists with a tool that
ranks together with the biotechnical developments of genetic and
protein engineering. Monoclonal antibodies provide researchers
with specific reagents so that they can now study questions that
previously had been impossible to consider.

Some of the early research, using monoclonal antibodies
focused within the medical community, attempted to recognize
antigens unique to cancer cells, with the hope of providing a
means for early diagnosis or even inducing remission. It was
thought that a "magic bullet" for cancerous growth could be
devised by conjugating radioactive isotopes to monoclonal
antibodies, thus directly destroying cancer cells. Although the
magic bullet has yet to be discovered, monoclonal antibodies are
being applied to many other areas of research. The medical
profession has been using them to purify drugs and desirable
biochemicals. Anti-interferon monoclonal antibodies are being
fixed onto solid-phase columns and used to isolate and
concentrate this highly valued substance for research and
clinical trials. The increased availability and reduced cost
for interferon have reflected the true value of this new
technology. Industry now provides kits with specific monoclonal
antibodies to aid in collection of various hormones, drugs, and
antibodies.

In comparison with polyclonal sera, monoclonal antibodies
have provided numerous, previously elusive, benefits to plant
and animal scientists. The traditional means for producing
polyclonal antiserum to antigen was replete with pitfalls
because of the inconstant immune responses of each sensitized
animal. With the same antigen, different laboratories produced
different titers and specifications. Even when excellent
antiserum production was achieved, the death of a prized animal
proved an inevitable setback for the immunologist.

Traditional means for optimizing industrial fermentations depended upon enhancing the output of a desired end product. This was achieved by testing for strain variability and encouraging conventional mutations to provide this desired end product. Monoclonal antibodies revolutionize both the standard immunological and fermentation strategies. Purification of proteins with monoclonal antibodies reduces undesirable immunological cross-reactions and allows for advanced study of target proteins, which previously had been impossible. Epitope mapping of these proteins will aid in the inevitable construction of synthetic proteins to mediate biochemical reactions of agricultural importance. Antibody reagents specific for particular antigens can now be made consistently and can be used to enhance and concentrate desirable end products of fermentations.

The applications of the technology to research problems are limited only by the questions we ask. Papers in this volume include the application of monoclonal antibodies to specify antigenic determinants allowing for molecular dissection of complex antigens. The precision and specificity of the technology enable researchers to characterize epitopes and delineate the specific functions of these moleculer sites. Also included are papers on the association between the inheritance of histocompatibility genes and susceptibility to certain types of disease. Distinctive molecules on cell surfaces may confer resistance to disease and may be genetically expressed. The characterization of the lymph system components by monoclonals is being used to further the study of animal disease and resistance to infection.

In horticultural sciences, the use of monoclonals for detection of viral pathogens in plants is of international interest. Because seed stocks are distributed worldwide, it is necessary to confine endogenous viral pathogens and prevent their spread. In addition, with this new technology detection of viruses before germplasm is disseminated now seems plausible. Monoclonal antibodies are also used in research that characterizes plant hormones, work that aims to increase productivity and to develop lines of plants that can withstand drought, insects, disease, and other environmental hazards.

In microbiology, this new approach is being used to characterize pathogens of livestock and humans. Immunological tests have the potential to provide diagnoses in shorter times than those provided by traditional cultural methods. The mechanisms of disease are being elucidated by research using monoclonal antibodies, and it is the hope of microbiologists to produce monoclonal antibodies that serve as vaccines. As we begin to characterize bacterial toxins and adhesion factors of pathogens, there appears the hope for neutralizing these disease-mediating factors.

The contents of this volume represent contributions presented at a seminar on Hybridoma Technology in Agricultural and Veterinary Research held in the Center of Adult Education, College Park, Maryland, on October 30 to November 1, 1983. A sense of excitement and enthusiasm existed among scientists pursuing this new biotechnology. Hybridoma technology in agricultural and veterinary fields is being translated from theoretical speculations to research and technical realities. Innovative research solutions, scientific principles, and applications were exchanged by the scientists who attended; researchers developed confidence in their abilities to recognize and apply this new approach in their research; new cooperative efforts were formulated; and further future advances were postulated through the use of hybridoma technology.

CHAPTER 1

FROM IMMUNOFANTASY TO MONOCLONAL REALITY

MATTHEW D. SCHARFF
Department of Cell Biology
Albert Einstein College of Medicine
Bronx, NY 10461

Much of the research done in biology requires the
identification, quantification, and localization of
macromolecules, either free in solution or as part of some more
complex structure. Different kinds of assays may be used for
different types of molecules: enzymes are identified by their
reaction with specific substrates, receptors are studied by
examining their interaction with ligands, microorganisms are
often cultured, etc. Although it is clearly possible to develop
a unique and specific assay for each sort of molecule,
scientists working in all areas of biology have recognized for
years that immunoassays provide a more general approach.
Antibodies can be generated that are specific for the substance
to be studied, and such antibodies can be employed in any of the
many standard immunoassays to carry out almost every conceivable
task. Further, if it is desirable, antibodies can be and have
been used in vivo for the diagnosis and therapy of some diseases.
Considering the relative technical simplicity of generating
and using antibodies, it is surprising that they have not been
used more widely. However, when one examines the problems that
are sometimes encountered during the conventional generation of
antibodies and their actual use, it becomes obvious why
non-immunologists have often been reluctant to use immunologic
approaches. Many are turned away by the unpredictability of the
immune response. Some macromolecules are only weakly
immunogenic, and it is difficult to obtain higher titer antisera
for their detection. Even substances that induce excellent
antibody responses must be available in 10- to 100-μg
quantities in relatively pure form. If impure antigens are
used, the antiserum must be absorbed extensively. Even when
large amounts of a highly purified and potent immunogen are
injected, the recipient animal will make antibodies that differ

widely in their affinity, specificity, and ability to carry out functions such as precipitation, fixation of complement, and binding to phagocytic cells. Each bleeding is therefore a unique serological reagent and the amounts of antibody available are often inadequate. This is especially true if weak or impure immunogens are used and absorption of contaminating antibodies is required.

These and other problems have led immunologists to seek techniques for producing more homogeneous and reproducible reagents in larger amounts. This was not a complete fantasy, for there have been some successes in producing homogeneous antibodies against bacterial polysaccharides (1, 2). Further, the disease multiple myeloma provided a model for obtaining monoclonal antibodies, for it revealed that a single antibody-forming cell can undergo malignant transformation, grow in an uncontrolled fashion, and secrete into the bloodstream large amounts of a single antibody molecule. This occurs because each of the antibody molecules produced by the clonal progeny of the myeloma cell has exactly the same amino acid sequence. Unfortunately, few myelomas produced antibodies with known specificities and it was not possible in practice to harness this model to produce serological reagents routinely.

In 1975, Kohler and Milstein revolutionized serology when they described a simple method for converting normal antibody-forming cells into myeloma tumors (3). They showed that mouse myeloma cells adapted to tissue culture could be fused to the spleen cells from immunized mice (or rats) and that some of the hybrids made antibodies that were specific for the antigen used to immunize the animal. Further, these hybridoma cells maintain the characteristics of the myeloma parent and grow continuously in culture, can be injected back into mice to produce tumors and large amounts of antibody, and can be frozen away and recovered from the freezer at will (3).

Although this dramatic discovery arose during somatic cell genetic experiments aimed at understanding the control of expression of immunoglobulin genes, it was immediately apparent that the technology could be used to produce the kinds of serological reagents that immunologists had fantasized about but had seldom been able to obtain. In fact, as the technology was improved and applied more widely, it became obvious that the benefits of monoclonal antibodies would be even greater than was initially appreciated. Monoclonal antibodies are indeed homogeneous chemical reagents that can be renewed whenever needed. Once a useful hybridoma has been obtained, it is relatively simple and inexpensive to produce hundreds of milligrams of that antibody. Although these practical benefits are crucial, the unique contributions of monoclonal antibodies may derive from two other aspects of the basic technique. First, because the identification of useful hybridomas involves the cloning out of individual antibody-forming cells, pure antibodies are routinely obtained from impure antigens. This is especially important, because many of the most interesting and useful antigens are weak immunogens and have not yet been purified. It is even possible to use the hybridoma technology to look for antigens whose existence is only suspected and whose structure and properties are completely unknown. For example, it has been possible to immunize mice with malignant human cells, screen for monoclonal antibodies that react with malignant but not normal cells, and identify those that react with oncofetal antigens on the surface of those cells even though they represent only a small percent of the foreign macromolecules in the human cells that were injected into the mouse (4).

A second major advantage of the technology is that it is often possible to obtain monoclonal antibodies that are ideally suited for a particular need. This can be done by selecting among a large battery of monoclonal antibodies for one that has the desired properties. For example, in most situations, high-affinity antibodies are needed. If many monoclonal antibodies have been identified with the desired specificity, then the one with the highest affinity can be chosen for future use. For some uses, it is essential to have a specific reagent, but in other situations an antibody that is very cross-reactive is needed. This is true where evolutionary relationships or general classes of molecules or microorganisms are being identified. It is also possible to select antibodies that can or cannot carry out certain effector functions. For example, mouse IgM antibodies bind complement effectively and are cytotoxic but IgG_1 antibodies are not.

Unfortunately, in many cases it is difficult to generate a large battery of monoclonal antibodies representing a wide range of affinities, specificities, and classes. In this situation, it is technically possible to obtain somatic mutants of a particular monoclonal that may be more useful than the original antibody. One recent example of this approach has been reported by Muller and Rajewsky (5). When newborn mice are injected with anti-idiotypic (anti-variable region) antibodies, they do not express that idiotype as adults (6). A series of studies had suggested that such neonatal suppression could be achieved only with some subclasses of antibodies (6). However, it was not clear whether the variable suppressive effect of different monoclonal antibodies was due to differences in the epitopes being recognized, variations in affinity, or the differences in the effector functions mediated by different subclasses. To resolve this question, Muller and Rajewsky (5) took one anti-idiotypic monoclonal antibody and generated from it class switch variants (7) in tissue culture so that they had a series of antibodies, all with exactly the same antigen binding site, that differed only in the IgG subclass. All of these were equally effective in neonatal suppression, indicating that the subclass of the antibody is irrelevant (5, 6).

Somatic cell mutants with changes in the structure of their constant region may also prove useful both as research tools and in _in vivo_ diagnosis and therapy. For example, the binding and phagocytosis of antibodies by macrophages and other cells having Fc receptors often result in nonspecific targeting of antibodies. This may be solved by digesting antibodies with papain or pepsin to make Fab fragments. However, it is also possible to generate somatic mutants of monoclonal antibodies that have deletions in their Fc region and will no longer bind to such receptors (8). These then provide a permanent source of monoclonal antibodies that behave like Fab fragments. Based on studies with mouse myelomas, it may also be possible to isolate somatic mutants with changes in affinity (9, 10). Ultimately, it is likely that the genes of useful monoclonal antibodies will be cloned and manipulated using recombinant DNA technology and site-specific mutagenesis to produce serological reagents that are custom made for each task.

The benefits of monoclonal antibodies have resulted in the application of the hybridoma technology to many basic and practical problems. Each month between 200 and 300 papers with the words monoclonal antibody or hybridoma in their titles or abstracts are published. These studies range from an examination of phage receptors on _Escherichia coli_ (11) to the treatment and diagnosis of human cancer (12). In spite of the extraordinary utilization of this technique, there are still some persisting technical problems; overcoming these problems should result in even wider use of antibodies in research and medicine. It is still difficult to generate monoclonal antibodies against weakly immunogenic molecules. No one has yet described a generally useful, simple, and effective technique for increasing the numbers of antibody-forming cells that can then be used to fuse to myeloma cells. The present solutions to this problem are to make weak immunogens more potent by chemical modification and attachment to immunogenic carriers (13). In addition, some increase in antibody-forming cells can be achieved by injecting recently immunized spleen cells into irradiated recipients along with antigen and fusing the repopulated spleen cells (14). It is also apparently feasible to increase antibody-forming cells by culturing them _in vitro_ in the presence of antigen (15).

A second major problem has been the difficulty of generating hybridomas from species other than mouse and rat. This has been particularly frustrating for those who wish to use monoclonal antibodies for _in vivo_ therapy or diagnosis in humans and other species. Part of the problem seems to be due to a lack of adequate cultured myeloma-like cells from these species. However, many seemingly appropriate cell lines have been used to obtain human X human hybrids, and none have been generally useful (16). New techniques such as electrofusion (see Zimmerman, this volume) or oncogene transfection (17) may help to overcome this problem. It is worth noting that many human monoclonal antibodies have been generated by transforming cells with EB virus (18) and that B-cell tropic viruses may be useful for transforming cells of other species.

In conclusion, the hybridoma technology has made it possible to routinely generate homogeneous antibodies against a wide variety of antigens. Because these antibodies are chemically defined reagents that can be produced in large amounts relatively inexpensively and the same antibody can be renewed whenever it is needed, antibodies are being employed as research tools in every area of biology. Mouse and rat monoclonal antibodies are also powerful tools in the diagnostic laboratory, where they can be used to identify foreign agents in blood and tissues of animals and in plants and their products. Once human monoclonal antibodies can be produced with ease, it is inevitable that they will be used in the diagnosis and therapy of diseases. Similarly, where monoclonal antibodies can be generated in other species, those antibodies should be useful in veterinary diseases.

REFERENCES

1. Haber, E. (1970) Fed. Proc. 29:66.
2. Krause, R.M. (1970) Adv. Immunol. 12:1.
3. Kohler, G., and Milstein, C. (1975) Nature 256:495.
4. Yelton, D.E., and Scharff, M.D. (1981) Ann. Rev. Biochem. 50:657.
5. Muller, C., and Rajewsky, K. (1983) J. Immunol. 131:877.
6. Rajewsky, K., and Takemuri, T. (1983) Ann. Rev. Immunol. 1:569.
7. Preud'homme, J.-L., Birshtein, B.K., and Scharff, M.D. (1975) Proc. Natl. Acad. Sci. USA 72:1427.
8. Yelton, D.E., and Scharff, M.D. (1982) J. Exp. Med. 156:1131.
9. Cook, W.D., Rudikoff, S., Giusti, A.M., and Scharff, M.D. (1982) Proc. Natl. Acad. Sci. USA 79:1240.
10. Rudikoff, S., Giusti, A.M., Cook, W.D., and Scharff, M.D. (1982) Proc. Natl. Acad. Sci. USA 79:1979.
11. Gabay, J. and Schwartz, M. (1982) J. Biol. Chem. 257:6627.
12. Levy, R., and Miller, R.A. (1983) Fed. Proc. 42:2650.
13. Sakato, N., and Eisen, H.M. (1975) J. Exp. Med. 141:1411.
14. Fox, P.C., Berenstein, E.H., and Siraganian, R.P. (1981) Eur. J. Immunol. 11:431.
15. Lubin, R.A., and Mohler, M.A. (1980) Mol. Immunol. 17:635.
16. Olsson, L., Kromstrom, H., Lambon-De Mouzou, A., Hansik, C., Broden, T., and Jakobsen, B. (1983) J. Immunol. Meth. 61:17.
17. Jonak, Z.L., Bramin, V., and Kennett, R.H. (1983) Hybridoma 2:124.
18. Steinitz, M., Klein, G., Koskimies, S., and Makela, O. (1972) Nature 269:420.

CHAPTER 2

BACKGROUND, METHODS, AND TECHNIQUES IN HYBRIDOMA TECHNOLOGY

CELL CULTURE AND THE ORIGINS OF HYBRIDOMA TECHNOLOGY

RICHARD A. GOLDSBY, S. SRIKUMARAN, and ALBERT J. GUIDRY*
Department of Biology, Amherst College, Amherst, MA 01002
and *Laboratory of Milk Secretion and Mastitis Research
Agricultural Research Service, U.S. Department of Agriculture
Beltsville, MD 20705

The successful conduct of hybridoma technology depends on
the ability to grow, maintain, and clone cells in vitro. It is
not surprising that tracing the roots of this powerful technique
leads us to consider the origins of cell culture. An
appreciation and statement of the central problem of somatic
cell culture has been precisely articulated by Barnes and Sato
(1):

> The most significant contribution of the technique of
> animal cell culture to...research was the
> demonstration that it was possible. The initial
> wonder of this achievement has been dulled by the
> passing of time....Yet it is still astonishing that we
> can reconstruct the in vivo environment of a cell well
> enough in vitro to allow life to go on outside the
> body.

A general approach to the solution of the problem of in
vitro culture of normal and malignant cells has been available
for only a little more than a generation. However, the first
successful effort to maintain cells outside the organism was
reported by Ross G. Harrison in 1907. Harrison (2) demonstrated
that pieces of embryonic frog tissue known to give rise to nerve
fibers sealed in a chamber with frog lymph survived for periods
of up to 4 weeks. Furthermore, in many cases these tissue
cultures underwent differentiation and produced nerve fibers.
However, the manipulations required to initiate these cultures
were extremely delicate and difficult and the cultures
inevitably died after 4 weeks. The belief quickly grew that
more than a few days or weeks of life outside the body could not
be sustained. This developing dogma was short-lived, for the
experiments of Alexis Carrel in 1912 using cultures of chick
embryo produced a solution to the problem of growing cells in

culture for prolonged periods of time: It was necessary to feed cells (3).

During the 1940s and 1950s, Sanford et al. (4) demonstrated that mammalian cells in culture could be cloned and Gey et al. (5) showed that human tumor cells could be kept alive and vigorously proliferating in culture for indefinite periods of time.   It was during the latter part of this period that Harry Eagle and his colleagues turned the field of cell culture away from reliance on highly complex and undefined growth media composed of embryo extracts, broths, and large amounts of serum toward chemically defined culture media.   Eagle demonstrated that mouse and human cell lines could be grown in a mixture of salts, amino acids, vitamins, cofactors, and glucose to which 1% dialyzed horse serum had been added (6).   The evolution of culture media toward ever greater definition culminated in the 1970s with the completely defined media devised in the laboratories of Ham and Sato (1).   The Ham laboratory assumed that the basis of the serum requirement lay in the large number of essential trace elements that might be supplied by serum. They composed a medium that included optimized levels of a number of amino acids, vitamins, and ions that supported the serum-free clonal growth of established cell lines.   Sato and his associates assumed that serum provided hormonal factors or transport factors, such as ferritin, that were indispensable to the growth of particular cell lines.   His laboratory has demonstrated that supplementation of chemically defined media with a small number of factors and hormones permits the growth in culture of a wide variety of cell lines.

Even by 1960, the routine growth of normal untransformed cells in culture was not routine.   The work of Peter Nowell (7) demonstrated that cell division could be initiated in human lymphoid cells by treatment with the plant lectin phytohemagglutinin (PHA).   This proved to be a discovery of considerable importance because it made possible the establishment of short-term proliferating cell cultures from the readily available "liquid" tissue, blood.   Furthermore, PHA and later many other mitogens made it possible to initiate cultures of T lymphocytes (PHA, concanavalin A) or B lymphocytes (lipopolysaccharide) in a wide variety of species.   The ability to mitogenically stimulate lymphoid cells in culture remains a major tool for the analysis of the immune system.

In 1961 we learned that the life of an untransformed normal cell in culture is finite.   Hayflick and Moorhead showed that normal human lung fibroblasts become senescent and die after about 50 generations in cell culture.   This work indicated what many subsequent studies of others have confirmed--the finite life-span is not a simple matter of culture conditions but appears to reflect the operation of a clock that counts cell divisions and, significantly, reflects the age of the donor.

The field of cell culture began with the efforts of Harrison to study development by the establishment of cell

cultures. The late 1960s and the decade of the 1970s saw the
development of general methodologies for establishing cell
cultures that expressed differentiated functions. The trick is
to begin with a tumor that expresses the desired function and to
adapt the tumor to growth in culture. Perhaps the most
successful practitioner of this art has been Gordon Sato (9, 10).

CELL FUSION AND THE DERIVATION OF SOMATIC CELL HYBRIDS

     Fusion between cells is not limited to the confines of the
research laboratory. Fertilization that involves the fusion of
gametes is obviously widely distributed in nature. In the
course of muscle development, individual cells known as
myoblasts align themselves and fuse to become the multinucleated
syncytia that are the fibers of striated muscle. These examples
notwithstanding, spontaneous fusion is, of necessity, not a
frequent occurrence but rather must be induced by the
application of specific protocols. Fusion, the union of
membranes to produce polykaryons (homokaryons if the nuclei are
the same; heterokaryons if they are different) from two or more
participating cells or organelles, is a necessary first step in
the production of somatic cell hybrids. Subsequent steps
involve the intracellular union of nuclei to yield a single
hybrid nucleus and produce a cell that is capable of division.
     The field of somatic cell hybridization was launched with
the demonstration by Barski, Sorieul, and Cornefert (11) that
the co-cultivation of cell cultures of two mouse
sarcoma-producing lines resulted in the occasional spontaneous
formation of hybrid cells. Although the hybrids occurred in low
frequency and were identified because they grew more vigorously
than the parental cell lines, their hybrid character was
confirmed by karyological analysis. However, such a
hybridization technology that simply mixes two different cell
types and hopes for the emergence of more rapidly growing hybrid
cells is likely to be a low yield "hit or miss" proposition that
was certain to miss less vigorous hybrids.
     Littlefield (12) introduced a rational selective strategy
for the selection of somatic cell hybrids. The method employed
the HAT (hypoxanthine, aminopterin, and thymidine) selective
system of Szybalski and his colleagues (13) and drug-resistant
parental cell lines. One of these lines lacked the purine
salvage enzyme hypoxanthine-guanine phosphoribosyl transferase
(HGPT) and the other lacked the thymidine salvage enzyme
thymidine kinase (TK). Because aminopterin blocks the
biosynthesis of purines and thymidine, only cells with salvage
enzymes for both purines and thymidine can utilize the
hypoxanthine and thymidine that are supplied in HAT medium.
This powerful selective strategy, based on complementation in
the hybrid, has in one modification or another, been a major
tool for the selection of an extremely broad range of hybrids.

The next step was the development of a procedure that increased the yield of hybrids. Because the formation of hybrids begins with the fusion of parental cells, the key element in a strategy designed to increase the number of hybrids is to devise a method for efficiently fusing cells. In 1958 Okada (14) demonstrated that the paramyxovirus, Sendai, efficiently fused cells. By 1965, three groups, Harris and Watkins (15) in England, Okada and Murayama (16) in Japan, and Ephrussi and Weiss (17) in the United States demonstrated that Sendai virus could be used to fuse cells from different species. Ephrussi and Weiss demonstrated that these interspecific hybrids were capable of proliferation and a year later, they (18) showed that at least some of the genes of both parental lines could be expressed in proliferating hybrid cells.

In reviewing the developments that have led to the present point in hybridoma technology, it is important to emphasize the landmark and pivotal discovery of Weiss and Green (19). These workers discovered that interspecific human X mouse hybrids preferentially lost human chromosomes. Furthermore, they pointed out that by concordance analysis of interspecific hybrids in which the chromosomes of one of the participating species are preferentially lost, it is possible to assign particular genes to particular chromosomes. This observation provided a general approach to the problem of assigning genes to chromosomes, thereby providing an alternative to assignment by classic genetic methods that involve the deliberate crossing of individuals and the subsequent analysis of their progeny to determine linkage by recombination analysis. These latter techniques, which are the stock in trade of mouse and drosophila geneticists, represent an approach to gene mapping that is unethical in humans and, because of generation time and the expense of rearing, impractical in many species of veterinary interest. One suspects that the more aggressive application of this approach to veterinary genetics would result in a rapid expansion in our knowledge of the genetic maps of species of veterinary interest.

Although the use of UV- or chemically inactivated Sendai virus when teamed with HAT selection made it possible to obtain large numbers of hybrids from a variety of parasexual matings in culture, it was not without its drawbacks. Foremost among these was the cost in material and especially time that had to be expended to prepare and inactivate enough Sendai virus to conduct a dozen or so fusion experiments. Another consideration was the fact that not all cells have receptors for Sendai virus and hence some fusions cannot be promoted by it. This is true of the cells of many invertebrates and protoplasts of prokaryotes. All of this was changed when the introduction from plant protoplast research of fusions mediated by inexpensive, stable, easily prepared polyethylene glycol (PEG)-based fusing solutions greatly simplified the task of making hybrids (20). Polyethylene glycol preparations can be used to induce not only

intra- and interspecific fusions but also to promote fusion
between cells of different phyla and even between the cells of
different kingdoms.

THE DELIBERATE CONSTRUCTION OF IMMUNOGLOBULIN-PRODUCING HYBRIDS

In 1973, Schwaber and Cohen used inactivated Sendai virus
to fuse human lymphocytes from normal donors to a mouse
plasmacytoma (21). These workers were the first to demonstrate
that one could immortalize the antibody production of a normal
cell of the B lymphocyte lineage by fusing it _in vitro_ with a
plasma cell tumor. It should be noted that these workers found
that some of their hybrids produced immunoglobulin of human as
well as mouse origin. Their report was therefore the first to
establish the feasibility of fusing mouse myeloma cells with
normal immunocytes of another species as an approach to the
production of nonmurine monoclonal antibodies. Other workers
have used this approach to produce human monoclonal antibodies
(22), rat monoclonal antibodies (23), and monoclonal bovine
immunoglobulin (24).

In 1975, Kohler and Milstein (25) devised and demonstrated
a strategy for the deliberate and rational construction of
continuous cell lines that secrete monoclonal antibodies of a
desired specificity. They fused a HAT-selectable mouse myeloma
line with spleen cells from mice that had been previously
immunized with sheep red blood cells. They then screened the
resulting hybrids for the production of antibody specific for
sheep red blood cells, the immunizing antigen. Their success,
which has been widely reproduced, revolutionized serology,
provided tools for the examination of basic immunological
mechanisms, and created an industry. Table 1 provides a partial
list of the antigens that have been successfully targeted by
monoclonal antibodies.

It is clear that in the short time since its inception,
monoclonal antibody technology has influenced an extraordinary
range of pure and applied biological problems. Increasingly,
monoclonal serology will replace the useful but undefined
polyclonal antisera that have for so many years been the
mainstay of immunology. The rapid development of this
technology has already provided useful tools for the pursuit of
research on veterinary immunology, animal parasitology, and
plant virology. Increasingly, standard reference and analytical
reagents for food technology, agriculture, and veterinary
medicine will come from hybridoma technology.

Table 1--Monoclonal antibodies have been produced against a wide
variety of antigens

Microbes and microbial products, i.e., bacteria, viruses, and
bacterial toxins

Parasites

Hormones and drugs

Tumor-associated cell surface and soluble antigens

Classes and subclasses of leukocytes

Plant proteins, hormones, and pathogens

Populations and subpopulations of neuronal cells

Major histocompatibility complex determined antigens

DNA (single and double stranded)

RNA

Numerous soluble proteins from immunoglobulin to tublin

Glycolipids

Carbohydrates

REFERENCES

1. Barnes, D., and Sato, G. (1980) Cell 22:649.
2. Harrison, R.G. (1907) Proc. Soc. Exp. Biol. Med. 4:140.
3. Carrel, A. (1912) J. Exp. Med. 15:516.
4. Sanford, K.K., Earle, W.R., and Likely, G.D. (1948) J.
    Natl. Cancer Inst. 9:229.
5. Gey, G.O., Coffman, W.D., and Kubicek, M.T. (1952) Cancer
    Res. 12:264.
6. Eagle, H. (1955) Science 122:501.
7. Nowell, P.C. (1960) Cancer Res. 20:462.
8. Hayflick, L., and Moorhead, P.S. (1961) Exp. Cell Res.
    25:585.
9. Buonassisi, V., Sato, G., and Cohen, A.I. (1962) Proc.
    Natl. Acad. Sci. USA 48:1184.
10. Augusti-Tocco, G., and Sato, G. (1969) Proc. Natl. Acad.
    Sci. USA 64:311.
11. Barski, G., Sorieul, S., and Cornefert, F. (1961) J. Natl.
    Cancer Inst. 26:1269.

12. Littlefield, J.W. (1964) Science 145:709.
13. Szybalski, W., Szybalska, E.H., and Regni, G. (1962) Natl.
     Cancer Inst. Monogr. 7:75.
14. Okada, Y. (1958) Bikens J. 1:103.
15. Harris, H., and Watkins, J.F. (1965) Nature 207:606.
16. Okada, Y., and Murayama, F. (1965) Exp. Cell Res. 40:154.
17. Ephrussi, B., and Weiss, M.C. (1965) Proc. Natl. Acad. Sci.
     USA 53:1040.
18. Weiss, M.C., and Ephrussi, B. (1966) Genetics 54:1095.
19. Weiss, M.C., and Green, H. (1967) Proc. Natl. Acad. Sci.
     USA 58:1104.
20. Pontecorvo, G. (1976) Somatic Cell Gen. 1:397.
21. Schwaber, J., and Cohen, E.P. (1973) Nature 244:444.
22. Nowinski, R., Berglund, C., Lane, J., Lostrom, M.,
     Bernstein, I., Young, W., Hakemori, S.I., Hill, L., and
     Cooney, M. (1980) Science 210:537.
23. Springer, T., Galfre, G., Secher, D., and Milstein, C.
     (1978) Curr. Top. Microbiol. Immunol. 81:45.
24. Srikumaran, S., Guidry, A.J., and Goldsby, R.A. (1983)
     Science 220:522.
25. Kohler, G., and Milstein, C. (1975) Nature 256:495.

MAKING HYBRIDOMAS

RICHARD A. VAN DEUSEN
National Veterinary Services Laboratories
P.O. Box 844, Ames, Iowa 50010

In this summary, I will touch only on key points that I
think a novice in this work should be aware of. The details of
how one goes about making a hybridoma have been published by
several authors (1-7). For the purposes of this presentation, I
will restrict my description to the methods I currently use (6,
8). You will find that there are more sophisticated methods,
such as electrofusion of cells, as described by Dr. Zimmerman in
his article. There are methods using various myelomas. Some
methods include the use of feeder cells. You can find differences in media preparations, and there are differences in cloning
methods. These are all equally good methods that I suggest you
explore once your needs for hybridomas are well defined.

The basis for most of the cell handling techniques used in
hybridoma production lies in the peculiarities of the mouse
myeloma cell lines that are used. I use the SP2/0 line (9). It
is a non-anchorage-dependent cell that lacks the hypoxanthine-
guanine phosphoribosyl transferase enzyme and thus is
susceptible to aminopterin or methotrexate. The SP2/0 line is
incapable of producing either heavy or light chain myeloma
protein either before or after cell fusion. Therefore,
hybridomas produced using this parent will secrete only
immunoglobulin protein coded for by the B cell parent. Another
important characteristic is that these cells are dependent upon
products of their own metabolism in order to grow, and therefore
may not be diluted far below 1 x 10^5 cells/ml in fresh media
and still survive. For this reason, either feeder cells or
conditioned medium must be used whenever one is attempting to
propagate hybrids beginning with very low numbers of live cells.

When you fuse cells by any of the methods available, the
growing hybridomas will arise from parent cells that are both in
metaphase at the time of fusion. For this reason, restimulating
the donor mouse with antigen 3 to 4 days prior to use is
necessary in order that the highest possible number of
antigen-specific B cells are actively dividing. Likewise, the

myeloma cell culture must be fed in such a way as to have it in log-phase at the time of cell fusion.

A newly fused cell will contain the 73 chromosomes from the SP 2/0 cell plus the 40 contributed from the B cell. During the first few cell divisions, the hybrids often lose chromosomes, making it possible to have a single colony of growing cells with only a portion of the cells secreting antibody. It is therefore necessary to clone every hybridoma of interest.

From my viewpoint, the really difficult part of producing monoclonal antibodies is the screening and selection of the primary hybrids, and later the clones, which are producing the desired product. It is imperative that a screening method be decided upon and be fully functional before beginning a cell fusion procedure. Otherwise, you will find that the capacity of your incubator and your technician's patience are both quickly surpassed by the demands of a very large number of growing hybridomas that cannot be culled intelligently. It is important also that decisions be made quickly and cloning be done as soon as possible to avoid the possibility that the antibody-secreting cells of a primary hybridoma will be overgrown by non-secreting cells in the same vessel.

The following protocol for cell fusion, hybridoma propagation, cell freezing, and ascites production is adapted from Van Deusen and Whetstone (6).

MATERIALS AND METHODS

Media Preparations

Dulbecco's minimal essential medium (DMEM) with 4500 mg of glucose/l (GIBCO Laboratories, Grand Island, NY) is the basic nutrient medium from which all of the preparations listed below are derived. All media are passed through a 0.22-μ filter for sterilization, and then tested for sterility prior to addition of any anti-microbial agent for at least 1 week before use. DMEM with no additives is used in the cell fusion procedure.

Serum added to the medium is either fetal bovine serum (FBS) or a combination of two parts gamma-globulin-free horse serum and 1 part specific-pathogen-free (SPF) calf serum. The latter is less expensive and has performed well in routine use in our laboratory. Each new lot of serum is filtered, tested for sterility, and then tested for suitability in the system by propagating SP2/0 cells (see below) for at least 2 weeks in media supplemented with the serum.

Before use, 1% L-glutamine is added to any medium that has been stored for more than 2 weeks after preparation.

DMEM-R, or regular growth medium, is prepared by adding 75 ml of serum to 925 ml of DMEM, then adding 50 mg gentamicin. This is bottled in 100-ml amounts for convenience. DMEM-R is used for all cells being grown in flasks.

DMEM-CF is the medium used in both cloning and cell-freezing procedures. It is prepared by adding 200 ml of serum to 800 ml of DMEM, adding 50 mg of gentamicin, and bottling in 100 ml amounts.

DMEM-F consists of DMEM-CF with 100 μg of amphotericin B (Fungazone, GIBCO)/l and bottled in 100-ml amounts. DMEM-F is used in the early steps of maintaining fused cells in tissue culture plates when the risk of contamination is greatest.

DMEM-8-aza is DMEM-R with 8-azaguanine (8-aza) added. This preparation is made by adding 2 mg of 8-aza to a liter of distilled water. This is left on a magnetic mixer in a 37°C incubator overnight. DMEM is then made from dry ingredients according to package instructions using the 8-aza solution in place of water. After filtering, 75 ml of serum is added to each 925 ml of medium, then 50 mg/l of gentamicin is added and the medium is bottled in 100-ml amounts.

Conditioned medium (CM) is prepared by harvesting medium in which SP2/0 cells have been growing for 2 to 3 days. (We keep at least two 150-cm^2 flasks, each containing SP2/0 cells, growing in 50 ml of DMEM-R at all times to assure a sufficient supply of CM.) The CM is harvested by centrifugation to remove cells and filtered to assure sterility.

HAT, or hypoxanthine, aminopterin, and thymidine solution, is made as follows, but with the more readily obtained analog, methotrexate, substituted for aminopterin. Hypoxanthine solution is made by dissolving 68 mg of the powder in 100 ml of distilled water using a heated magnetic stir plate set at 60°C. Methotrexate solution consists of 0.1 ml (0.27 mg) of the pharmaceutical preparation added to 9.9 ml of distilled water. Thymidine solution contains 2.42 mg of the powder dissolved in 10 ml of distilled water at room temperature. The three solutions are mixed together, filtered (0.22 μm), bottled in 1.5-ml amounts, and stored at −70°C.

H-T, or hypoxanthine and thymidine solution, is made by omitting methotrexate from the HAT solution protocol above.

HAT medium is prepared by mixing together 50 ml each of DMEM-F and CM, then adding 1.2 ml of HAT solution. H-T medium is prepared by mixing together 50 ml each of DMEM-CF and CM, and then adding 1.1 ml of H-T solution.

Myeloma Cells

The SP2/0 cell line established by Shulman, Wilde, and Kohler (9) is maintained in continuous culture by passage two to three times weekly at a 1:10 ratio in DMEM-R. The cells are grown in plastic flasks with loosened caps kept at 37°C with 5 to 7% CO_2. Selection of hybridized cells following fusion is dependent on susceptibility of the SP2/0 cell to methotrexate. Susceptibility is maintained by two or three passages in DMEM-8-aza at 4- to 8-week intervals.

Mice

In all steps involving mice, we usually use the BALB/c strain, because the SP2/0 cell line is of BALB/c origin. Immunizations to produce antibody-secreting B lymphocytes for cell fusions are carried out in 16- to 18-g mice of either sex. Implantation of cells for ascites production or cell rescue is done in females 5 or more months of age that are primed 3 or more weeks prior to use by intraperitoneal (IP) administration of 0.5 ml of pristane (2,6,10,14-tetramethylpentadecane).

Preparation for Cell Fusion

Before considering a cell fusion, a suitable method for screening hybridomas for antibody production must be available and ready for repeated routine use. Preparations should be made for performing a large number of tests (up to 300 to 400) beginning about 10 to 14 days following the cell fusion date. The screening test method and the antigen to be used for immunizing mice will depend to some degree on the laboratory test method in which the monoclonal antibody product is to be used.

Once a suitable antigen has been prepared, at least 4 mice are inoculated by a route appropriate for the antigen preparation used. When mouse IgG is the desired product, we allow 14 to 18 days to elapse, after which a 0.1- to 0.2-ml blood sample is taken from each mouse and serum is collected by centrifugation of the clotted blood using Sure-Sep II (General Diagnostics, Morris Plains, NJ) to maximize serum yield. The serum samples are tested for antibodies against the immunizing agent. If necessary, the mice are re-inoculated (repeatedly if required) to obtain a positive serum antibody response.

When one or more mice are suitably "turned on" to the antigen, a date for the cell fusion procedure--at least 3 weeks following the last administration of antigen to the mice--is set.

Three days prior to the cell fusion date the SP2/0 cells are passed in DMEM-R and one or two of the immunized mice are hyperimmunized, usually by intravenous administration of antigen. These steps assure a high percentage of actively dividing B cells and SP2/0 cells for fusion. This is also the appropriate time to pristane-prime a number of mice as future recipients of cloned hybridomas.

If mouse IgM-producing hybridomas are desired, we reinoculate the mice 4 days after initial exposure and perform the cell fusion 3 days later.

Polyethylene glycol (PEG) 1540 (MCB Chemicals, Gibbstown, NJ) can be measured, 2.5 g per vial, in 2-dram vials for convenience. A diluent for staining and counting viable SP2/0 cells and spleen lymphocytes is prepared by mixing 1 part of 0.4% trypan blue stain with 9 parts of 0.1 M phosphate-buffered saline (pH 7.2) (PBS). DMEM with no additives can be bottled in

25- to 50-ml amounts for use during the cell fusion procedure.
"Sawed-off pipets" are prepared by scoring and snapping off the
tips of 1-ml disposable glass pipets, and fire-polishing the cut
ends. These pipets are used when handling cells during the PEG
fusion steps to reduce shearing the cell clumps.

<u>Cell Fusion</u>

Here is a stepwise procedure for cell fusion along with a
list of the necessary supplies for each step:

Step		Supplies
1.	Prepare HAT medium by mixing 50 ml DMEM-CF with 50 ml filtered CM and 1.2 ml HAT solution.	- HAT solution - DMEM-CF - CM
2.	Prepare a fresh PEG solution by adding 2.7 ml PBS to 2.25 g PEG-1540 and warming in a 60°C water bath.	- PBS - PEG-1 vial - (1) 5-ml pipet - 60°C water bath
3.	Prepare 4 tubes for making cell counts by adding 0.4 ml of diluent to each of 2 tubes and 0.9 ml to each of 2 tubes.	- (4) 12 x 75 mm tubes - PBS diluent with trypan blue - (1) 1-ml pipet
4.	Prepare SP2/0 cell suspension by combining cells from 2 25-cm^2 flasks in a 50-ml centrifuge tube.	- (1) 50-ml centrifuge tube
5.	Harvest spleen: kill mouse by exsanguination. Save the blood to prepare a positive control serum for screening tests. Soak the mouse with 70% ETOH. Cut away skin and muscle over left flank. Use new instruments to take out the spleen with a minimum of attached fat and blood vessels. Place spleen in Petri dish and add 3.5 ml of DMEM.	- (1) 1-cc syringe with a 25-ga 5/8 inch needle - 70% ETOH - (2) forceps: 1 rat-toothed, 1 smooth-edge - (2) pairs sterile scissors - (1) 60-mm plastic Petri dish - DMEM - (1) 5-ml pipet
6.	Extract cells: Inject 1 ml of DMEM (from Petri dish) into spleen. Set spleen on screen	- (1) "Cellector" tissue sieve with an 80-mesh screen

<u>Cell Fusion Protocol (continued)</u>

over Petri dish. (Cut spleen into 3-4 pieces with scissors.) Push spleen pulp through screen by <u>gently</u> massaging with gloved finger. Rinse through screen with DMEM from Petri dish. Remove cell suspension from Petri dish with a pipet and place in a 15-ml centrifuge tube.

- (Bellco Glass, Inc., Vineland, NJ)
- (1) pr. disposable sterile gloves
- (1) pr. sterile forceps (smooth)
- (1) 1-cc tuberculin syringe with a 25-ga 5/8-inch needle
- (1) 15-ml centrifuge tube
- (1) pr sterile scissors

7. Prepare to count SP2/0 cells and spleen cells by adding 0.1 ml of each cell suspension to one of the tubes containing 0.4 ml of diluent, mixing and transferring 0.1 ml to a tube containing 0.9 ml of diluent. (Final dilution of each is 1:50.) Record volume in mls of SP2/0 cell and spleen cell suspensions. Begin centrifugation of the SP2/0 cell suspension (200 X g for 8 min.) Fill one hemacytometer chamber with each trypan blue stained suspension using a clean Pasteur pipet for each. Count the 4 corner squares and center square of each chamber. Calculate the total number of cells of each type: Total counted cells x 10^5 x ml of cell suspension.

- (1) hemacytometer
- (2) Pasteur pipets
- (2) 1-ml pipets
- tubes containing pre-measured diluent

Calculate cell mixture. The desired ratio of SP2/0 to spleen cells is between 1:1 and 2:1. (SP2/0 cells will be resuspended in 5 ml of medium.)

8. Pour supernatant off of the SP2/0 cells and resuspend in 5 ml of DMEM. Add the calculated volume of spleen cell suspension to the SP2/0 cell suspension, mix and centrifuge the mixture at 200 X g for 8 min.

- DMEM
- (2) 5-ml pipets

<u>Cell Fusion Protocol (continued)</u>

9. Remove supernatant from cell
    mixture by pouring off.

10. Disperse pellet by flicking the
    tip of the centrifuge tube several
    times.

11. Mix the dissolved PEG by   - PEG solution
    inverting several times, then   - (1) 3-ml syringe
    let it cool and draw 3 ml of the   - (1) 0.22-μm
    PEG into a syringe with needle.   filter
    Discard the needle and place filter   - (1) 18-ga 1.5 inch
    on the syringe. Start filtering   needle
    until first drop appears.

12.* Total time 45 sec. Slowly   - timer
    add 1 ml PEG to the dispersed
    cells (about 15 sec). <u>Gently</u>   - 1000-ml beaker half
    mix for 30 sec by swirling with   filled with 37°C
    the tube in water bath.   bath

13.* Place tube in 37°C water for 90
    sec. <u>Do not disturb</u>.

14.* Total time 1 min. Add 1 ml of   - (1) 1-ml pipet
    DMEM slowly (about 30 sec).   - DMEM
    Tilt tube and rotate <u>gently</u>
    to mix.

15.* Total time 45 seconds. Add   - (1) 10-ml pipet
    10 ml DMEM. Swirl tube <u>gently</u>   - DMEM
    to mix. Let stand <u>undisturbed</u>
    for 5 minutes in 37°C H_2O.

16.* Centrifuge at 200 X g for 8
    min. Pour off supernatant.

17. The pellet may be first dispersed by   - HAT medium
    flicking the tube gently. Dilute   - sawed-off pipet
    cells to the equivalent of 5 x 10^5
    myeloma cells/ml by first adding 1 ml
    HAT medium and dispersing pellet <u>very</u>
    <u>gently</u> with a sawed-off pipet. Then
    add 4 to 5 ml HAT medium slowly.
    Invert tube slowly to mix. Pour
    this suspension into vessel containing
    sufficient HAT medium to attain final
    desired concentration of cells.

Cell Fusion Protocol (continued)

18.   Dispense 2/3 of the cell suspension   - 96-well plates
      into wells by delivering 0.2 ml   - sawed-off pipets
      to each well with a sawed-off pipet.

19.   Dilute the remaining cell   - HAT medium
      suspension with an equal
      amount of HAT medium and
      dispense as in 18, above.

20.   Incubate at 37°C, 5% CO_2,
      high humidity.

*** TIME THESE STEPS CAREFULLY:**

Note: Total running time of steps 12-15 is
        9 min. Lay out all necessary supplies
        so you can work without interruptions.

Nurturing New Hybrids

You will note in the following detailed description that we keep back-up cultures of all hybrids, or clones, until each is safely frozen.

During the first 3 days after cell fusion, plates are examined only for unusual color changes or signs of contamination. When the medium begins to change to yellow-orange it is time to feed the cells. In any case, the cells should be fed on the fourth or fifth day after fusion. Feeding is done by carefully withdrawing 0.1 ml of medium from near the top of each well. At this stage a single 5-ml pipet may be used to draw medium off of all the wells of one plate. Two drops (0.1 ml) of HAT medium are then added to each well.

An inverted microscope is used to examine each well for the appearance of hybrid colonies at about the tenth day after fusion. Healthy hybrids appear as clusters of round, bright, translucent cells that are easily distinguished from the debris and dead cells that are left by the selective HAT treatment. From this point, each hybrid is considered unique and is handled as a separate cell culture. This is necessary to prevent carrying cells or antibody product from one well to another, which would confuse the results of screening tests.

Once a new hybrid, or primary, hybridoma has grown sufficiently to cause the medium in its well to turn yellow-orange, it should be screened for antibody production. This is accomplished by carefully withdrawing 0.1 ml of the medium for use in the screening test. The well is then re-fed with 0.1 ml of HAT medium.

Each primary hybridoma positive for antibody production is moved into one well of a 48-well plate (Costar, Cambridge, MA). A Pasteur pipet or disposable plastic pipet may be used to suspend and aspirate the cells from the original well for the transfer. The cells are fed at this time by adding 0.3 ml of H-T medium to each of the new wells and 0.1 ml to the original well (back-up #1).

The primary hybridomas are examined regularly. When the growing cells have caused the medium to become yellow-orange, the cells are moved to a single well of 24-well culture plate with 0.3 ml of H-T medium. The 48-well plate wells are each re-fed with 0.2 ml of H-T medium (back-up #2). When the color of the medium again begins to change, the new well is fed by the addition of 0.5 ml of DMEM-F. At the next change of medium color, the cells are moved to a 25-cm^2 flask with 2 ml of DMEM-R. The 24-well plate well is re-fed with 0.5 ml of DMEM-F (back-up #3). The original plate (back-up #1) may be discarded at this time.

The flask is incubated in an upright position with loosened cap until the medium changes color. Three milliliters of DMEM-R is added and the flask then incubated in the horizontal position. Another feeding of 4 ml of DMEM-R should result in sufficient cells to allow a 1:3 split of the primary hybridoma, resulting in 3 flasks, each containing 10 ml of cell suspension. Back-up #2 may be dispensed with once these cells are actively growing.

When the media begin to turn orange in the three flasks, the primary hybridoma is ready to be cloned and frozen for storage. Two of the flasks are shaken vigorously to suspend the cells. The suspension is poured into a tube and centrifuged at 200 X g for 8 to 10 minutes. The cells are resuspended in 2.5 ml of DMEM-CF for freezing (see cell freezing).

Cloning Primary Hybrids

Dilute the primary hybrid cells to 300/ml in DMEM-R. Prepare a 20-ml suspension of 5 x 10^5 SP2/0 cells/ml in HAT medium. Add 0.2 ml of the primary hybrid suspension of the SP2/0 suspension and mix thoroughly. Dispense 0.2 ml of this mixture into each well of a 96-well plate. Incubate as for a new cell fusion plate.

Beginning on the fourth day after seeding, examine each well daily and identify all wells in which a growing colony of cells appears. Do not attempt to feed these wells. Only wells containing a single colony on the tenth day after seeding are to be sampled, fed, and propagated further.

Useful clones are "moved up" in gradually increasing amounts of medium and sampled for antibody production as outlined for primary hybridomas. Non-secreting clones are discarded and secreting clones are expanded and frozen as quickly as possible. All spent media from each of the secreting

clones are saved for determining the specificity and usefulness
of the antibody product.

Cell Freezing

We prepare a small quantity of DMEM-CF with 20% dimethyl
sulfoxide (DMSO) in a sterile container and allow it to cool to
no more than 37°C. Cells to be frozen are counted and then
concentrated to 2 x 10^6 cells/ml by centrifugation and
resuspension in an appropriate volume of DMEM-CF. The volume is
then doubled by adding an equal amount of the 20% DMSO medium
preparation. The cell suspension is then dispensed into ampules
and sealed for freezing.

Maximum cell viability is maintained if freezing is
accomplished by a constant 1°C per minute temperature reduction
until -80°C is achieved. The ampules may then be placed
directly into liquid N_2 for storage.

To recover frozen cells, thaw quickly, dilute to 15 ml with
DMEM, centrifuge, resuspend in 3 to 4 ml of equal parts DMEM-CF
and CM and incubate in an upright 25-cm^2 flask. Add DMEM-R
medium and lay the flask on its side when cells are actively
growing.

Ascites Fluid Production

A large amount of single determinant-specific monoclonal
antibody is easily produced by implanting cloned hybrid cells in
a pristane-primed mouse. Cells are first washed once and
resuspended in DMEM at approximately 2 x 10^6 cells/ml. Inject
0.5 ml into the peritoneal cavity of the recipient mouse.
Observe the mouse three times weekly for abdominal swelling.
When swelling becomes pronounced, aspirate the ascites fluid
with a disposable myelography needle (Becton-Dickinson, Itasca,
IL) and a 10-ml syringe. Repeated harvests are made at 2- to
3-day intervals until the mouse dies. Each fluid harvest is
clarified by centrifugation using Sure-Sep II[TM] to maximize
yield. We re-use the myelography needles after washing them
with a protein detergent and sterilizing them with ethylene
oxide.

Alternate, In Vitro, Monoclonal Antibody Production Method

Some cloned hybrids do not induce ascites production in
recipient mice. If the antibody product from such a clone is
desirable, we have expanded the clone sufficiently to seed one
or more 150-cm^2 flasks, each with 50 ml of cell suspension.
The cells are allowed to grow in the medium until it is
exhausted (usually 7 to 10 days). The cells are removed by
centrifugation and the fluid frozen for storage.

Cell Rescue

We have experienced a few instances when SP2/0, primary
hybridoma, or cloned hybrid cells have become contaminated with
bacteria or yeast. Cells have been successfully rescued by
following the ascites fluid production protocol, allowing the
mouse to overcome the contaminant, and reculturing the cells
from harvested ascites fluid. The method was suggested to us by
Hinshaw (10). Once recovered from the mouse, about 0.5 ml of
ascites fluid is added to 9.5 ml of DMEM-R. Red blood cells and
macrophages are, for the most part, eliminated upon one passage
of a 1:20 dilution. We then clone the rescued cells.

REFERENCES

1. Chang, T.H., Zenon, S., and Koprowski, H. (1980) J.
    Immunol. Meth. 39:369-375.
2. de St. Groth, S.F., and Scheidegger, D. (1980) J. Immunol.
    Meth. 35:1-21.
3. Galfre, G., and Milstein, C. (1981) Meth. Enzymol. (Part B)
    70:3-45.
4. Goding, J.W. (1980) J. Immunol. Meth. 39:285-308.
5. Shulman, M., and Kohler, G. (eds.)(1979) Cells of
    Immunoglobulin Synthesis. Academic Press, New York,
    pp. 275-293.
6. Van Deusen, R.A., and Whetstone, C.A. (1981) AAVLD 24th
    Annual Proceedings; pp. 211-228.
7. Yelton, D.E., and Scharff, M.D. (1981) Ann. Rev. of
    Biochem. 50:657-680.
8. Van Deusen, R.A. (1984) Submitted for Publication.
9. Shulman, M., Wilde, C.D., and Kohler, G. (1978) Nature,
    Vol. 276, 16 November.
10. Hinshaw, V.I. (1981) St. Jude Children's Research
    Hospital. Memphis, Tennessee (Personal communication).

SCREENING HYBRIDOMAS FOR ANTIBODY PRODUCTION

H. R. GAMBLE
Animal Parasitology Institute, Agricultural Research Service
Helminthic Disease Laboratory
U.S. Department of Agriculture, Beltsville, Maryland 20705

A good screening assay is a key element in the successful
production of monoclonal antibodies. The assay should select
only those hybrids of interest, so that unnecessary clones are
not retained, and additionally, it should detect all hybrids of
interest. Establishing a valid screening procedure before a
fusion is attempted will avoid problems and prevent the loss of
potentially important hybrids.

The requirements for the assay are a) sensitivity, the
ability to detect even weakly secreting clones; b) specificity,
the ability to detect only those clones secreting antibodies
directed toward the antigens of interest; c) rapidity, the
capability of screening several hundred potentially positive
hybrids in a few hours; and d) reproducibility, low background
values that allow positive samples to be selected reliably and
accurately.

A variety of assay systems are available to screen for
monoclonal antibody production (Table 1). The particular assay
chosen depends on the specific requirements of the investigation.

BINDING ASSAYS

The most widely used and generally applicable screening
procedures are binding assays, including enzyme-linked
immunosorbent assay (ELISA) (1), radioimmunoassay (RIA) (2), and
indirect fluorescent antibody (IFA) assay (3, 4). These assays
are generally performed with antigen linked to a solid phase
such as the plastic surface of a microtiter plate. Binding of
the primary, monoclonal antibody is determined indirectly by use
of a labeled second or even third antibody.

The steps involved in an indirect binding assay are
depicted in Figure 1. Antigen that is bound to a solid surface
is reacted with hybridoma culture supernatant, unbound

Table 1--Common screening assays for monoclonal antibody
production

Binding assays

- Enzyme-linked immunosorbent assay (ELISA)
- Radioimmunoassay (RIA)
- Indirect fluorescent antibody (IFA)

Lytic or cytotoxic assays (direct or indirect)

- Hemolytic assays (plaque assays)
- ^{51}Cr release, dye exclusion

Hemagglutination (direct or indirect)

Biological assays

---

antibodies are removed, and then a second, labeled antibody is
added. Bound label is visualized by a) a color change after the
addition of substrate in the case of an enzyme assay, b) an
increase in bound radioactivity, or c) an increase in
fluorescence.

Different techniques can be used for fixing an antigen to
the solid surface, depending on the nature of the antigen.
Insoluble antigens such as cell surface proteins or intact cells
can be fixed to plastic by treatment with glutaraldehyde (5).
Alternatively, assay plates containing cells can be centrifuged
between reagent steps (6). Soluble antigens may be bound to
plastic by passive means, generally in the presence of a high pH
carbonate buffer, or to plates pretreated with poly-L-lysine (7).

Binding assays can be modified based on the second, or
detecting, antibody. As mentioned, the second antibody may be
labeled with an enzyme, radionuclide, or fluorescein.
Additionally, the second antibody may be specific for a
particular immunoglobulin class or subclass, or it may be
rendered isotype independent through the use of an anti-light
chain reagent. As an alternative to second antibody, a protein
A conjugate may be used to detect bound monoclonal antibody (8);
however, not all immunoglobulin classes will be detected equally
well (9). Protein A may be preferred if its use is planned for
subsequent antibody or antigen purification. Finally, an
unlabeled second antibody may be combined with a labeled third
antibody (10) or protein A (11). This modification generally
will increase the sensitivity of the assay because of the
additional amplification generated by multiple binding sites at
each step.

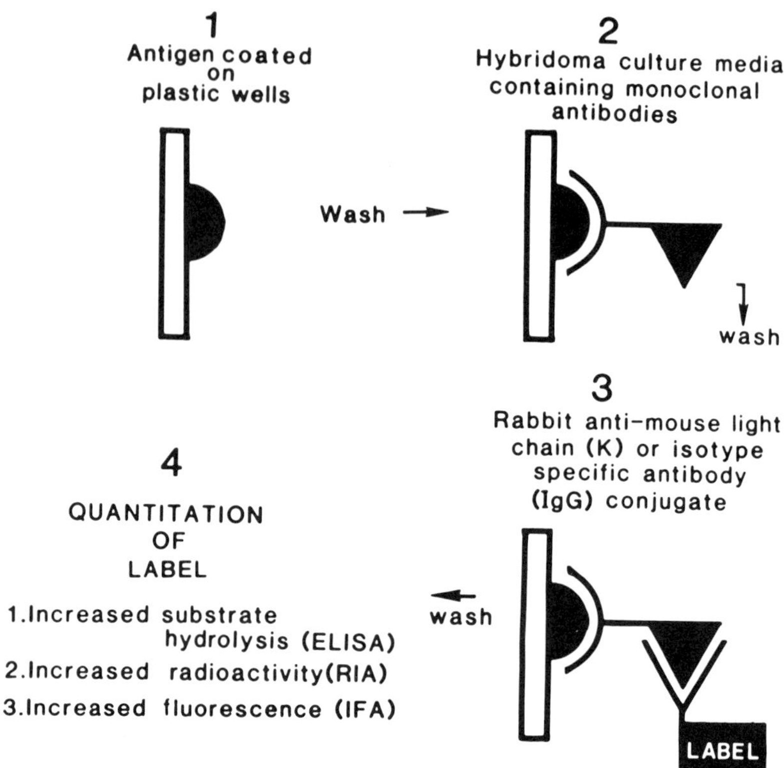

Figure 1--Double antibody indirect binding assay for detection
of monoclonal antibodies.

Some common antibody labels and coupling procedures are
listed in Table 2. Glutaraldehyde is commonly used in a
one-step procedure for enzyme conjugation (12). For indirect
binding assays with radiolabel, ^{125}I is generally used and
conjugated to antibody by the chloramine-T method (13). For
direct binding assays (see below), monoclonal antibodies are
labeled metabolically with ^{3}H, ^{14}C, or ^{35}S (14, 15).
Monoclonal, second or third antibodies can also be labeled with
fluorescein (3) or biotin for use with avidin (16).

LYTIC ASSAYS

In cytotoxic assays, the presence of specific antibodies in
a hybridoma cell supernatant is measured by complement-mediated

Table 2--Common antibody labels and labeling methods

Label	Method
Enzyme - Peroxidase - Alkaline phosphatase - β-galactosidase - Penicillinase	Glutaraldehyde conjugation
Radiolabel - ^{125}I - ^{35}S, ^{3}H, ^{14}C	Chloramine-T Metabolic labeling
Biotin	Biotin-N-hydroxysuccinimide
Fluorescein	Fluorescein isothiocyanate

cell lysis. Cell lysis in turn is measured by methods such as ^{51}Cr release (17) after preloading of cells and dye exclusion (18). These methods are particularly applicable to cell surface antigens. Specificity for soluble antigens is measured by similar techniques after the antigen has been attached to sheep red blood cells. Lysis of red blood cells (fixed in agar) by antibody and complement is determined visually (14, 19). Direct cytotoxic assays are limited to the detection of IgM and certain subclasses of IgG antibodies (those that bind complement). This limitation can be overcome by the use of a second complement-fixing antibody in an indirect cytotoxic assay.

A convenient modification of these lytic assays allows for the simultaneous cloning and selection of antibody-producing cells (14). Hybrids are cloned in agarose that contains antigen-coated sheep red blood cells. After the hybrids are allowed to grow, clones are overlaid with a second antibody and a complement source. Areas of lysis are determined visually. Clones are then allowed to grow and can be picked from the agarose and expanded.

HEMAGGLUTINATION ASSAYS

Hemagglutination assays (14, 20), in which red blood cells are coated with antigen, have the advantage of being extremely simple and rapid and give visually readable results. Assays are performed in 96-well microtiter plates. However, the assays have disadvantages: higher concentrations of antibody might inhibit agglutination due to prozone effects and lower

concentrations of antibody might go undetected because the
agglutination is too weak to be detected. The sensitivity of
the assay can be enhanced by the use of a second antibody in an
indirect hemagglutination assay. Inhibition of hemagglutination
might also be used as a simple method for determining antibody
specificity.

BIOLOGICAL ASSAYS

     In some cases, biological assays may be useful in screening
for the production of a specific antibody. Supernatants from
hybrid cultures are added to any antigen preparation that is
defined by a specific biological activity; a subsequent decrease
in that activity is considered preliminary evidence for the
presence of specific antibody. Examples of such assays range
from inhibition of an enzyme reaction to the inhibition of
penetration of a host cell by a parasite (21).

COMPETITIVE ASSAYS

     Determining whether antibodies secreted by different hybrid
clones have common epitope specificities is often of interest to
researchers. Such an assay might be included as a primary
screening assay. Competitive binding assays, used to determine
common epitope specificities, are performed with a labeled monoclonal antibody (22-24). Labels may include enzymes, radio-
nuclides, fluorescein, or biotin. For this particular
application, biotin labels are particularly useful. In a
competitive assay, antigen is bound to a solid surface and a
mixture of biotinylated monoclonal antibody and a potentially
competing monoclonal antibody is added. When both antibodies
recognize the same antigen epitope, the number of biotinylated
antibody molecules bound will be reduced. The sequential
addition of avidin-enzyme and substrate will allow this reduced
binding to be quantified. Maximum binding is determined from a
control, and competition is reflected as a percent of inhibition.

CONCLUSION

     I have been able to present only a portion of the vast
number of immunological assays that can be used to screen hybridomas for antibody production. New, simpler, and more sensitive
assays are constantly being developed. The importance of the
screening step in the production of monoclonal antibodies should
not be underestimated. Careful planning and preparation of
screening procedures will save time and additional work and
greatly aid in the successful production of highly specific
antibodies.

REFERENCES

1. Engvall, E., and Perlmann, P. (1971) Immunochemistry
    8:871-874.
2. Tsu, T.T., and Herzenberg, L.A. (1980) In: Selected
    Methods in Cellular Immunology. Mishell, B.M., and Shiigi,
    S.M. (eds.) W.H. Freeman, San Francisco, pp. 373-397.
3. Goding, J.W. (1976) J. Immunol. Meth. 13:215-226.
4. Micheel, B., Karsten, U., and Fiebach, H. (1981) J.
    Immunol. Meth. 46:41-46.
5. Heusser, C.H., Stocker, J.W., and Gisler, R.H. (1981)
    Meth. Enzymol. 63:406-417.
6. Posner, M.R., Antoniou, D., Griffin, J., Schlossman, S.F.,
    and Lazarus, H. (1982) J. Immunol. Meth. 48:23-31.
7. Douillard, J.Y., and Hoffman, T. (1983) Meth. Enzymol.
    92:168-174.
8. Jonsson, S., and Kronvall, G. (1974) Eur. J Immunol.
    4:29-33.
9. Ey, P.L., Prowse, S.J., and Jenkin, C.R. (1978)
    Immunochemistry 15:429-436.
10. Gamble, H.R., and Graham, C.E. (1984) Am. J. Vet. Res.
    45:67-73.
11. Brown, J.P., Hellstrom, K.E., and Hellstrom, I. (1983)
    Meth. Enzymol. 92:160-167.
12. O'Sullivan, M.J., and Marks, V. (1981) Meth. Enzymol.
    63:147-165.
13. Klinman, D.M., and Howard, J.C. (1980) In: Monoclonal
    Antibodies: Hybridomas--A new dimension in biological
    analysis. Kennett, R.H., McKearn, T.J., and Bechtol, K.B.
    (eds.) Plenum Press, New York, pp. 401-402.
14. Galfre, G., and Milstein, C. (1981) Meth. Enzymol. 63:1-45.
15. Ahmed-Zadeh, C., Piguet, J.D., and Colli, L. (1971)
    Immunology 21:1065-1071.
16. Warnke, R., and Levy, R. (1980) J. Histochem. Cytochem.
    28:771-776.
17. McKearn, T.J. (1980) In: Monoclonal antibodies:
    Hybridomas--A new dimension in biological analysis.
    Kennett, R.H., McKearn, T.J., and Bechtol, K.B. (eds.)
    Plenum Press, New York, pp. 393-394.
18. Kennett, R.H. (1980) In: Monoclonal Antibodies:
    Hybridomas--A new dimension in biological analysis.
    Kennett, R.H., McKearn, T.J., and Bechtol, K.B. (eds.)
    Plenum Press, New York, pp. 391-392.
19. Jerne, N.K., and Nordin, A.A. (1963) Science 140:405.
20. Coombs, R.R.A. (1980) In: Immunoassays for the 80's.
    Voller, A., Bartlett, A., and Bidwell, D.E., (eds.)
    University Park Press, Baltimore, pp. 17-34
21. Epstein, N., Miller, L.H., Kaushel, D.C., Udeinya, I.J.,
    Rener, J., Howard, R.J., Asofsky, R., Aikawam, M., and
    Hess, R.L. (1981) J. Immunol. 127:212-217.

22. Stahli, C., Miggiano, V., Stocker, J., Staehlin, T., Haring, P., and Takacs, B. (1983) Meth. Enzymol. 92:242-253.
23. Lampson, L.A. (1980) In: Monoclonal Antibodies: Hybridomas--A new dimension in biological analysis. Kennett, R.H., McKearn, T.J., and Bechtol, K.B. (eds.) Plenum Press, New York, pp. 398-400.
24. Springer, T., Galfre, G., Secher, D.S., and Milstein, C. (1978) Eur. J. Immunol. 8:539-551.

CHARACTERIZATION OF MONOCLONAL ANTIBODIES

TERRY W. PEARSON
Department of Biochemistry and Microbiology
University of Victoria, Victoria, British Columbia
Canada V8W 2Y2

SPECIAL PROPERTIES OF MONOCLONAL ANTIBODIES

In order to fully comprehend the properties of monoclonal
antibodies that make them "special" it is important to
understand a) the structure of the antibodies manufactured by
the hybridoma cell and b) the conditions that govern their
selection. It has been clearly shown by Milstein and his
collaborators (1-2) that antibodies secreted by myeloma hybrids
can contain light chains and heavy chains from both parental
cells and that these can associate in a random fashion. For
this reason, only some of the antibodies are functional (bind
antigen) and thus parental myeloma cells were derived which make
no immunoglobulin chains of their own (e.g., SP2/0,
P3X63-Ag8.653). Thus it is now possible to derive hybridomas
that secrete monoclonal antibodies that are 100% active. It was
also shown by Milstein and collaborators that mixing of the
variable and constant regions of the heavy and light chains does
not occur in hybridomas. Thus, antibodies derived from fusions
between nonsynthesizing myeloma parent cells and immune
lymphocytes have the properties of the antibodies from the
immune lymphocytes only. That is, the dual role of recognition
(antigen-binding specificity) and effector function (fixing of
complement components, binding to mast cells, etc.) is found in
one polypeptide chain that is synthesized under the control of
independent genetic loci of the parental spleen cell. This is
true for both heavy and light chains. It is therefore clear
that the type of antibody (as defined by its specificity and
effector function) obtained in cell fusion experiments depends
on the immune status of the animal used as lymphocyte donor.
For our purposes, the important point is that hybridomas do
not represent samplings from all the lymphoid cells of the
immune donor but a representation from the antibody-producing
cells or precursors. It is believed therefore that by deriving

hybridomas, one is able to dissect the antibody response of an animal.

The advantages of monoclonal antibodies have been discussed extensively (3-4). Essentially, they are chemically homogeneous reagents that bind to single antigenic sites on a molecule and can be produced in unlimited quantities without using pure antigens for immunization. The result is that each monoclonal antibody is a unique protein with its own biochemical and immunochemical characteristics unlike conventional antisera, which are composed of antibodies of different classes, affinities, and specificities. This biochemical uniformity of the monoclonal reagents results in their exquisite specificity and high titres obtained after growth of cloned hybridomas. In addition their individuality presents several other advantages and disadvantages, some of which will be outlined here.

DETECTION OF SPECIFIC MONOCLONAL ANTIBODIES

Because monoclonal antibodies will be of a certain class (isotype) and will have their own distinct effector function (in addition to antigen-binding specificity and affinity), it is important to consider carefully the methods used for their detection. Usually, screening assays depend upon the use of second antibodies (rabbit or goat anti-mouse immunoglobulin) that are labeled with ^{125}I (radioimmunoassays), coupled with an enzyme (enzyme-linked immunosorbent assays), or with fluorescent compounds (immunofluorescence). It is important for many applications that the second antibodies detect all isotypes of mouse Ig in order that all monoclonals are detected. Many commercially available antibodies are made to IgG fractions and detect some Ig classes better than others and some not at all. Light-chain-specific antisera can be made, however, that ensure that all Ig classes will be detected (5). Of course heavy-chain-specific reagents should be used for screening if certain antibody isotypes are desired. Usually binding assays are used for detection of specific monoclonal antibodies. Proven methods are established for cell surface antigens (4), for surface and internal antigens of parasites (6), and for soluble antigens (7). In all cases, it is the quality of the second antibody that determines success in detection of high numbers of antibodies. In addition, the type of assay used determines whether or not all specificities will be detected (see section on specificity analysis).

DETECTION OF SPECIFIC EFFECTOR FUNCTIONS

Individual monoclonal antibodies will differ in their effector function; thus, if a specific functional antibody is required, assays other than binding assays are necessitated.

For example, complement-fixing antibodies can be detected in lytic assays using [51]chromium-release as an indicator of cell lysis (3). Immunoglobulin M reagents are often detected in these assays as they are usually excellent at fixing complement. IgG subclasses are detected also but it should be kept in mind that _each_ monoclonal reagent is unique and should be tested independently. "Rules of thumb" often break down when monoclonal antibodies are examined for effector functions that in the past were ascribed to certain isotypes. A further consideration is that some monoclonal antibodies complement each other or act synergistically in lysis of cells (8).

For certain purposes it may be desirable to obtain monoclonal reagents that bind to protein A from _Staphylococcus aureus_, for example, to facilitate purification of the antibody. Screening assays that employ ^{125}I-labeled protein A have been successfully used to detect monoclonal antibodies of various isotypes (9). Again the dogma that only certain subclasses bind protein A should not be believed since monoclonals of most isotypes have been shown to bind protein A (10; unpublished data).

Antibodies that cause immunoprecipitation are often desired for certain experimental purposes. Monoclonal antibodies once again show their special characteristics as they often fail to precipitate antigens (for example in Ouchterlony gel diffusion). Only when the antigen has several identical or closely related antigenic sites will precipitation occur. By adding two different monoclonal antibodies (each specific for different antigenic sites on the same molecule) precipitation can be effected since a "lattice" can be formed. Incidentally, this observation is the first formal test for the lattice theory, as until the advent of monoclonal reagents the experiment was extremely difficult or impossible to perform (11). By using a second antibody or _Staphylococcus aureus_ protein A (as whole organism or coupled to Sepharose) it is possible to "immunoprecipitate" a monoclonal antibody and its bound antigen, even when the monoclonal fails to effect immunoprecipitation on its own (see section on specificity analysis).

IMMUNOCHEMISTRY

An antiserum is a collection of antibodies of different specificities, classes and affinities whereas a monoclonal antibody is biochemically homogeneous. This is most clearly shown by two-dimensional gel analysis of polyvalent antisera or monoclonal antibodies. The procedure allows resolution of individual protein gene products in a mixture (the gels are usually run under reducing conditions) enabling visualization of discrete heavy and light chain spots from the antibodies in the starting material (12-13). An example of a two-dimensional gel

of monoclonal antibody-containing ascites fluids is shown in
Figure 1. Monoclonal antibodies show on the gels as discrete
heavy and light chain spots. Each monoclonal heavy chain or
light chain has its own characteristic isoelectric point, again
indicating the fact that each antibody is a unique species that
has been selected by the fusion process from the vast repertoire
of antibodies that can be made by an animal. Figure 2 shows a
portion of a two-dimensional gel on which was run mouse
antibodies that bound to protein A-Sepharose. Only the light
chain spots are shown and these reveal the great heterogeneity
of the light chain species. Simply by running two-dimensional
gels it is possible to assess the antibody repertoire of an
animal at any time (14).

The gel analyses clearly highlight the central point of
this paper, that is, that each monoclonal antibody is an
individual, unique biochemical species. The implications of
this biochemical individuality are many:

Purification of Monoclonals

Standard column chromatographic procedures for purification
of mouse or rat antibodies are often inadequate for monoclonal
antibodies. Elution from DEAE cellulose columns should be
performed using a salt gradient (0.01 to 0.1 M NaCl) in order to
ensure removal of all monoclonals, some of which bind well even
in high salt. A single step elution is usually adequate for Ig
purification from an antiserum.

Stability of Monoclonals

Monoclonal antibodies show a wide range of stability or
fragility. For example, some are easily rendered inactive after
iodination using the oxidizing agent chloramine T. Thus, other
iodination procedures such as that described by Fraker and Speck
(15) should be tried. Similarly, some monoclonal reagents are
extremely sensitive to freezing-thawing and lose activity
quickly upon storage. Generally, storage of the antibodies at
high concentrations (5 to 25 mg/ml) at 4°C or -20°C in the
presence of 0.01% sodium azide is adequate, even for IgM
antibodies which are often fragile.

Monoclonal Antibody Immunoadsorbents

If a monoclonal antibody binds well to protein A, it can
easily be used as a solid-phase immunoadsorbent for antigen
purification or analysis (12). Monoclonal reagents can also be
used to make immunoadsorbents by coupling them directly to
solid-phase supports such as Sepharose even though their
activity may be severely reduced by the coupling procedure
(16). Figure 3 shows the two-dimensional gel profile of the
eluate from a monoclonal antihuman albumin immunoadsorbent and

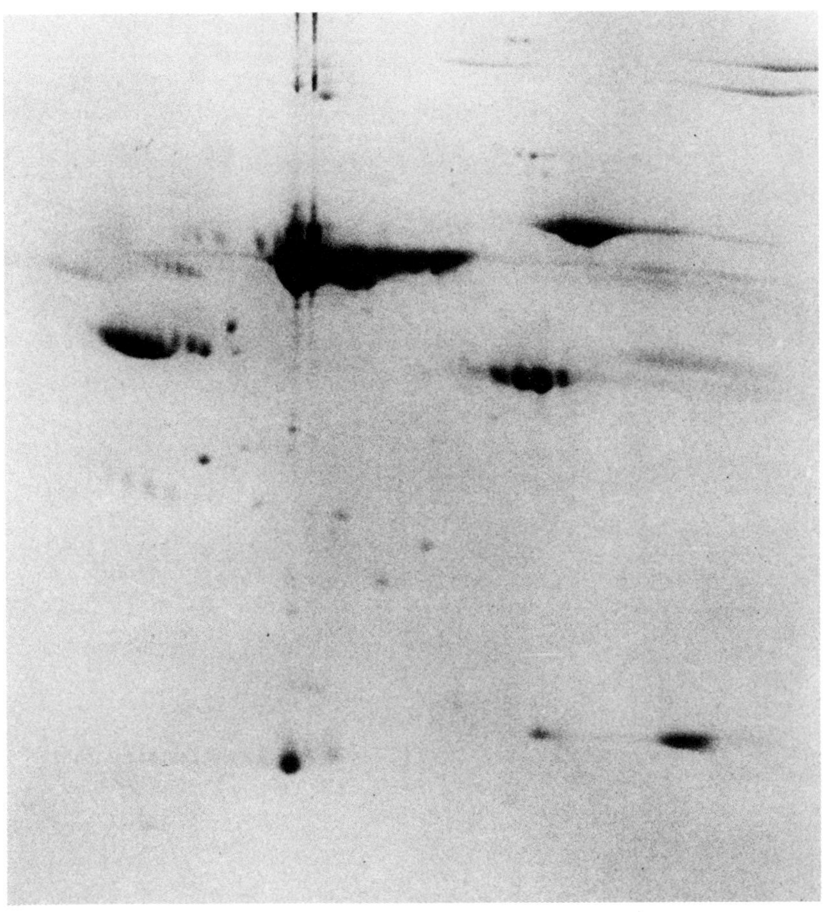

Figure 1--Two-dimensional gel profile of ascites fluid from a
BALB/c mouse bearing hybridoma cells (HSA1/25.1.3).   IgG1 heavy
chain spots, kappa (κ) light chain spots. The acid end of the
gel is to the left. The pH range is ≈4-8.

clearly demonstrates the specificity of the monoclonal reagent
in removing the molecules from human serum. Not all monoclonal
reagents are of sufficient stability or of high enough affinity
to be useful as immunosorbents and this should be kept in mind
if difficulties are encountered. The methods are generally
applicable, however, and when suitable monoclonal reagents are
found the power of the highly specific immunoadsorbents is
apparent (12).

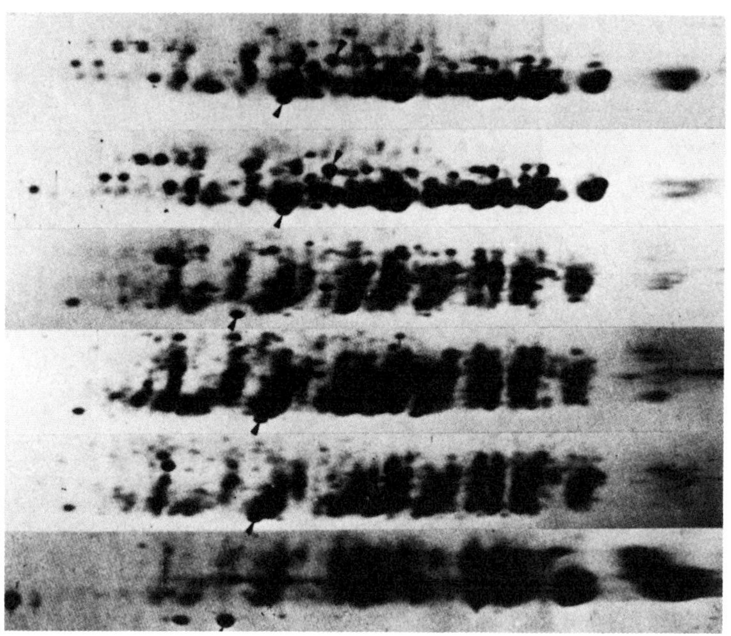

Figure 2--Two-dimensional gel profile of the light chains of
normal BALB/c mouse serum eluted from protein A Sepharose. Each
spot represents a different light chain species. The acid end
of the gel is to the left.

"BIOENGINEERED ANTIBODIES"

It is clear that for many purposes monoclonal antibodies
will in the future be "tailor-made" not only in terms of binding
specificity but for effector function, isotype stability, and
protein A-binding and affinity. Indeed, several commercial
firms are already advertising "bioengineered" monoclonal
reagents and to the extent that the engineering is currently
possible, this is a significant advance in antibody technology.

ANALYSIS OF SPECIFICITY OF MONOCLONAL ANTIBODIES

One of the most powerful features of hybridoma technology
is the fact that pure antigens are not required in order to
derive specific monoclonal antibodies. By enriching for the
antigen, however, the specificity analyses are made simpler.

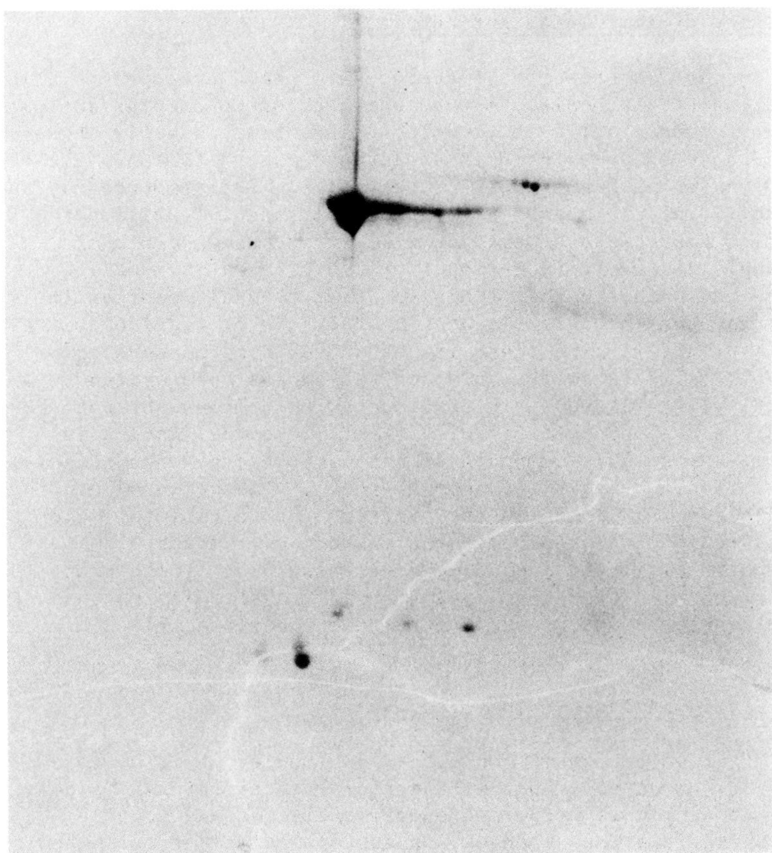

Figure 3--Two-dimensional gel profile of the eluate from a
monoclonal antibody Sepharose immunoadsorbent column. The
monoclonal antibody αHSA1/25.1.3 was coupled to Sepharose and
human serum was passed over the column. After washing, the
bound material was eluted and run on two-dimensional gel. The
major spot is human albumin. The spots to the right and
slightly above the albumin are human transferrin, which always
sticks nonspecifically to Sepharose.

For example, if monoclonal antibodies to a pure protein are
desired, it is a straightforward procedure to detect antibodies
directed against the protein. In contrast, when an antigen
mixture is used, the task of specificity analysis is much more
complex. For determination of specificities of individual
monoclonal reagents, it is perhaps convenient to outline three
main steps to definition of specificity:

Screening of Monoclonal Antibodies

In many laboratories, methods have been developed to allow
detection of antibodies to pure antigens or to complex antigens
either as whole cells or as soluble mixtures. Usually these are
solid-phase enzyme-linked immunosorbent assays (ELISA) (17) or
solid-phase immunoradiometric assays (7). Radioimmunoassay and
immunofluorescence tests are also widely used. I will not
describe these assays here but will offer a few points of
caution. First, it is necessary to have the assay working
before beginning fusion. This is not a trivial point as it is
important that decisions regarding the fate of hybridomas are
made straightaway and there is not enough time to develop
suitably reliable assays once the fusion has been started.
Secondly, it is frequently observed (with complex antigen
mixtures) that nonspecific binding of monoclonal antibodies
occurs. Presumably components in the mixture have an affinity
for antibodies (via Fc receptors, etc.). "Background" binding
of this type can sometimes be minimized by suitable dilution of
the antigen or by passage of the mixture over mouse
Ig-Sepharose. It is a good idea (although not always practical
due to the size of the assays) to screen hybridoma supernatants
on an unrelated antigen to ensure that specific antibodies are
being selected.

Specificity Analysis (Non-molecular)

After detection of monoclonal antibodies it is often
necessary to choose those with certain reactivities. To do
this, screening is performed on cells/tissues and soluble
molecules of various species, on subpopulations of cells and on
subcellular fractions (e.g., membranes, nuclei, cytosol) of
cells, etc. At this point it is not necessary to know which
antigen molecules are involved, only whether or not the
substance being examined contains the antigen or not. Again,
the many methods will not be described here. As examples,
methods for detecting antibodies to whole organisms or their
membrane or cytosol fractions are described by Naot and
Remington (18) and Pearson et al. (12).

Specificity Analysis (Molecular)

Determination of the molecular species bound by a
monoclonal antibody is usually performed by immunoprecipitation
or immunoadsorbent binding of the antigen and analysis by one-
or two-dimensional acrylamide gels. Detailed methods for this
type of analysis are described by Jones (19) and by Pearson and
Anderson (12). This last method utilizes protein A
microimmunoadsorbents and allows rapid analysis of monoclonal
antibody heavy and light chains and the bound antigen, all
simultaneously.

A promising approach for analyzing antigen-binding specificity is that of gel transfer technology. In this technique, for example, two-dimensional gel patterns are transferred electrophoretically to nitrocellulose paper and the proteins are then localized using antibodies and immunoperoxidase staining (20). Although conventional antisera usually react with transferred antigens and allow their localization on the gel transfers, monoclonal antibodies often do not react with the SDS-denatured, transferred antigens. This is probably due to denaturation of the protein antigens after SDS-electrophoresis and disruption of conformation-dependent antigenic sites. Derivation of many monoclonal reagents will allow selection of those that bind to transferred antigens, especially when SDS-denatured antigens are used as the original immunogen and for hybridoma screening (Pearson and Anderson, unpublished).

Many of the methods for analysis of monoclonal antibodies and their specificities are described in detail in Part E of Immunochemical Techniques (Methods in Enzymology Vol. 92, 1983), Van Vunakis, H., and Langone, J.J. (eds.), Academic Press, New York, p. 196.

It is clear that monoclonal antibodies are easily derived, usually in large numbers. In order to maximize the utility of monoclonal antibodies it is important to select the particular antibody desired for its ultimate use. For this reason it is good strategy to eliminate unwanted antibodies, even if they have the desired specificity. It is recommended that several demands be placed on the antibodies as early as possible during their selection. These include: a) Selection for specificity, b) Selection for effector function, c) Selection in the assay of ultimate use, d) Selection for immunochemical/biochemical characteristics (stability, ease of purification, etc.). In many instances, only by being ruthless in this selection will one be successful in finding the monoclonal reagents which will satisfy all requirements.

ACKNOWLEDGMENTS

I thank Ms. Linda Saya and Ms. Herma Neyndorff for running and photographing the two-dimensional gels.

REFERENCES

1. Cotton, R.G.H., and Milstein, C. (1973) Nature 244:42-43.
2. Köhler, G., and Milstein, C. (1975) Nature 256:495-497.
3. Pearson, T.W., Galfre, G., Ziegler, A., and Milstein, C. (1977) Eur. J. Immunol. 7:684-688.
4. Williams, A.F., Galfre, G., and Milstein, C. (1977) Cell 12:663-673.

5. Jensenius, J.C., and Williams, A.F. (1974) Eur. J. Immunol.
    4:91-97.
6. Pearson, T.W, Pinder, M., Roelants, G.E., Kar, S.K.,
    Lundin, L.B., Mayor-Withey, K.S., and Hewett, R.S. (1980)
    J. Immunol. Meth. 34:141-149.
7. Tsu, T.T., and Herzenberg, L.A. (1980) In: Selected
    Methods in Cellular Immunology. Mishell, B.M., and Shiigi,
    S.M. (eds.) W.H. Freeman & Co., San Francisco, p. 373.
8. Howard, J.C., Butcher, G.W., Galfre, G., Milstein, C., and
    Milstein, C.P. (1979) Immun. Rev. 47:139-173.
9. Nowinski, R.C., Lostrom, M.E., Tam, M.R., Stone, M.R., and
    Burnette, W.N. (1979) Virology 93:111-126.
10. McKenzie, M.R., Gutman, G.A., and Warner, N.L. (1978)
    Scand. J. Immunol. 7:367-370.
11. Milstein, C. (1981) Proc. R. Soc. Lond. B221:393-412.
12. Pearson, T.W., and Anderson, N.L. (1980) Anal. Biochem.
    101:377-386.
13. Anderson, N.G., and Anderson, N.L. (1979) Behring Inst.
    Mitt. 63:169-210.
14. Anderson, N.L. (1981) Immunol. Lett. 2:195-199.
15. Fraker, P.J., and Speck, J.C. (1978) Biochem. Biophys. Res.
    Commun. 80:849-859.
16. McMaster, R., and Williams, A. (1979) Immunological Rev.
    47:117-137.
17. Engvall, E., and Perlmann, P. (1971) Immunochemistry
    8:871-876.
18. Naot, Y., and Remington, J.S. (1981) J. Immunol. Meth.
    43:333-341.
19. Jones, P.P. (1980) In: Selected Methods in Cellular
    Immunology. Mishell, B.B., and Shiigi, S.M. (eds.)
    W.H. Freeman & Co., San Francisco, p. 398.
20. Anderson, N.L., Nance, S.L., Pearson, T.W., and Anderson,
    N.G. (1982) Electrophoresis 3:135-142.

CHAPTER 3

MOLECULAR DISSECTION OF ANTIGENS BY MONOCLONAL ANTIBODIES

M.H.V. VAN REGENMORTEL
Institut de Biologie Moléculaire et Cellulaire du C.N.R.S.
15, rue Descartes, Strasbourg, France

The antigenic reactivity of a substance describes its
capacity to undergo specific binding with antibodies. This
reactivity resides in restricted parts of the molecule known as
antigenic determinants or epitopes. An epitope possesses a
three-dimensional structure complementary to that of the binding
site of the antibody molecule. The binding sites of antibodies
are constituted of six hypervariable loops that form a pocket at
the distal end of the Fab fragments of the immunoglobulin
molecule. The size of the cleft formed by the hypervariable
loops is about 15 Å x 20 Å x 10 Å (1), and it is generally
accepted that the site is able to accommodate protein epitopes
consisting of about seven amino acid residues. It has been
shown by Janin (2) that the surface area of globular proteins of
molecular weight 6000 to 35,000 is proportional to the
two-thirds power of their molecular weight, i.e., ASA = 11.1 $M^{2/3}$
where ASA is the accessible surface area, in $Å^2$. This relationship makes it possible to calculate the total surface area of an
antigen that is available for binding to antibody. In the case
of lysozyme, for instance, the accessible surface area of the
native structure is 6500 $Å^2$, whereas that of the unfolded
molecule is 20,500 $Å^2$ (3).
     As a first approximation, the number of antibody molecules
capable of binding simultaneously to a protein defines the
minimum number of epitopes present on the antigen. This number,
which corresponds to the antigenic valence, is proportional to
the accessible surface area of the molecule. However, steric
hindrance puts an upper limit to the number of antibody
molecules that can bind simultaneously to the surface of the
antigen, and the number of epitopes present on an antigen is
usually larger than the valence. This situation is clearly
illustrated in the case of viruses, for which the antigenic

valence is usually smaller than the number of identical subunits
constituting the viral capsid.

Proteins are multideterminant antigens, and they give rise,
in immunized animals, to a variety of heterogeneous antibody
families, each family comprising members able to recognize, with
various degrees of fit, one particular epitope of the protein.
This double heterogeneity of polyclonal antisera raised against
proteins is responsible for the fact that unambiguous
delineation of the epitopes of proteins has been difficult.
Even when the three-dimensional structure of a protein is known,
as in the case of myoglobin and lysozyme, attempts to localize
the epitopes by means of polyclonal antisera have led to
considerable disagreement between several groups of workers
(4-9). Discrepancies in the results by different workers may be
ascribed to the uniqueness of each antiserum, to differences in
methods of immunization and animal species used, and to the fact
that different immunochemical assays may emphasize different
subsets of the total antibody population.

The use of monoclonal antibodies for elucidating the
complex antigenic structure of proteins clearly represents an
ideal way to obviate the difficulties inherent in the use of
heterogeneous antibody mixtures. Although, compared to
polyclonal antisera, monoclonal antibodies often possess a
superior discriminatory capacity for revealing small differences
in the structure of epitopes, they also suffer from several
disadvantages that limit their use as analytical reagents. The
advantages and limitations of monoclonal antibodies for
dissecting the antigenic structure of proteins can be fully
appreciated only when this new tool is placed in the general
context of our knowledge of protein immunochemistry. There is a
major operational element present in the notion of epitope, and
this is clearly perceived only through an in-depth analysis of
the different methods used for the elucidation of antigenic
structures.

TYPES OF ANTIGENIC DETERMINANTS IN PROTEINS

It has been customary, for many years, to distinguish two
types of epitopes, the so-called _sequential_ and _conformational_
determinants (10, 11). Whereas a sequential determinant is
believed to correspond to a sequence of about seven amino acid
residues in its unfolded random coil form, a conformational
determinant is defined by a number of residues that are kept in
a particular conformation, usually within the confines of a
particular macromolecular structure. It is generally assumed
that antibodies directed to conformational determinants do not
react with the unfolded peptides derived from the corresponding
part of the native molecule. Because monoclonal antibodies
directed to globular proteins usually do not bind to peptide
fragments of the antigen (12-14), it has been suggested that

monoclonal antibodies are specific mainly for conformational determinants. It should be emphasized that the distinction between conformation-dependent and independent determinants is somewhat artificial because any sequence of residues within a native globular protein possesses a particular conformation. It has been suggested (15) that a more satisfactory distinction would be to differentiate between <u>continuous</u> and <u>discontinuous determinants</u>. A continuous determinant is defined as a continuous sequence of residues exposed at the surface of a native protein and possessing distinctive conformational features. A discontinuous determinant consists in the juxtaposition in space of residues that are not contiguous in the primary structure. Distant residues could become contiguous through the folding of the polypeptide chain or the juxtaposition of two separate peptide chains. The capacity of antibodies to recognize conformational features of protein molecules has led to many studies of protein folding and polymerization (9, 16-23). There is also good evidence that additional amino acids situated outside the epitope region that is in direct contact with the antibody binding site are able to modulate the conformation of the determinant (7, 24). The term <u>topographic determinant</u> has been used (25) to describe the situation in which an epitope gains its structural uniqueness by long-range interactions at the level of secondary and tertiary structure. It is now generally accepted (6, 26, 27) that certain amino acid substitutions outside antigenic determinants are able, by some kind of allosteric mechanism, to produce conformational changes that alter antigenic reactivity. Because the conformational distortions arising from the excision of an epitope region from the surrounding polypeptide environment are likely to be at least as important as those induced in the native epitopes by distant substitutions, there is little prospect of reproducing exactly the structure of an epitope by fragmentation of the protein or by synthesis of short peptides.

The fact that antibodies raised by immunization with denatured proteins often do not react with the corresponding native molecule has given rise to the concept of <u>hidden epitopes</u> or <u>cryptotopes</u> (28). Cryptotopes are determinants that become antigenically active only after breakage, depolymerization, or denaturation of the antigen. In the case of viral capsids, cryptotopes are found on the surfaces of the protein subunits that are turned inward and become buried after polymerization. There is also evidence that polymerized proteins possess epitopes specific for the quaternary structure; such epitopes, which have been called <u>neotopes</u> (29), are found in most virions and are absent in the constituent monomeric subunits of the capsid (9, 30). Neotopes owe their existence either to conformational changes of the protein induced by intersubunit bonds or to the juxtaposition of residues from neighboring subunits.

METHODS USED IN THE LOCALIZATION OF EPITOPES

Fragmentation of the Protein

In this method a series of peptides obtained by chemical or enzymatic cleavage of the protein are screened for antigenic reactivity. The main shortcoming of this approach is that most of the isolated peptides do not maintain the conformation they possessed in the parent molecule. When the fragments are tested by means of polyclonal antiserum directed to the native molecule, large molar excesses of peptide over intact protein are needed to demonstrate the capacity of peptides to inhibit the reaction between antibodies and the native antigen. This is usually ascribed to the fact that peptides exist in solution in a variety of random conformations in equilibrium with the native one, and that only conformations that approximate to the native form will bind to antibody. Classic experiments performed with fragments of the staphylococcal nuclease molecule led to the suggestion that 1 in 5000 molecules of peptide fragment was correctly folded (17, 31). This means that the affinity constant for antibody binding to the peptide is reduced more than a thousandfold compared to binding to the native molecule. In spite of the low residual antigenic reactivity that is usually present on short peptide fragments, the study of cleavage products has nevertheless allowed a number of different protein epitopes to be identified (32-34).

Instead of testing the fragments for their capacity to react with antibodies prepared against the intact protein, one can also use the fragments for immunization. If the resulting antibodies react with the native molecule, it is assumed that the isolated peptide approximates to an epitope of the protein. This approach has been successfully used for identifying epitopes of lysozyme (5) and β-galactosidase (35).

Use of Synthetic Peptides

Once antigenic reactivity has been localized on a particular peptide fragment of a protein, more precise delineation of the antigenic region is usually attempted by synthesizing a series of short peptides corresponding to the putative sequence of the epitope. A large number of epitopes have been defined in this way, for instance in myoglobin (34), hemoglobin (36), cytochrome c (37), and the coat proteins of tobacco mosaic virus (38, 39) and turnip yellow mosaic virus (40). In addition to the procedures of classic solid-phase peptide synthesis (41), it is also possible to test the antigenic reactivity of a growing peptide at each stage of synthesis without cleaving it from the resin support (42, 43). Recently, hydrophilic resins that swell readily in aqueous solution and admit antibody molecules have been introduced;

after each step of synthesis the binding of antibodies can be
tested directly (44).

When no information besides the amino acid sequence of the
protein is available, it is possible to make an educated guess
as to the likely location of epitopes by identifying the regions
of greatest local hydrophilicity in the sequence. This is based
on the assumption that hydrophilic regions are predominantly
surface-oriented and therefore more likely to be antigenic.
Hopp and Woods (45, 46) have developed a prediction method for
locating epitopes that uses twelve test antigens of partially
known antigenic structure. Each amino acid was assigned a
hydrophilicity value, and these values were then averaged along
the peptide chain, yielding an optimal result with a six-residue
average. This led to the suggestion that peptides of at least
12 residues should be synthesized, including three residues on
either side of the six amino acids that yield the highest
average hydrophilicity in the protein (46). Although the
prediction success rate is not very high, this method represents
a good basis on which to build additional predictions of
secondary structure and surface orientation (47). It should be
emphasized, however, that current methods for predicting protein
secondary structure are not very satisfactory, since none of
them predict more than 56% of the residues correctly (48, 49).
In recent years, the information on amino acid sequence of
proteins has been increasingly obtained from nucleotide sequence
analysis (50-52). The antigen is often not available in
sufficient quantity to allow a conventional immunochemical study
to be carried out. It may also be necessary to verify the very
existence of the antigen by synthesizing one of its putative
antigenic regions; after raising antibodies against such a
synthetic epitope, it may then be possible to reveal the
presence of the complete molecule by means of an appropriate
immunoassay (53, 54).

Ultimately, the success to be reached in mimicking
antigenic regions of proteins by synthesizing linear peptides
depends on the extent to which such linear sequences truly
represent protein epitopes. When the surface of a globular
protein is represented by space-filling models, it becomes
apparent that only relatively small areas of the surface
correspond to a linear array of contiguous residues in the
protein sequence. Exposed surface patches of the protein are
frequently made up of side chains contributed by residues
distant in the sequence but brought in close proximity by
folding of the peptide chain. These surface patches provide
multipoint contacts for antibody binding, and only some of these
contact points will be supplied by the residues that are in
linear sequence. A linear synthetic peptide may therefore
correspond to only part of the epitope surface, and in addition
it may be largely folded incorrectly. It is therefore not
astonishing that the antigenic reactivity of synthetic peptides
is usually very low. One type of peptide that possesses higher

than average antigenicity corresponds to the C-terminal part of
different proteins, e.g., tobacco mosaic virus (TMV) coat
protein (55-57), histone H3 (58, 59), and a polypeptide of
foot-and-mouth disease virus (FMDV) (60). Synthetic peptides
corresponding to C-terminal sequences were able to induce
antibodies that neutralized the infectivity of TMV and FMDV,
respectively. It seems likely that the surface orientation of
the C-terminal residues reduces the conformational constraints
of the peptide in the native structure and provides for a better
than average correspondence between the synthetic and native
structures.

In order to mimic regions of the protein surface that are
made up of spatially adjacent residues that are not in direct
peptide bond linkage, Atassi et al. (61) developed a method
called surface simulation synthesis. A series of discontinuous
epitopes of lysozyme were constructed by linking residues that
are held apart in the sequence by means of glycine spacers
(62). This type of synthesis is likely to be used increasingly
in the future as it becomes apparent that simple linear
sequences of residues cannot mimic adequately the complex
surface topography of proteins.

In recent years, the immunogenicity of synthetic peptides
has received considerable attention. A series of 20 partially
overlapping peptides that cover 75% of the 330 residues of the
HA1 chain of influenza virus were synthesized, coupled to
keyhole limpet hemocyanin, and used to immunize rabbits. Of the
20 peptide antisera that were obtained, 18 were found to
recognize the intact hemagglutinin molecule. Anti-hemagglutinin
antibodies present in influenza virus antisera did not react
with any of the synthetic peptides (63). These 18 immunogenic
regions were scattered throughout the entire HA1 chain and did
not correspond to the epitopes of HA mapped by studies on virus
mutants (64). It seems that more regions of the HA are
immunogenic when fragments of the molecule are used for
immunization instead of the whole molecule.

It has been suggested that the reason why anti-peptide
antibodies so often react with the native structure is that they
can adopt a number of conformations in solution, one of which
approximates that adopted in the native molecule (65). This
so-called stochastic model is based on the assumption that a
rare native-like conformation of the peptide is responsible for
generating the antibodies that recognize the intact protein.
This model was tested by estimating the frequency with which
monoclonal antibodies directed against several peptides
recognize the corresponding intact protein (66). Of 21
hybridomas that synthesized antibody against an hemagglutinin
peptide, 16 produced antibodies that reacted with the complete
molecule. This high frequency contradicts the view that the
immunogenic form of the peptide corresponds to a rare
conformational occurrence and thus refutes the stochastic
model. Several other models have been proposed to explain the

high frequency with which anti-peptide antibodies react with
intact proteins, but none is entirely satisfactory. One
interpretation is that the immune system is able to recognize
preferentially native-like conformations; another suggestion is
that local disorder occurs on short segments of the protein,
allowing reaction with anti-peptide antibodies (66). At the
present time, there is no adequate explanation for the following
contradiction: on the one hand monoclonal antibodies raised
against a native protein are so conformation-specific that they
usually do not bind to peptide fragments of the molecule, while
on the other hand, the majority of monoclonal antibodies raised
against a peptide fragment recognize also the native molecule.

Recently, several laboratories have reported another
puzzling observation, namely that it is possible to immunize
animals with short synthetic peptides without conjugating them
to macromolecular carriers, and that the antibodies raised in
this manner react with epitopes on the intact molecule (67-69).

By immunizing mice with uncoupled synthetic peptides
corresponding to the five epitopes of myoglobin delineated by
Atassi (34), it was possible to produce monoclonal antibodies
that reacted with intact myoglobin (70). Furthermore it was
also possible, when the appropriate synthetic peptides were used
as immunogens, to produce monoclonal antibodies to regions that
are not immunogenic in myoglobin (for instance regions 1-6 and
121-127). Some of these observations are difficult to reconcile
with our presently held views on the molecular basis of
immunogenicity. However, the ability to produce monoclonal
antibodies that bind to protein regions that are nonimmunogenic
when the protein is the immunogen constitutes a phenomenon of
great heuristic value. Further studies using monoclonal
antibodies obtained in this way will no doubt contribute greatly
to our understanding of epitope specificity.

Study of Immunological Cross-Reactivity Between Closely Related Proteins

This type of study requires a series of closely related
proteins possessing a limited number of amino acid
substitutions. When mutants of the protein with single amino
acid exchanges are available, it is possible to ascertain
whether the particular substitution leads to a change in
antibody binding. This approach has been used for instance with
mutants of cytochrome c (71), hemoglobin (72, 73), and TMV (74,
75). If the substitution leads to an antigenic change, it is
concluded that the residue contributes to the structure of an
epitope. However, without additional information, for instance
about the three-dimensional structure of the molecule, one
cannot be sure that the mutated residue is located within the
boundaries of an epitope. It is now well established that
substitutions occurring outside the surface domain of epitopes

are able to produce conformational changes that alter the
antigenic reactivity of proteins (6, 24, 26, 27).

Immunochemical comparisons within several families of
related proteins of known amino acid sequence have demonstrated
that there is a strong correlation between degree of sequence
difference and degree of antigenic difference (76, 77). From
the extent of correlation, it has been inferred that about 80%
of the amino acid substitutions that have accumulated during the
evolution of monomeric globular proteins are immunologically
detectable (24). Such a result is in keeping with the view that
the whole surface of proteins is antigenic, and that internal
substitutions influence the reactivity of antigens by long-range
interactions at the level of secondary and tertiary structure.

The superior discriminatory capacity of monoclonal
antibodies compared to polyclonal antisera is especially
relevant in cross-reactivity studies. In the case of
multideterminant antigens, those antibodies in an antiserum that
recognize an alteration in one particular epitope are often
swamped by the large number of antibodies that continue to react
in the normal way with unchanged epitopes. Furthermore, not all
antibodies specific for one epitope will recognize a particular
substitution equally well, and those antibodies in the antiserum
that retain sufficient reactivity may blur the discrimination
achieved by some others. The molecular homogeneity of a
monoclonal antibody preparation ensures that only one epitope is
investigated at one time, and the influence of a substitution on
antibody binding is therefore more readily detectable. It is,
in fact, in the area of cross-reactivity studies that monoclonal
antibodies have been most useful for elucidating the structure
of epitopes; in contrast, the use of hybridomas has been less
rewarding in the analysis of protein fragments and synthetic
peptides.

ANTIGENIC STRUCTURE OF MYOGLOBIN

Sperm-whale myoglobin is a protein composed of a single
polypeptide chain comprising 153 residues folded in a highly
helical compact structure. The antigenic structure of this
molecule has been studied extensively by Atassi and
collaborators, using chemical modifications of specific side
chains, isolation of antigenic fragments and peptide synthesis
(34, 78). The results led to the claim that five sequences
corresponding to residues 15-22, 56-62, 94-99, 113-119, and
145-151 represent the totality of the antigenic structure of
myoglobin. This conclusion was based on the findings that
serial elution with the five peptides displaced about 81% of the
total elutable antibody from a myoglobin-Sepharose
immunoabsorbent (79), and that the five peptides could achieve
89 to 94% inhibition of the precipitin reaction between
myoglobin and its antibodies (4). The location of the five

epitopes delineated by Atassi in the three-dimensional structure
of myoglobin is illustrated in Figure 1. All epitopes are
prominently exposed at the surface of the molecule and four of
them are situated on bends linking two adjacent helices. Some
of the residues of epitope 4 (residues 113-119) are part of an
α-helix.

The claim that the entire antigenic structure of myoglobin
can be described in terms of these five epitopes has been
challenged by several other workers who studied the
immunochemical properties of myoglobin.

For instance, it has been shown that complete sequence
homology within one of the five epitopes does not guarantee that
two myoglobins will cross-react serologically. In the case of
sperm-whale and cattle myoglobins, which possess one identical
epitope (residues 56-62), the two molecules did not cross-react
when tested by a radioimmunoassay procedure (7, 26). It was
also found that the presence of an identical epitope (residues
15-22) in the three myoglobins of beef, sheep, and pig did not
lead to the same extent of binding with antibodies specific for
the fragment 1-55 of myoglobin. This fragment contains only one
of the five epitopes of Atassi (i.e., residues 15-22) and
differs by only one exchange, at position 9, between beef and
sheep myoglobins. Because antibodies specific for the fragment
1-55 of beef myoglobin (isolated on an immunoadsorbent column)
could discriminate between beef, pig, and sheep myoglobins, it
was deduced that the antigenic reactivity of fragment 1-55
cannot reside solely in the sequence 15-22 (7). The
three-dimensional structure of myoglobin shows that the only
exchanges between beef and sheep myoglobins in the vicinity of
residues 15-22 are at positions 9 and 124 (7, 80). It seems
likely therefore that these two residues (which lie outside the
five epitopes of Atassi) contribute to the topography of one or
more epitopes recognized by antibodies isolated on the fragment
1-55 immunoadsorbent column. There is, in fact, independent
evidence that peptides corresponding to residues 25-55 and 72-89
are able to bind to anti-beef myoglobin antibodies (44).

Additional regions of the molecule, for instance residues
1-6 and 121-127, have also been implicated in the antigenic
properties of myoglobin. Peptides corresponding to the five
epitopes of Atassi as well as residues 1-6 and 121-127 injected
in their free form (without coupling to a carrier) were found to
elicit antibodies capable of binding to native myoglobin (68).
Monoclonal antibodies were also raised against the same seven
peptides in their free form (70, 81).

The antigenic structure of myoglobin has also been studied
by testing the ability of monoclonal antibodies to distinguish
between several mammalian myoglobins (13, 82). Each different
monoclonal antibody against myoglobin had a different spectrum
of affinities for related myoglobins. By identifying which
residues are conserved among the cross-reactive myoglobins while
differing for the non-cross-reactive myoglobins, it was possible

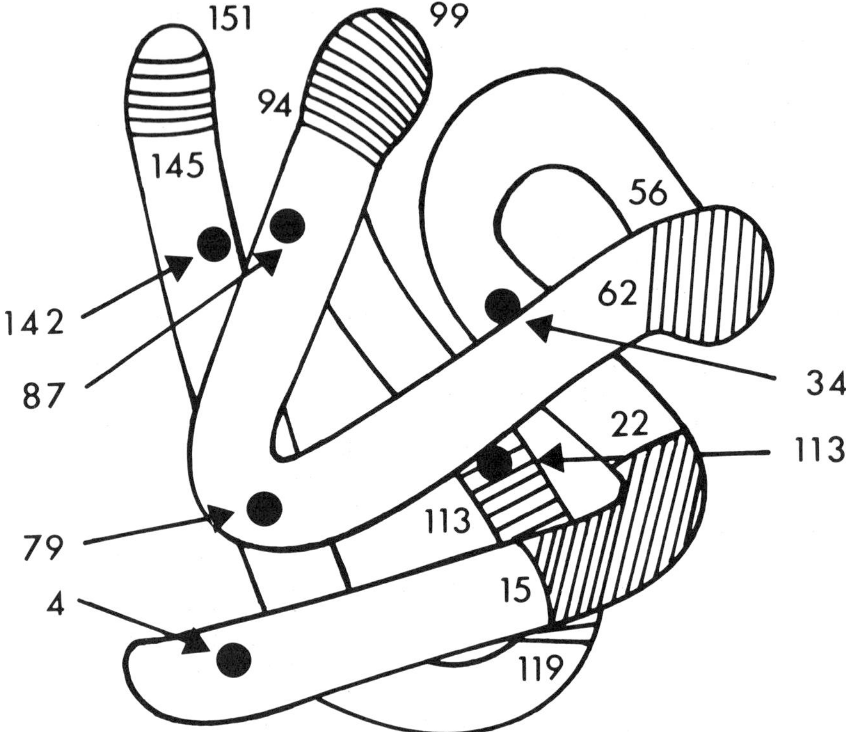

Figure 1--Drawing of the three-dimensional structure of
myoglobin showing the five epitopes delineated by Atassi (34) as
hatched regions. Pairs of numbers designated by arrows refer to
residues that have been implicated together in the binding of
individual monoclonal antibodies (13, 82).

to implicate a number of residues in the formation of certain
epitopes. It was suggested that residues 34 and 113, which are
spatially close to each other, contributed to one epitope, and
the neighboring residues 87 and 142 contributed to another
(82). Similar considerations led Berzofsky et al. (13) to
conclude that two other monoclonal antibodies recognized two
groups of residues that are far apart in primary structure but
close together in the three-dimensional structure (i.e.,
residues 4 and 79 for the one epitope and residues 83, 144 and
145 for the other). As shown in Figure 1, nearly all of the
residues implicated in the binding to several monoclonal
antibodies lie outside the five epitopes studied by Atassi
(34). Furthermore, pairs of residues implicated in the binding
to one particular antibody are located in regions far apart in
the sequence. These results clearly establish the presence of
topographic epitopes in myoglobin and refute the notion that the

antigenic structure of this protein can be described solely in terms of five linear sequences each consisting of 6-7 residues.

It has been argued (68) that immunochemical cross-reaction studies between related myoglobins lead to erroneous assignments of epitopes because these studies allegedly disregard the possibility that the conformation of the protein and that of distal epitopes may be affected by the substitutions. It is of course true that when a substitution that alters the antigenicity is present in a highly exposed position at the surface of the protein, this does not by itself establish that the residue is an integral part of an epitope. However, if the myoglobin substitutions acted by altering the conformation of epitopes elsewhere in the molecule, this effect nevertheless did not completely abolish the ability of monoclonal antibodies to recognize related myoglobins (13). On the other hand, cleavage of sperm-whale myoglobin into three large fragments altered the conformation of the epitopes to such a degree that they could no longer recognize monoclonal antibodies specific for the homologous intact molecule (12, 13). These findings demonstrate that fragmentation of a protein can be even more detrimental to the conformational integrity of its epitopes than the alterations brought about by amino acid substitutions. It seems, therefore, that there is little ground for the belief that the study of protein fragments and synthetic peptides provides more reliable information on the nature of epitopes than the study of immunochemical cross-reactions.

ANTIGENIC STRUCTURE OF LYSOZYME

Hen egg-white lysozyme consists of a single polypeptide chain made up of 129 amino acid residues and internally cross-linked by four disulfide bonds. Cleavage of the disulfide bonds leads to the unfolding of the molecule and to a total loss of antigenic reactivity as measured by antibodies directed to native lysozyme. Immunization with the unfolded molecule leads to the production of antibodies that do not cross-react with native lysozyme in the usual <u>in vitro</u> tests. However, when tests that measure cell-mediated immunity are used, the native and unfolded forms do exhibit cross-reactivity (5).

Tryptic digestion of the reduced lysozyme molecule produced fragments devoid of antigenic reactivity (83). However, when the molecule was digested with trypsin without rupturing the disulfide bonds, three antigenically reactive disulfide-containing peptides were obtained (84). On the basis of precipitation inhibition tests, it was claimed that these three peptides (see Figure 2) were capable of inhibiting the reaction of lysozyme with its specific antiserum by 85 to 89%, and that they accounted for almost the entire antigenic reactivity of lysozyme (85, 86). Independent studies (87, 88) have demonstrated that the N- and C-terminal regions of the molecule

(i.e., residues 1-27 and 123-129 linked by the 6-127 disulfide
bridge) encompass one of the major epitopes of lysozyme. Other
investigators have shown that the so-called loop region, which
consists of residues 64-80 (Figure 2), also possesses antigenic
reactivity (89, 91, 97). The reactivity of this loop peptide
was shown to be drastically reduced when the disulfide bridge
joining residues 64-80 is cleaved (5). Another region of
lysozyme that has been shown to possess antigenic reactivity
corresponds to residues 38-54 (92).

A more precise localization of the epitopes within the
three disulfide-containing peptides isolated by Atassi et al.
(84) was attempted by the method of surface-simulation synthesis
(61, 62). The inhibitory activity of several peptides that
simulate parts of the surface topography of the molecule was
measured. The results led to the claim that three peptides
(Figure 2) accounted for 90 to 95% of the antigenic reactivity
of native lysozyme (93). When these three synthetic structures
are compared with the three disulfide-containing peptides
obtained by cleavage (Figure 2), it is obvious that in spite of
their similar total inhibitory capacities, the two sets of
peptides are very different in structure. In particular, when
the two synthetic peptides made up of residues 5-7-13-14-125 and
residues 33-34-113-114-116 are compared with the corresponding
cleavage peptides, it appears that in each case, three of the
five residues forming the synthetic epitopes are not present in
the original structures obtained by cleavage. This discrepancy,
as well as the large excesses of peptides required to achieve
maximum inhibition, casts doubt on the claim that the three
surface-simulation peptides account quantitatively for the total
antigenic reactivity of lysozyme.

The conclusions of Atassi and Lee (93) are also at variance
with the results of cross-reactivity studies of a large number
of avian and mammalian lysozymes of known primary structure. It
was found, for instance, that several bird lysozymes that
present exchanges exclusively outside the presumed epitopes were
nevertheless distinguishable antigenically (6).

Several attempts have been made to identify some of the
contact residues of the lysozyme epitopes by means of monoclonal
antibodies (94-96). By comparing the relative ability of
several avian lysozymes with known substitutions to inhibit the
binding of a particular monoclonal antibody to chicken lysozyme,
it was established for instance that residue 68 is part of an
epitope (95). The boundaries of this epitope were defined by
examining space-filling models of chicken lysozyme and noting
the position of residues that vary in lysozymes showing
antigenic identity to chicken lysozyme. This led to the
proposal that the boundary residues of the epitope comprising
residues 68 and 45 include residues 41, 62, 63, 102, 103, and
114. The determining residues to two other epitopes recognized
by monoclonal antibodies also did not correspond to any of the
three epitopes of lysozyme defined by Atassi and Lee (93).

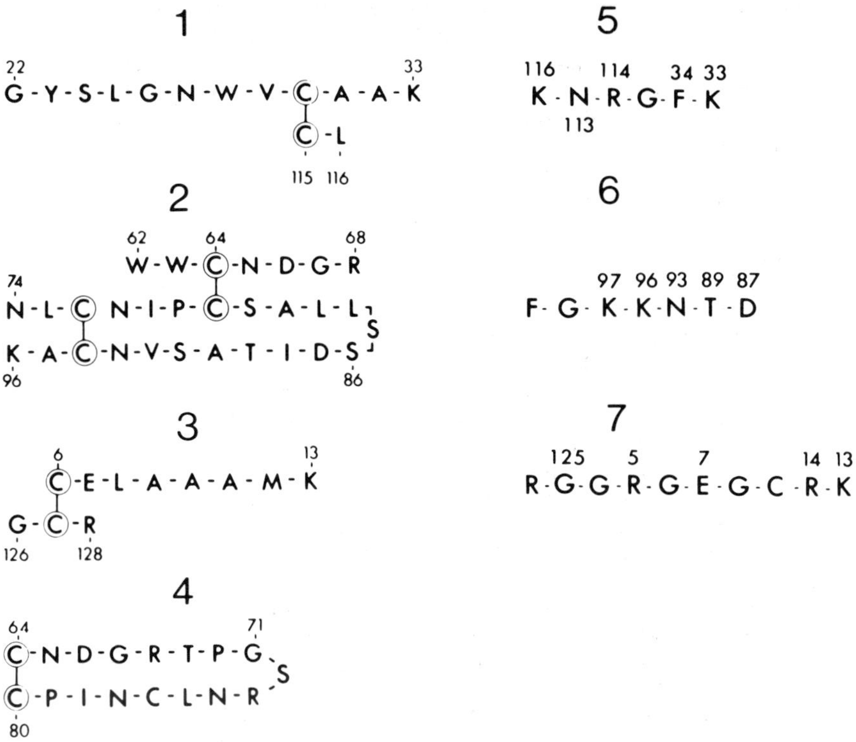

1

22   33
G - Y - S - L - G - N - W - V - C - A - A - K
                             C - L
                           115 116

5

116   114   34 33
K - N - R - G - F - K
113

2

      62   64   68
     W - W - C - N - D - G - R
74
N - L - C - N - I - P - C - S - A - L - L
K - A - C - N - V - S - A - T - I - D - S
96   86

6

97 96 93 89 87
F - G - K - K - N - T - D

3

   6   13
 C - E - L - A - A - A - M - K
G - C - R
126   128

7

125   5   7   14 13
R - G - G - R - G - E - G - C - R - K

4

64   71
C - N - D - G - R - T - P - G
                        S
C - P - I - N - C - L - N - R
80

Figure 2--Structure of some of the peptides implicated in the
antigenic structure of lysozyme. Peptides 1, 2, and 3 were
obtained by tryptic digestion and were capable of inhibiting (85
to 89%) the reaction of lysozyme with its specific antiserum
(84, 86). Peptide 4 is the loop peptide studied by Arnon et al.
(90) and Ibrahimi et al. (97). Peptides 5, 6, and 7 are the
surface simulation synthetic structures which together possess
95% inhibitory activity. These synthetic peptides are alleged
to represent the complete antigenic structure of lysozyme (93).
The numbers of the constituent residues correspond to their
position in the lysozyme sequence.

STRUCTURE OF INFLUENZA VIRUS HEMAGGLUTININ

The membrane of the influenza virus particle contains two
glycoproteins, hemagglutinin (HA) and neuraminidase (NA), that
play important roles in the process of infection. The HA
mediates attachment of the virus to host cells as well as the
fusion of viral and host membranes, while the NA effects the
release of virus from infected cells. The three-dimensional
structure of both proteins has been determined by X-ray
crystallography (98, 99).

The HA is a triangular, rod-shaped molecule composed of three pairs of disulfide-linked polypeptide chains, which are called HA1 and HA2. The HA1 chain comprises about 330 amino acid residues. The NA is a tetrameric protein made up of four roughly spherical heads and of a stalk that anchors the molecule in the membrane.

Before the development of monoclonal antibodies, only two groups of antigenic determinants, the strain-specific and cross-reacting epitopes, could be distinguished on the HA (100); the heterogeneity of the antisera used in these studies prevented further elucidation of the antigenic structure of the influenza glycoproteins. Studies initiated at the Wistar Institute in Philadelphia (101, 102) led to the production of a large number of hybridomas that were used to delineate the structure of some of the epitopes of the HA molecule of influence viruses A/Aichi/68 (H3N2) and A/PR/8/34 (H1N1). Monoclonal antibodies were employed to select non-neutralized virus mutants in which the epitope recognized by the selecting antibody was altered. Antigenic variants of influenza virus selected after a single passage of the virus grown in the presence of individual monoclonal antibodies were shown to have single amino acid substitutions in the sequence of the HA polypeptide. Variants that grew in the presence of monoclonal antibodies existed in the wild-type stock at a frequency of about 1 in 10^5. No variants grew in the presence of a mixture of two monoclonal antibodies. Because the infectivity titer of the wild-type virus was about 10^8 EID_{50}, this finding is in agreement with the expected frequency of 1 in 10^{10} for variants possessing changes in two independent sites.

The loss in the ability of the variants to bind to the monoclonal antibody used in their selection was caused by single amino acid exchanges in the sequence of the HA1 polypeptide. These exchanges were localized by comparing the composition of tryptic peptides of the variants with the known sequences of these peptides in the wild-type virus. From the location of these exchanges on the surface of the molecule (64), four clusters of exchanges were identified as potential antigenic determinants. There is no direct evidence that all the variable residues represent actual contact amino acids, for the possibility that certain substitutions induce conformational changes in epitopes located elsewhere in the molecule cannot be excluded.

Most single substitutions totally abolished the ability of the monoclonal antibody used to select the variant to bind to the hemagglutinin; at the same time, the binding to other monoclonal antibodies usually was not affected. Although the majority of single point mutants selected under the immunological pressure of monoclonal antibodies could not be differentiated from wild-type virus by means of polyclonal antisera (103, 104), a certain number of them were antigenically distinct when analyzed by antisera. This was the case, for instance, for variants showing a change at positions 143, 144, or 145. Because residues 140-146 correspond to a protruding loop of the HA molecule (98), it seems likely that this region does in fact correspond to an epitope of the hemagglutinin molecule (64, 105).

By locating the substitutions causing antigenic variation in the three-dimensional structure of HA, it was possible to identify three other clusters of residue exchanges. These regions probably also correspond to independent epitopes of the hemagglutinin molecule (64).

A similar study of mutants of influenza virus A/PR/8/34 (subtype H1) selected by monoclonal antibodies has been reported (106, 107). Although the H1 and H3 subtypes of hemagglutinin show only a 35% homology in sequence in the HA1 subunit (108), it is probable that they possess a similar tertiary structure. Many structurally important amino acids are conserved in both HA subtypes; for instance the cysteines and many prolines are conserved as well as the distribution of hydrophilic and hydrophobic regions (109).

Substitutions in the H1 and H3 hemagglutinin mutants that altered the reactivity to monoclonal antibodies are summarized in Table 1. Nearly all changed residues are found in accessible positions at the surface of the HA molecule, and there is considerable agreement between the two antigenic maps. Some of the differences between the proposed locations of epitopes in the two subtypes have been ascribed to differences in sites of carbohydrate attachment in the two molecules (107).

Supporting evidence for the identity of the epitopes labeled by Wiley et al. (64) A, B, and D has been presented by Underwood (105). This author identified, in the HA of 16 virus strains, groups of variable amino acids that affected the binding of a panel of 125 monoclonal antibodies raised against the hemagglutinin of influenza A/NT/60/68. However, instead of defining separate discrete epitopes with precise boundaries, Underwood (105) interpreted his data in terms of a continuous surface area showing antigenic reactivity. The variable residues studied by Underwood (105) and the location of some of the suggested epitopes of hemagglutinin are illustrated in Figure 3.

Four distinct antigenic regions have also been detected in HA by a different experimental approach, namely competitive radioimmunoassay (110). This method measures the capacity of unlabeled monoclonal antibodies to block the binding of heterologous radiolabeled hybridoma antibodies; it is assumed that if the epitopes recognized by two monoclonal antibodies are physically close, the binding of one antibody will sterically block the binding of the other. If the epitopes are sufficiently distant, no blocking is expected to occur. Only two of the antigenic regions defined in this way (110) corresponded to sites identified by the analysis of antigenic variants.

If some antibodies bind monovalently to viral subunits
(111), the blocking effect in a competition experiment is likely
to be less than if they bind bivalently to separate copies of
the same epitope on different monomers (112). The extent of
competition measured in reciprocal experiments with two
antibodies binding monovalently and bivalently, respectively,
could thus be markedly different.

As pointed out by Yewdell and Gerhard (102) and Friguet et
al. (113), if monoclonal antibodies are observed to compete for
binding to antigen, several explanations are possible. First,
the antibodies may delineate structurally overlapping epitopes.
Second, binding of one antibody may prevent access of the second
antibody to a physically distinct site; in this case the
proximity of the two epitopes creates steric hindrance between
the antibody molecules. Furthermore, the binding of some
monoclonal antibodies may induce conformational changes in the
antigen, thereby altering the epitope recognized by the
competing antibody (114). There is also another possibility,
namely that the antigen oscillates between two conformations:
if the first antibody species binds exclusively to one

Table 1--Residues involved in epitopes of influenza virus
hemagglutinin (H1 and H3 subtypes). Substitutions at the
indicated positions abolish the binding of the hemagglutinin to
individual monoclonal antibodies. See Figure 3 for the location
of these residues in the three-dimensional structure of the
virus.

Epitope	H1 Subtype [a] Sequence positions	Epitope	H3 Subtype [b] Sequence positions
Sa	124, 125, 154, 156, 158, 161, 162, 163	A	133, 137, 140-146
Sb	152, 155, 188, 189, 192, 194	B	155-160, 187-197
Ca 1	165, 169, 203, 236, 270	C	53, 54, 275, 278 (CYS 52 - CYS 277)
Ca 2	136, 139, 141, 220, 221	D	201, 205, 207, 217, 220
Cb	70, 71, 73, 74, 75, 115		

[a]Caton et al. (107).
[b]Wiley et al. (64).

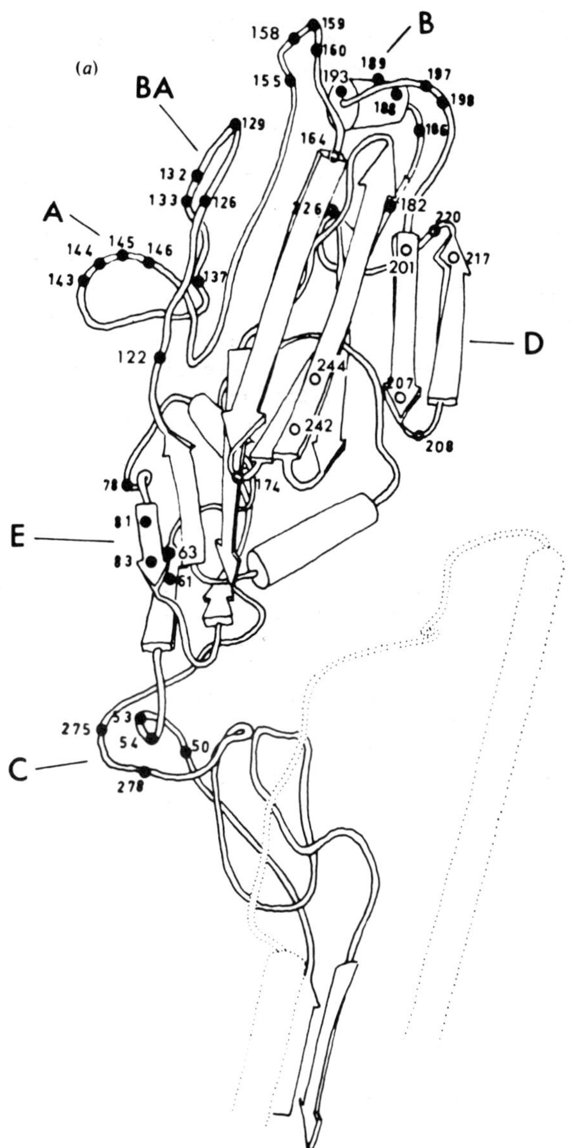

Figure 3--Drawing of the three-dimensional structure of the upper part of a monomer of influenza virus hemagglutinin. Some of the variable amino acids marked by numbers are involved in the epitopes marked by letters (from Underwood, 105).

conformation while the second one binds exclusively to the
other, it will be impossible for the two antibodies to bind
simultaneously to the antigen even if they are specific for
distinct epitopes (113).

Recently, a study of the epitopes of the neuraminidase of
influenza virus has been undertaken, using the same approach as
in the case of the hemagglutinin (115). Antigenic variants of
neuraminidase were selected after a single passage of the virus
in the presence of monoclonal antibodies. The variants did not
bind to the antibody used for their selection, and each showed a
single substitution in the neuraminidase sequence. Seven
variable segments were identified in the three-dimensional
structure of the molecule, but these did not segregate into
spatially distinct non-overlapping areas (115). It seems,
therefore, that the epitopes of the neuraminidase molecule form
a nearly continuous antigenic surface across the top of the
monomer.

ANTIGENIC STRUCTURE OF TOBACCO MOSAIC VIRUS COAT PROTEIN

Tobacco mosaic virus (TMV) is a rod-shaped particle, 300 nm
long, that consists of 2130 identical protein subunits arranged
as a helix around an RNA molecule of molecular weight 2×10^6.
The protein subunit (TMVP) contains 158 amino acid residues
(Table 2), and its three-dimensional structure was established
by X-ray crystallography (116, 117). The central part of the
subunit consists of two pairs of α-helices, which are composed
of about 60 residues. In the assembled virus, the RNA fits
between the protein subunits in a groove at a radius of 4 nm
(118).

Table 2--Amino acid sequence of tobacco mosaic virus protein

```
SYSITTPSQFVFLSSAWADPIELINLCTNALGNQFQTQQARTVVQRQFSEVWKT
    |         |         |         |         |
    10        20        30        40        50


SPQVTVRFPDSDFKVYRYNAVLDPLVTALLGAFDTRNRIIEVENQANPTTAETL
    |         |         |         |         |
    60        70        80        90        100


DATRRVDDATVAIRSAINNLIVELIRGTGSYNRSSFESSSGLVWTSGPAT
 |         |         |         |         |         |
 110       120       130       140       150       158
```

Analysis of TMV Epitopes by Means of Polyclonal Antisera

Several methods have been used to locate the epitopes of
TMVP. The first approach used by Anderer (55) was to test the
capacity of different cleavage peptides of TMVP to inhibit the
precipitin reaction between virus and antibody. Regions
corresponding to residues 18-23, 62-68, 123-134, and 153-158 in
TMVP were found to possess the highest inhibitory activity.
Antisera were also prepared by using as the immunogen the
C-terminal hexa-, penta-, tetra-, or tripeptides of TMVP coupled
to bovine serum albumin. The resulting antibodies were found to
precipitate the virus and to neutralize its infectivity (56,
57). The presence of antibodies reactive with the C-terminal
peptides could also be detected in TMV antisera by a sensitive
passive hemagglutination test (119, 120).

The second approach used to locate the epitopes of TMVP
consisted of studying the binding of cleavage peptides to
antibodies prepared against depolymerized subunits instead of
against the virus. Using the technique of inhibition of
complement fixation, Benjamini et al. (121) found that only the
region 93-113 of TMVP possessed inhibitory activity. The
antigenic reactivity of this region was then studied extensively
with synthetic peptides ranging in length from dipeptide to
decapeptide (122, 123). The shortest peptide that possessed
demonstrable binding activity was the pentapeptide 108-112. The
binding of this peptide of sequence LDATR was greatly enhanced
by the addition of five alanine residues at its N-terminal end.
Furthermore, the inactive tripeptide 110-112 (ATR) acquired
binding activity by N-octanoylation (124). This finding showed
that increased hydrophobicity of the modified peptide could
enhance its overall binding energy in a nonspecific manner. It
seems that, in such a case, the dissection of an epitope has
reached its ultimate limit. This finding illustrates the truism
that antigen-antibody binding is mediated only through the usual
forces of protein-protein interactions such as electrostatic and
hydrophobic bonds, and that these are brought into play by a
conformational element that brings contact residues of the
antigen into close apposition with complementary residues of the
antibody binding site. In the case of short peptides, the
conformational component in the binding may be nearly
insignificant, and the addition at a nonspecific location of
hydrophobic or charged residues may increase the reactivity of
the peptide above the threshold value needed for detection. It
has been shown, for instance, that the addition of a lysine
residue to the N-terminal of a peptide can greatly increase the
binding of antibodies, whether from normal or immune sera (44).
Control experiments with unrelated peptides and unrelated
antisera are thus required to establish the specificity of the
observed reaction.

From the three-dimensional structure of TMVP, it is now apparent that the epitope studied exhaustively by Benjamini and colleagues is located in the central hole of the assembled virus particle. Because this epitope is not reactive in the virus, it corresponds to a cryptotope of the assembled virion. Recently the whole question of how many epitopes are expressed on the TMVP monomer and how many are found at the outer surface of the capsid was again investigated by means of enzyme-linked immunosorbent assays (ELISA) and complement fixation tests (27, 125, 126). In complement fixation assays, tryptic peptide 1 corresponding to residues 1-41 did not inhibit complement fixation in the TMVP anti-TMVP system, but on the contrary, enhanced the amount of fixation. However, when antiserum to the coat protein was absorbed beforehand with the virus in order to remove antibodies specific for the capsid, peptide 1 was found to inhibit complement fixation. This suggested that the region 1-14 contains two epitopes, one that is a cryptotope expressed only in the protein monomer and the other an epitope common to the virus and monomer. By means of inhibition assays with synthetic peptides it could be shown that the viral epitope is located in the N-terminal 10 residues of TMVP whereas the cryptotope is located in residues 34-39, a region that corresponds to a tight turn between two helices (39). From the analysis of cleavage and synthetic peptides, the following conclusions regarding the antigenic structures of TMVP have been drawn (39): (1) The dissociated TMVP molecule possesses seven epitopes situated in the vicinity of residues 1-10, 34-39, 55-61, 62-68, 80-90, 108-112, and 153-158. (2) Three of these epitopes located in the vicinity of residues 1-10, 62-68, and 153-158 are also expressed at the surface of the virus particle. (3) The four epitopes located in the vicinity of residues 34-39, 55-61, 80-90, and 108-112 are cryptotopes, i.e., epitopes that are only reactive in the dissociated viral subunits.

The third approach used to analyze the antigenic structure of TMV consisted in studying the immunological cross-reactivity between wild-type TMV and mutants showing 1-3 substitutions in the coat protein. By means of precipitin tests, it was shown that substitutions at positions 65, 66, 107, 136, 138, 140, 148, and 156 altered the antigenic reactivity of the virus (74, 75); on the contrary exchanges at positions 20, 21, 25, 33, 46, 59, 63, 81, 97, 99, 126, and 129 did not alter the antigenic properties. Some of the substitutions, at positions 20 (P→T) and 63 (P→S) which did not affect the epitopes of the virus, clearly altered the antigenic reactivity of the dissociated subunits (27). Both proline substitutions at positions 20 and 156 seemed to affect the antigenic reactivity by changing the conformation of the polypeptide chain. The exchange P→L in position 156 allowed the mutant virion to react with heterospecific antibodies present in TMV antisera (29, 75, 127, 128). Such heterospecific or heteroclitic antibodies, which seem to be elicited in all animals immunized with TMV, are

antibodies that are better recognized by another antigen than
the one used for immunization. Depending on the sensitivity of
the serological test, it may happen that the weak reactivity of
heterospecific antibodies with the immunogen is below detection
level. In such a case it may appear as if the antibody had been
elicited by an antigen with which it is unable to react. The
reactivity of heterospecific antibodies found in TMV antisera is
illustrated in Figure 4. In this experiment, a TMV antiserum
was repeatedly absorbed with homologous TMV and then tested by
ELISA for residual activity towards several mutants and
dissociated coat proteins.

Analysis of the Epitopes of TMV by Means of Monoclonal Antibodies

The availability of many single point mutants (129) and
natural strains of known primary structure (130) has allowed the
antigenic structure of TMV to be mapped by means of monoclonal
antibodies (131). Because the three-dimensional structure of
the common strain of the virus is known, it was possible to
predict which amino acids are potentially capable of acting as
contact residues in epitopes of TMV. A series of TMV mutants
showing substitutions either at the outer surface of the virion
or far away from the surface were selected for study. Fifteen
mutants showing single or double amino acid substitutions were
compared by means of nine monoclonal antibodies. The
substitutions present in each of the mutants are listed in Table
3. Mutants NI 430, Fu 27, Ni 1688, 414, Ni 116, Ni 109, and CP
415 were derived from wild-type TMV, whereas mutants Ni 458, Ni
1045, Ni 630, Ni 725, and Ni 1927 were derived from strain A 14,
which has an exchange I→T at position 129 compared to
wild-type virus (133). This exchange has been shown not to
alter the antigenic reactivity of the virus with respect to both
antisera and monoclonal antibodies. Mutant OM is the common
Japanese strain of TMV (134), whereas mutant 06 is an orchid
strain (135). Two other naturally occurring strains of TMV were
also included in the comparison--the Y-TAMV strain, which shows
18% difference in sequence compared to TMV, and strain U2, which
shows 26% difference in sequence (130). The ability of
monoclonal antibodies to differentiate between TMV and the
mutants was measured by an indirect double sandwich type of
ELISA test. Typical results showing the comparative binding of
monoclonal antibody 19 to nine mutants are illustrated in
Figure 5. The antibody reactivity towards six mutants was
reduced to various degrees compared to TMV, but it was
significantly higher in the case of four other mutants. A
series of ELISA tests showing the binding of monoclonal
antibodies 19, 20, and 22 to different viral strains are
illustrated in Figure 6. The variability in reactivity of
different antibodies is clearly seen in the case of strain 06
which, compared to TMV, reacted equally well with antibody 22,
better with antibody 19, and not at all with antibody 20

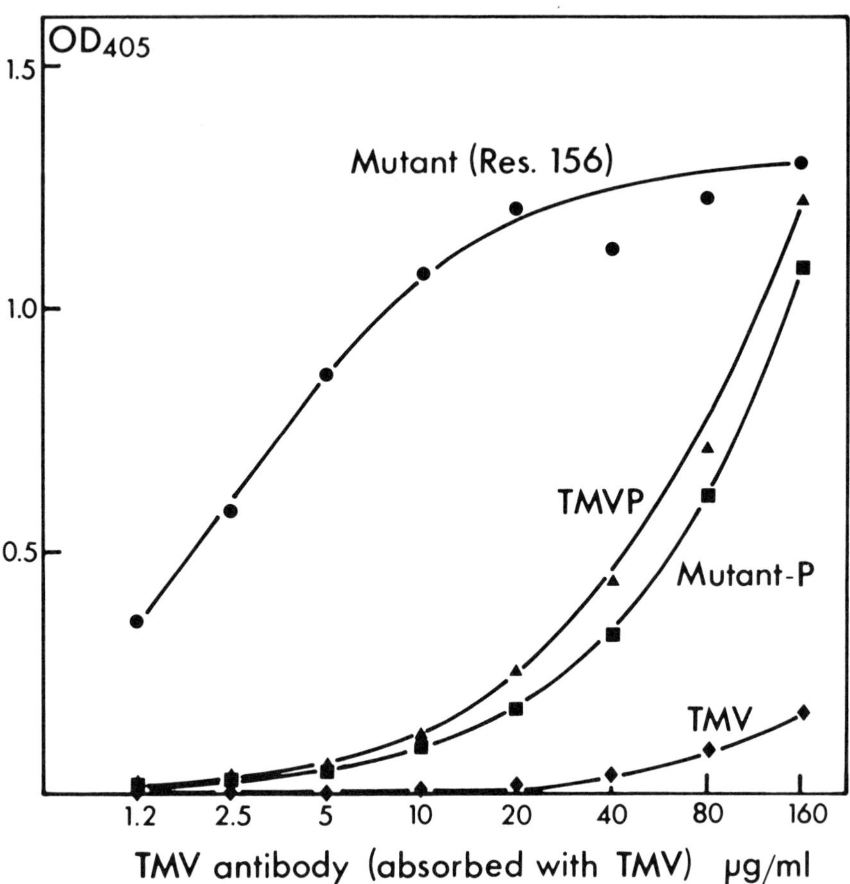

Figure 4--Demonstration of the presence of heterospecific
antibodies in TMV antiserum by means of ELISA. The TMV
antiserum was absorbed twice with the homologous TMV antigen and
centrifuged to remove virus and bound antibodies. Before
absorption, 1 µg/ml of TMV globulin gave rise to an OD$_{465}$ =
1.6 with TMV tested under the same conditions. After
absorption, 100 µg/ml of TMV globulin produced an OD$_{405}$ of
only 0.1 with TMV. The mutant possessing an exchange at
position 156 produced an OD$_{405}$ = 1.1 with 10 µg/ml TMV
globulin. The ELISA procedure used with the viral capsids
consisted of the steps AB[G], AG, AB[R] (absorbed), anti-R[G]-E
(see Table 5); in the case of the viral proteins, the procedure
was AG, AB[R] (absorbed) anti-R[G]-E.

Table 3--Amino acid substitutions of TMV mutants studied with monoclonal antibodies (131, 132)

	1	9	20	25	58	59	63	65	66	97	107	129	140	153	156	158
TMV	S	Q	P	N	V	T	P	S	D	E	T	I	N	T	P	T
06		H						R				M				
430			T													
OM												V		N		
FU 27					A											
Ni 458						I						T				
Ni 1045 -							S					T				
Ni 1688							S								L	
414								G								
Ni 116									G							
Ni 109										G						
Ni 630											M	T				
Ni 725											M	T				
A 14												T				
CP 415													K			
Ni 1927												T			L	

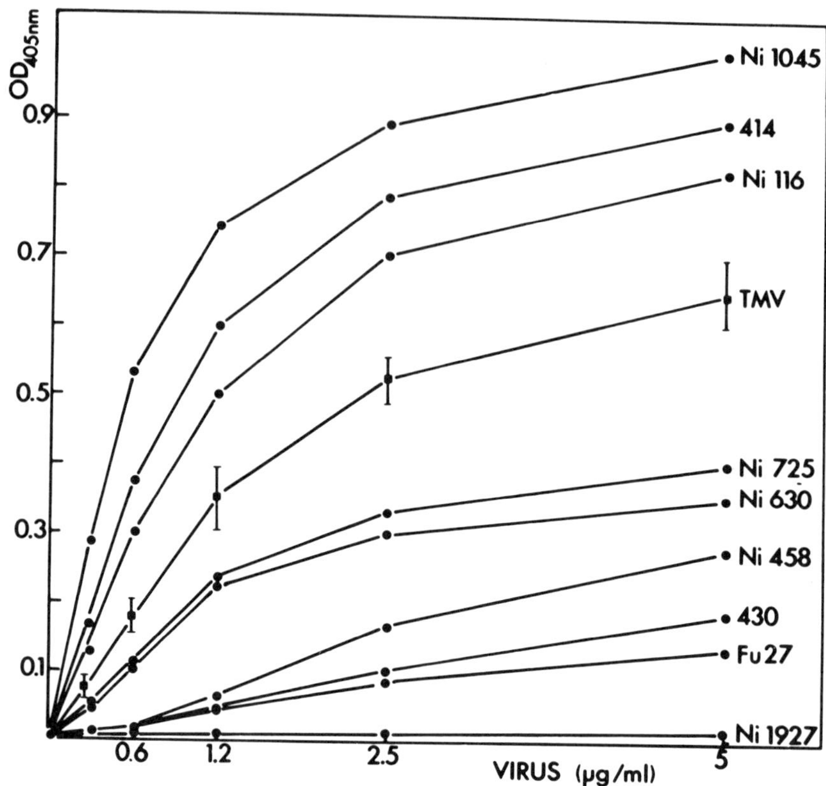

Figure 5--Comparison between nine mutants and wild-type TMV by indirect ELISA, using procedure 3 (see Table 5) and monoclonal antibody 19 diluted 1:2000. The enzyme conjugate was prepared with alkaline phosphatase (from Al Moudallal et al., 131).

(Figure 6). Compared to TMV, strain Y-TAMV reacted equally well with antibody 19 and not at all with antibody 20. The fact that strain Y-TAMV could not be differentiated from TMV by two monoclonal antibodies (Nos. 18 and 19), in spite of an 18% difference in sequence between the two viruses, is of considerable importance. Because these two antibodies obviously recognize an identical epitope in strains Y-TAMV and TMV, it shows that monoclonal antibodies do not always possess a greater discriminatory power than polyclonal antisera. Strain Y-TAMV is a member of the tomato mosaic group of strains and it is always clearly distinguishable from TMV by means of heterogeneous antisera (130, 136). In contrast, strain U2, which shows a 26% difference in sequence with TMV, was not recognized by any of the monoclonal antibodies.

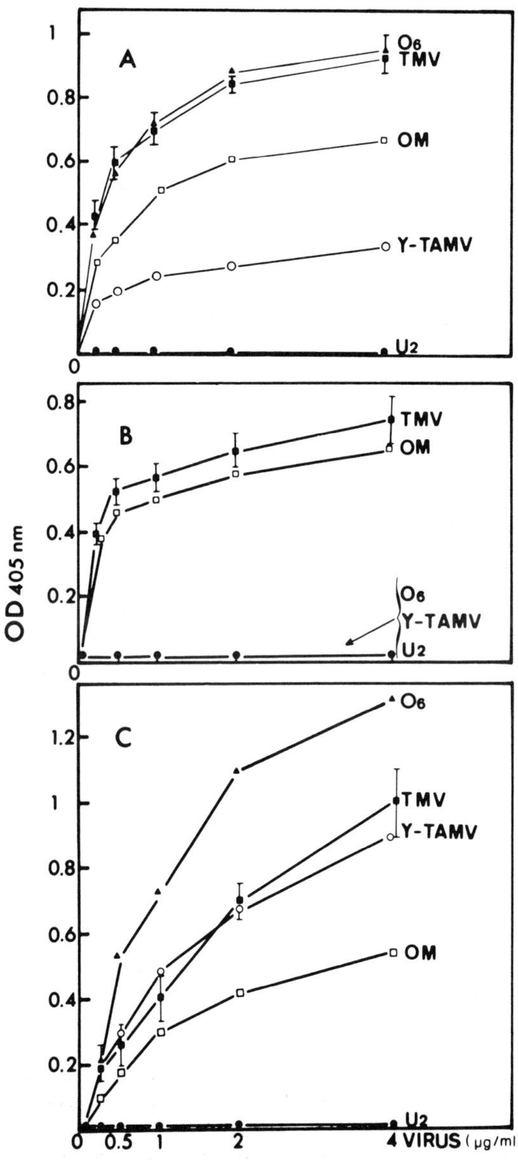

Figure 6--Detection of TMV mutants and strains by indirect ELISA using procedure 3 (Table 5) and monoclonal antibodies 22, 20, and 19 in A, B, and C respectively (from Briand et al., 132).

Table 4--Comparative binding of monoclonal antibodies to wild-type TMV and to mutants and strains of the virus

Substitution position	Antibody clone no.									No. of clones able to detect the exchanges
	3	8	18	19	20	21	22	25	28	
20	=	=	=	−	=	=	=	=	=	1
58	=	=	=	−	=	=	=	=	=	1
59	=	=	0	−	=	=	=	=	=	2
63	=	=	=	+	=	+	=	=	=	2
65	=	=	−	+	=	=	=	−	=	3
66	=	=	=	+	0	=	=	=	=	2
97	=	=	=	=	=	=	=	=	=	0
107	=	=	=	−	=	−	=	=	=	2
129	=	=	=	=	=	=	=	=	=	0
140	=	=	=	=	−	=	=	−	−	3
156	=	=	0	0	0	=	−	−	=	5
9 + 65	=	=	−	+	0	=	=	−	=	4
25 + 153	−	−	−	−	=	−	−	−	=	7
63 + 156	−	0	0	0	0	−	0	−	0	9
No. of single substitutions detected by each antibody	0	0	3	8	3	2	1	3	1	

+, Higher binding than with wild-type virus; =, equal binding as with wild-type virus; −, lower binding than with wild-type virus; 0, no binding at all.

The results of all the comparisons between TMV and 15 mutants are summarized in Table 4. Seven of the single substitutions that led to altered binding by monoclonal antibodies are situated near the outer surface of the viral capsid (i.e., residues 58, 59, 63, 65, 66, 140, and 156). Exchanges at positions 97 and 129, situated 3 to 5 nm from the surface, did not alter antibody binding. On the other hand, two substitutions that altered antibody binding (at positions 20 and

107) are located 2 and 5 nm away from the viral surface, respectively, and they must have influenced the reactivity of an epitope through long-range conformational changes. As far as strains 06 and OM are concerned, it seems likely that substitutions at positions 65 and 153 are responsible for their altered antigenic reactivity (132). It is of interest that although strains 06 and OM (as well as mutants having an exchange at position 156) were clearly distinct from TMV when compared by means of any polyclonal antiserum, not all monoclonal antibodies recognized the alterations at positions 65, 153, and 156.

Attempts were made to locate some of the TMV epitopes recognized by different monoclonal antibodies by means of a graphics modeling program using the atomic coordinates of TMVP (117). Substitutions that affected the binding of monoclonal antibodies were identified by color changes in the C^α-skeleton of the viral proteins by means of the Frodo graphics computer program designed by A. Jones. Stereo pairs of the backbone of TMV protein obtained by this program are shown in Figure 7.

Six antigenic regions of TMVP that were identified by inhibition studies with short peptides (39) are shown in Figure 7A. Substitutions that affect the accessible outer tip of the viral subunit and are recognized by monoclonal antibody 19 are shown in Figure 7B (residues 58, 59, 63, 65, 66, and 156). Because antibody 19 could not differentiate between Y-TAMV and TMV, additional residues that were altered in the exposed region of these two proteins were also identified in the same stereo pair (Figure 7C). These altered residues 6, 64, 153, and 156 are likely to be situated outside the epitope recognized by monoclonal antibody 19, and thus define approximate boundaries of the epitope. These considerations lead to the suggestion that antibody 19 recognizes a region approximating either to residues 1-5 or to residues 150-154, and that exchanges at positions 63, 65, 66, and 156 alter the antibody reactivity by an indirect conformational effect. Additional evidence for the location of the epitope in residues 150-154 was obtained by direct binding experiments between different monoclonal antibodies and various tryptic peptides of TMVP coupled to a carrier protein. The only combination of peptide and monoclonal antibody that showed some weak binding was the reaction between tryptic peptide 12 (residues 142-158) and monoclonal antibody 19 (143).

The complexity of the task of delineating the epitope recognized by antibody 19 is further illustrated by the fact that this antibody is heterospecific, i.e., it recognizes mutants with substitutions at positions 63, 65, and 66 better than wild-type virus (Table 4). Furthermore, in different ELISA tests, antibody 19 recognized the viral monomeric subunit only weakly or not at all. This last finding requires some further explanation, as it exemplifies the importance of selecting carefully the method used for measuring the activity of monoclonal antibodies.

In the first report that described the binding specificity of nine monoclonal antibodies raised against TMV (131), it was stated that none of them reacted with the monomeric subunit of the viral coat protein. This result was obtained by testing the monoclonal antibodies by an ELISA procedure in which the microtiter wells had been coated directly with the protein subunits. Because the positive binding of the same antibodies to the virus particles had been observed by a different ELISA in which the wells were first coated with anti-TMV globulin, it was decided to verify whether the type of ELISA used could alter the apparent binding specificity of monoclonal antibodies. Seven different ELISA procedures were compared to establish their suitability for detecting binding activity in monoclonal antibodies (137). The different ELISA procedures tested are presented schematically in Table 5.

When the binding of the nine monoclonal antibodies was tested by ELISA procedure 1, not a single antibody was found to react with either TMV or viral subunits (Figure 8). However, when the monoclonal antibodies were replaced by mouse antiserum (to TMV or to viral subunits), procedure 1 (Table 5) was found to be adequate for detecting antibody binding.

When procedure 2 was used, most of the monoclonal antibodies showed some binding to TMV, but the level of activity was low. When tested by procedure 3, all nine antibodies reacted strongly with TMV and four antibodies (Nos. 21, 22, 25 and 28) reacted with TMV subunits. The most sensitive ELISA tests were found to be procedures 4 and 7 which rely on the use of avian antibody to keep background readings low (Figures 9 and 10). The absence of cross-reaction between mammalian and avian antibodies is responsible for the fact that such multilayered sandwich procedures can be used in ELISA without producing high

---

Figure 7--Computer stereo pairs of the backbone of TMV protein based on the atomic coordinates of the molecule kindly provided by A.C. Bloomer and A. Klug. A--Six antigenic regions of the molecule identified by inhibition studies with short peptides are identified by color changes (residues 1-10, 34-39, 55-61, 62-68, 80-90, and 153-158). B--Substitutions affecting the accessible tip of the viral subunit and which are recognized by monoclonal antibody 19 are identified by color changes (residues 58, 59, 63, 65, 66, and 156). C--Substitutions between TMV and Y-TAMV affecting the tip of the viral subunit (residues 6, 64, 153, and 156) identified separately from the residue exchanges recognized by monoclonal antibody 19 (residues in red).
(see opposite page)

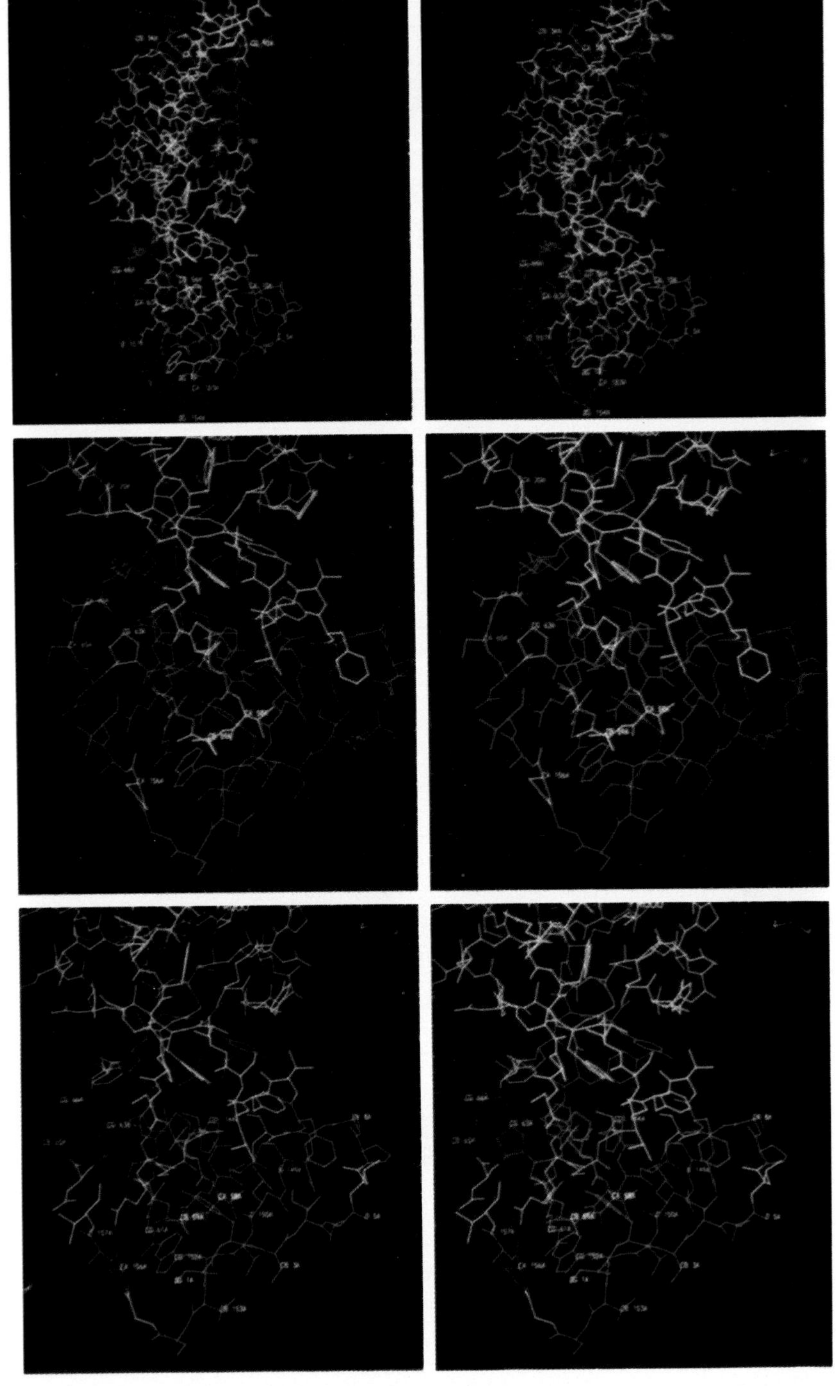

Table 5--ELISA procedures used for detecting binding
activity in monoclonal antibodies.

Procedure	Summary of successive steps of assay
1	AG, AB^M, anti M^R-E
2	AG, AB^M, anti M^R, anti R^G-E
3	AB^G, AG, AB^M, anti M^R-E
4	AB^C, AG, AB^M, anti M^R, anti R^G-E
5	AB^M, AG, AB^R, anti R^G-E
6	anti M^R, AB^M, AG, AB^R-E
7	anti M^S, AB^M, AG, AB^C, anti C^R-E

AG, antigen; AB, antibody; M, mouse; R, rabbit;
anti M^R, rabbit anti-mouse globulin; E, enzyme-label;
G, goat; C, chicken; S, sheep.

background readings. Procedures 3, 5, and 6 were less
sensitive than procedures 4 and 7 but were nevertheless
capable of detecting anti-TMVP activity in four of the nine
monoclonal antibodies. These results demonstrate the need to
use several different immunoassays in the screening of
monoclonal antibody activity.

EVOLUTION OF THE CONCEPT OF EPITOPE

     The classic studies of the antigenic structure of
myoglobin and lysozyme described by Atassi and collaborators
led to the view that the entire antigenic reactivity of
proteins could be described in terms of a few epitopes
situated in highly accessible locations at the surface of the
molecule. According to this view, protein surfaces possess
only a small number of immunodominant regions that account for
most of the antigenic reactivity of the molecule, and a large
portion of the surface is assumed to be devoid of apparent
antigenicity (see Figure 1). The epitopes of myoglobin and
lysozyme defined by Atassi (34, 78, 86) comprise five to eight
amino acid residues and constitute discrete patches on the
protein surface that retain their antigenic individuality even
when excised from the molecule. Although attempts have been
made to reconcile this definition of epitopes with the fact
that certain substitutions occurring outside antigenic
determinants are able to abolish antigenic reactivity, there
is no satisfactory explanation for this contradiction.
Conformational distortions arising from the excision of an
epitope from the molecule are likely to be at least as
important as those induced in native epitopes by distant

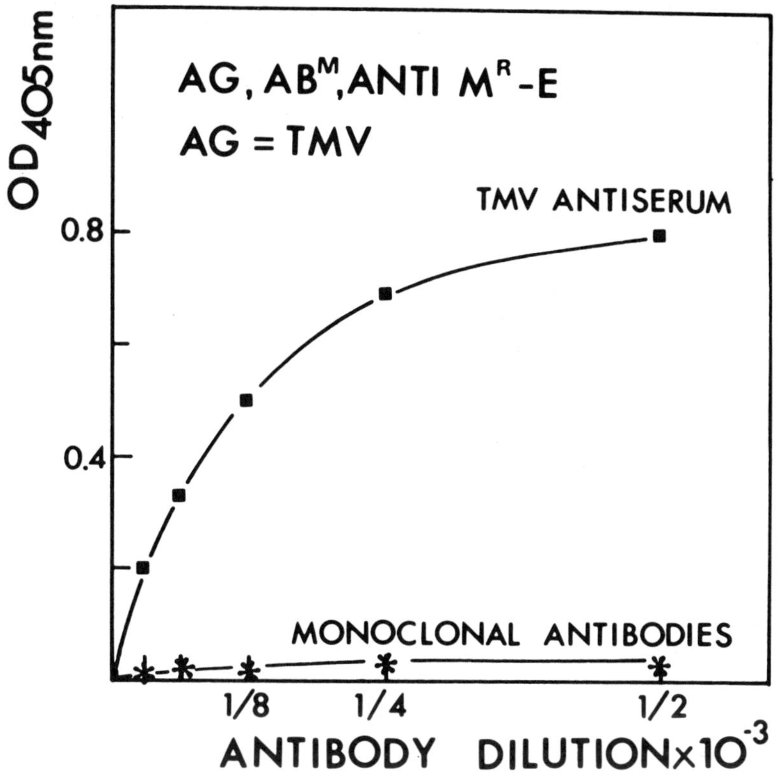

Figure 8--Detection of TMV by indirect ELISA using procedure 1
(see Table 5). Coating of wells was done with 4 µg/ml TMV for
3 hr at 37°C. Mouse antiserum to TMV and monoclonal antibodies
(in ascitic fluid) were allowed to interact for 2 hr at 37°C.
The enzyme conjugate (rabbit anti-mouse globulin diluted 1:1500)
was incubated for 18 hr at 4°C. Substrate hydrolysis time was 3
hr at 37°C. No reactivity could be detected in monoclonal
antibodies 3, 8, 18, 19, 20, 21, 22, 25, and 28 (from Al
Moudallal et al., 137).

substitutions; furthermore, an excised epitope is unlikely to
retain a memory of the conformational directives received from a
distal part of the intact molecule before the cleavage occurred.
     Studies of the immunological cross-reactivity between
related proteins carried out by means of both antisera and
monoclonal antibodies have given rise to a completely different
view of the nature of epitopes. Because about 80% of the
substitutions that have accumulated during the evolution of a
protein are detectable serologically (24), it has been inferred
that virtually the whole surface of the protein is capable of
combining with appropriate antibodies. Epitopes are then no

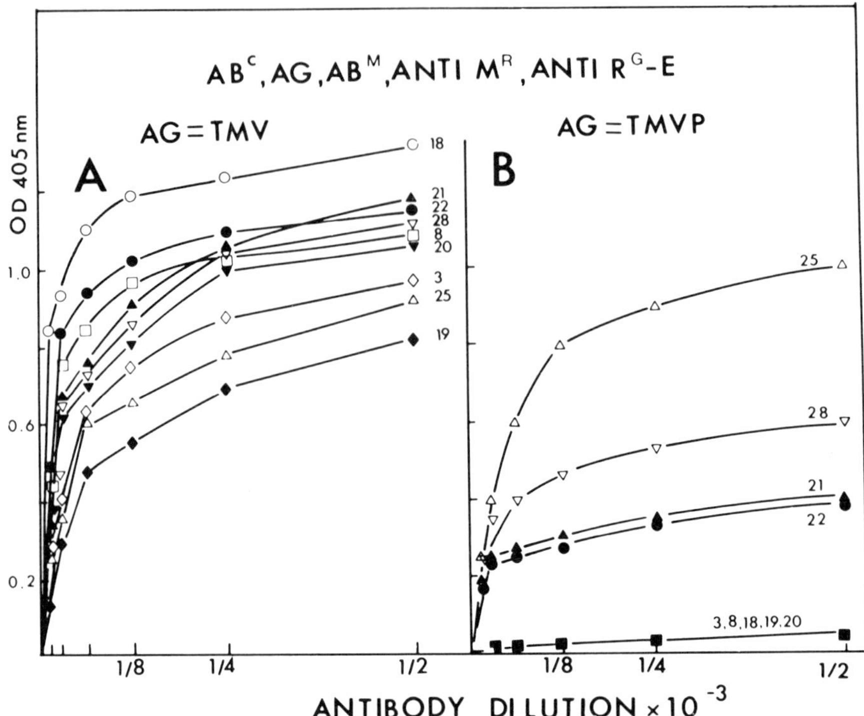

Figure 9--Detection of TMV (A) and TMV coat protein (B) by
indirect ELISA using procedure 4 (Table 5). Coating of wells
was done with 10 µg/ml chicken anti-TMV protein globulins for
1 1/4 h at 37°C. The antigen (2 µg/ml) as well as the
monoclonal antibodies, the rabbit anti-mouse globulins
(3 µg/ml) and the alkaline phosphatase conjugate (goat anti-
rabbit diluted 1:1500) were incubated for 2 h at 37°C.
Hydrolysis time was 30 min.

longer conceived of as particular regions of a protein
possessing antigenic reactivity but rather as surfaces
complementary to the binding sites of a series of immunoglobulin
molecules. The process of characterization of a protein antigen
is then transformed into an analysis of the specificity of a
series of antibody molecules. As is often the case in
experimental science, the analysis of the object under study
becomes in fact an analysis of the tool used in the study. In
this particular case, the antibody tool is also a mirror image
of the antigenic object under study. Instead of defining with
increasing precision a series of overlapping regions that make
up a particular antigen, the analysis of antigenic structure
becomes an enumeration of the specific binding sites, that are
present in the antibody repertoire of a particular animal,

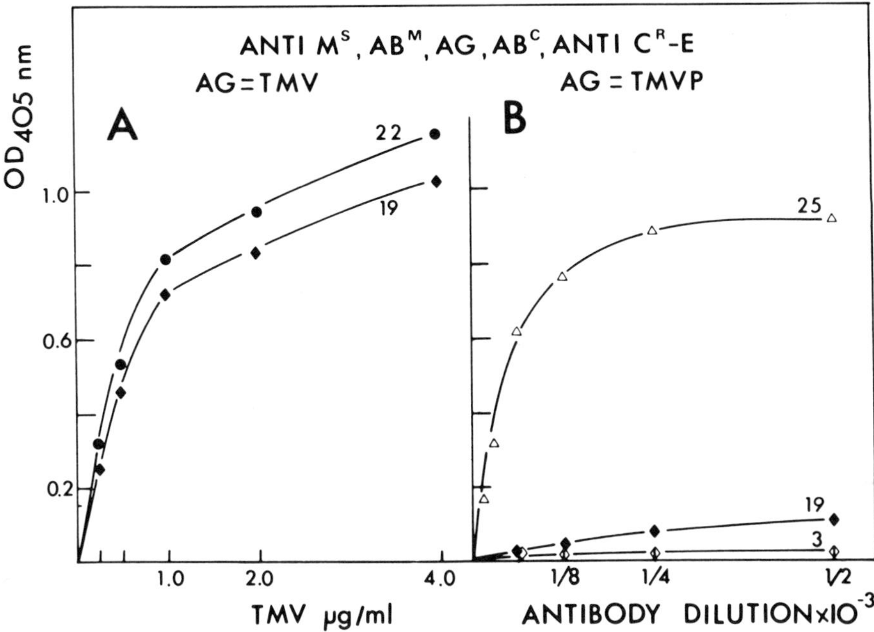

Figure 10--Detection of TMV (A) and TMV protein (B) by
indirect ELISA using procedure 7 (Table 5). Coating of wells was
done with sheep anti-mouse serum diluted 1:1000 for 1 1/2 h at
37°C. Monoclonal antibodies 22 (1:8000) and 9 (1:4000) were
incubated for 2 h at 37°C. TMV was incubated at 3 h at 37°C
followed by chicken anti-TMVP globulins (2 μg/ml) for 2 h at
37°C and rabbit anti-chicken alkaline phosphatase conjugate
(1:1000) for 2 h at 37°C. Hydrolysis time was 40 min.

endowed with a minimum threshold affinity for the antigen. It
is now well-established that the binding sites of antibodies are
polyfunctional (138-142) and that the specificity of polyclonal
antisera is a population phenomenon that masks the potential
cross-reactivity of each individual antibody (144). According
to this view, the heterospecific behavior of antibodies is in
fact the norm. Heterospecificity is only more conspicuous in
the case of monoclonal antibodies because the accompanying
antibodies normally found in an antiserum are not available to
dilute or mask the cross-reactivity. As a result, monoclonal
antibodies may actually be perceived as less specific for an
antigen than the average heterogeneous antiserum made up of a
population of antibodies sharing a common cross-reactivity for
the immunogen. Clearly, there is no reason to expect that the
clonal selection that triggers antibody production by a specific

B-cell clone must correspond necessarily to a situation where
the best possible fit between epitope and antibody binding site
occurs with the immunogen. Antigens with a better fit than the
immunogen usually are not looked for by the investigator and it
is no surprise, therefore, that they are rarely found. When
structural relatives of the immunogen, such as the TMV mutants
discussed in the previous section, are tested for their ability
to react with monoclonal antibodies, it is common to detect
heterospecific activities among them. The degree of fit between
antigen and antibody may be compared to the parallel situation
in human couples: a sufficient threshold affinity is required
for a reaction to occur, but a prolonged search may uncover
individuals endowed with a much better degree of fit. Calling
antibodies by the antigen's name is the immunochemist's
equivalent of modern marriage: there is no guarantee that
bearing the partner's name testifies of a unique physical
relationship.

ACKNOWLEDGMENTS

The author is grateful to Drs. A. Klug and A.C. Bloomer
(Cambridge) for providing the atomic coordinates of tobacco
mosaic virus protein.

REFERENCES

1. Poljak, R.J. (1978) In: Immunoglobulins. Litman, G.W.,
    and Good, R.A. (eds.), Plenum Press, New York, pp. 1-36.
2. Janin, J. (1976) J. Mol. Biol. 105:13-14.
3. Janin, J. (1979) Nature 277:491-492.
4. Atassi, M.Z. (1977) In: Immunochemistry of Proteins, Vol.
    2. Atassi, M.Z. (ed.) Plenum Press, New York, pp. 77-176.
5. Arnon, R. (1977) In: Immunochemistry of Enzymes and Their
    Antibodies. Salton, M.R.J. (ed.), John Wiley and Sons,
    New York, pp. 1-28.
6. Ibrahimi, I.M., Prager, E.M., White, T.J., and
    Wilson, A.C. (1979) Biochemistry 13:2736-2744.
7. East, I.J., Todd, P.E., and Leach, S.J. (1980) Molec.
    Immunol. 17:519-525.
8. Kabat, E.A. (1980) Meth. Enzymol. 70:3-49.
9. Van Regenmortel, M.H.V. (1982) Serology and
    Immunochemistry of Plant Viruses, Academic Press,
    New York, pp. 302.
10. Sela, M., Schechter, B., Schechter, I., and Borek, F.
    (1967) Cold Spring Harbor Symp. Quant. Biol. 32:537-545.
11. Sela, M. (1969) Science 166:1365-1374.
12. Berzofsky, J.A., Hicks, G., Fedorko, J., and Minna, J.
    (1980) J. Biol. Chem. 255:11188-11191.

13. Berzofsky, J.A., Buckenmeyer, G.K., Hicks, G., Gurd, F.R.N. Feldmann, R.J., and Minna, J. (1982) J. Biol. Chem. 257:3189-3198.

14. Jackson, D.C., Murray, J.M., Brown, L.E., and White, D.O. (1981) In: Genetic Variation among Influenza Viruses. Nayak, D.P., (ed.) ICN-UCLA Symposia on Molecular and Cellular Biology, Vol. 21, Academic Press, New York, pp. 355-372.

15. Atassi, M.Z., and Smith, J.A. (1978) Immunochemistry 15:609-610.

16. Celada, F., Ullmann, A., and Monod, J. (1974) Biochemistry 13:5543-5547.

17. Furie, B., Schechter, A.N., Sachs, D.H., and Anfinsen, C.B. (1975) J. Mol. Biol. 92:497-506.

18. Lazdunski, C.J., Pages, J.M., and Louvard, D. (1975) J. Mol. Biol. 97:309-335.

19. Teale, J.M., and Benjamin, D.C. (1976) J. Biol. Chem. 251:4603-4608.

20. Teale, J.M., and Benjamin, D.C. (1976) J. Biol. Chem. 251:4609-4615.

21. Hurrell, J.G.R., Smith, J.A., and Leach, S.J. (1977) Biochemistry 16:175-185.

22. Habeeb, A.F.S.A. (1977) In: Immunochemistry of Proteins, Vol. 1. Atassi, M.Z. (ed.), Plenum Press, New York, pp. 163-229.

23. Johnson, E.R., Anderson, W.L., Wetlaufer, D.B., Lee, C.L., and Atassi, M.Z. (1978) J. Biol. Chem. 253:3408-3414.

24. White, T.J., Ibrahimi, I.M., and Wilson, A.C. (1978) Nature 274:92-94.

25. Urbanski, G.J. and Margoliash, E. (1977) J. Immunol. 118:1170-1180.

26. Hurrell, J.G.R., Smith, J.A., Todd, P.E., and Leach, S.J. (1977) Immunochemistry 14:283-288.

27. Milton de L., R.C., Milton, S.C.F., Von Wechmar, M.B., and Van Regenmortel, M.H.V. (1980) Molec. Immunol. 17:1205-1212.

28. Jerne, N.K. (1960) Ann. Rev. Microbiol. 14:341-358.

29. Van Regenmortel, M.H.V. (1966) Adv. Virus Res. 12:207-271.

30. Neurath, A.R., and Rubin, B.A. (1971) Viral Structural Components as Immunogens of Prophylactic Value, S. Karger, Basel, 87 pp.

31. Sachs, D.H., Schechter, A.N., Eastlake, A., and Anfinsen, C.B. (1972) Proc. Natl. Acad. Sci. USA 69:3790-3794.

32. Crumpton, M.J. (1974) In: The Antigens, Vol. 2. Sela, M. (ed.), Academic Press, New York, pp. 1-78.

33. Goodman, J.W. (1975) In: The Antigens, Vol. 3, Sela, M., (ed.), Academic Press, New York, pp. 127-187.

34. Atassi, M.Z. (1975) Immunochemistry 12:423-438.

35. Celada, F., Fowler, A.V., and Zabin, I. (1978) Biochemistry 17:5156-5160.

36. Reichlin, M., and Noble, R.W. (1977) In: Immunochemistry
    of Proteins, Vol. 2. Atassi, M.Z. (ed.), Plenum Press,
    New York, pp. 311-351.

37. Jemmerson, R., and Margoliash, E. (1979) J. Biol. Chem.
    254:12706-12716.

38. Benjamini, E. (1977) In: Immunochemistry of Proteins,
    Vol. 2. Atassi, M.Z., (ed.), Plenum Press, New York, pp.
    265-310.

39. Altschuh, D., Hartmann, D., Reinbolt, J., and
    Van Regenmortel, M.H.V. (1983) Molec. Immunol. 20:271-278.

40. Quesniaux, V., Briand, J.P., and Van Regenmortel, M.H.V.
    (1983) Molec. Immunol. 20:179-185.

41. Stewart, J.M., and Young, J.D. (1969) Solid Phase Peptide
    Synthesis. W.H. Freeman, San Francisco, p. 103.

42. Smith, J.A., Hurrell, J.G.R., and Leach, S.J. (1977)
    Immunochemistry 14:565-568.

43. Hurrell, J.G.R., Smith, J.A., and Leach, S.J. (1978)
    Immunochemistry 15:297-302.

44. Leach, S.J. (1983) Biopolymers 22:425-440.

45. Hopp, T.P., and Woods, K.R. (1981) Proc. Natl. Acad. Sci.
    USA 78:3824-3828.

46. Hopp, T.P., and Woods, K.R. (1983) Molec. Immunol.
    20:483-489.

47. Pfaff, E., Mussagay, M., Bohm, H.O., Schulz, G.E., and
    Schaller, H. (1982) EMBO J. 1:869-874.

48. Sternberg, M.J.E., and Thornton, J.M. (1978) Nature
    271:15-20.

49. Kabasch, W., and Sander, C. (1983) FEBS Lett. 155:179-182.

50. Sutcliffe, J.G., Shinnick, T.M., Green, N., Liu, F.T.,
    Niman, H.L., and Lerner, R.A. (1980) Nature 287:801-805.

51. Lerner, R.A., Green, N., Alexander, H., Liu, F.T.,
    Sutcliffe, J.G., and Shinnick, T.M. (1981) Proc. Natl.
    Acad. Sci. USA 78:3403-3407.

52. Lerner, R.A., Sutcliffe, J.G., and Shinnick, T.M. (1981)
    Cell 23:309-310.

53. Walter, G., Scheidtmann, K.H., Carbone, A., Laudano, A.P.,
    and Doolittle, R.F. (1980) Proc. Natl. Acad. Sci. USA
    77:5197-5200.

54. Sutcliffe, J.G., Shinnick, T.M., Green, N., and Lerner,
    R.A. (1983) Science 219:660-666.

55. Anderer, F.A., (1963) Z. Naturforsch. 18b:1010-1014.

56. Anderer, F.A. (1963) Biochim. Biophys. Acta 71:246-248.

57. Anderer, F.A., and Schlumberger, H.D. (1965) Biochim.
    Biophys. Acta 97:503-509.

58. Absolom, D, and Van Regenmortel, M.H.V. (1977). FEBS
    Lett. 81:419-422.

59. Muller, S., Himmelspach, K., and Van Regenmortel, M.H.V.
    (1982) EMBO J. 1:421-425.

60. Bittle, J.L., Houghten, R.A., Alexander, H.,
    Shinnick, T.M., Sutcliffe, J.G., Lerner, R.A., Rowlands,
    D.J., and Brown, F. (1982) Nature 298:30-33.

61. Atassi, M.Z., Lee, C.L., and Pai, R.C. (1976) Biochim. Biophys. Acta 427:745-751.
62. Lee, C.L., and Atassi, M.Z. (1976) Biochem. J. 159:89-93.
63. Green, N., Alexander, H., Olson, A., Alexander, S., Shinnick, T.M., Sutcliffe, J.G., and Lerner, R.A. (1982) Cell 28:477-487.
64. Wiley, D.C., Wilson, I.A., and Skehel, J.J. (1981) Nature 289:373-378.
65. Lerner, R.A. (1982) Nature 299:592-596.
66. Niman, H.L., Houghten, R.A., Walker, L.E., Reisfeld, R.A., Wilson, I.A., Hogle, J.M., and Lerner, R.A. (1983) Proc. Natl. Acad. Sci. USA 80:4949-4953.
67. Jackson, D.C., Murray, J.M., White, D.O., Fagan, C.N., and Tregear, G.W. (1982) Virology 120:273-276.
68. Young, C.R., Schmitz, H.E., and Atassi, M.Z. (1983) Molec. Immunol. 20:567-570.
69. Atassi, M.Z., and Webster, R.G. (1983) Proc. Natl. Acad. Sci. USA 80:840-844.
70. Schmitz, H.E., Atassi, H., and Atassi, M.Z. (1983) Immunol. 12:161-175.
71. Nisonoff, A., Reichlin, M., and Margoliash, E. (1970) J. Biol. Chem. 245:940-946.
72. Reichlin, M. (1972) J. Mol. Biol. 64:485-496.
73. Reichlin, M. (1975) Advan. Immunol. 20:71-123.
74. Sengbusch, P. (1965) Z. Vererbungsl. 96:364-386.
75. Van Regenmortel, M.H.V. (1967) Virology 31:467-480.
76. Champion, A.B., Soderberg, K.L., Wilson A.C., and Ambler, R.P. (1975) J. Mol. Evol. 5:291-305.
77. Prager, E.M., Welling, G.W., and Wilson, A.C. (1978) J. Mol. Evol. 10:293-307.
78. Atassi, M.Z. (1979) CRC Crit. Rev. Biochem. 337-369.
79. Atassi, M.Z., and Koketsu, J. (1975) Immunochemistry 12:741-744.
80. Todd, P.E.E., East, I.J., and Leach, S.J. (1982) Trends Biochem. Sci. 7:212-216.
81. Schmitz, H.E., Atassi, H., and Atassi, M.Z. (1982) Molec. Immunol. 19:1699-1702.
82. East, I.J., Hurrell, J.G.R., Todd, P.E.E., and Leach, S.J. (1982) J. Biol. Chem. 257:3199-3202.
83. Young, J.D. and Leung, C.Y. (1970) Biochemistry 9:2755-2762.
84. Atassi, M.Z., Habeeb, A.F.S.A., and Ando, K. (1973) Biochim. Biophys. Acta 303:203-209.
85. Atassi, M.Z., and Habeeb, A.F.S.A. (1977) In: Immunochemistry of Proteins, Vol. 2. Atassi, M.Z. (ed.), Plenum Press, New York, pp. 177-264.
86. Atassi, M.Z. (1979) CRC Crit. Rev. Biochem. 371-400.
87. Fujio, H., Imanishi, M., Nishioka, K., and Amano, T. (1968) Biken's J. 11:207-218.
88. Ha, Y.M., Fujio, H., Sakato, N., and Amano, T. (1975) Biken's J. 18:47-60.

89. Arnon, R., and Sela, M. (1969) Proc. Natl. Acad. Sci. USA
     62:163-170.
90. Arnon, R., Maron, E., Sela, M., and Anfinsen, C.B. (1971)
     Proc. Natl. Acad. Sci. USA 68:1450-1455.
91. Teicher, E., Maron, E., and Arnon, R. (1973)
     Immunochemistry 10:265-271.
92. Takagaki, Y., Hirayama, A., Fujio, H., and Amano, T.
     (1980) Biochemistry 19:2498-2505.
93. Atassi, M.Z., and Lee, C.L. (1978) Biochem. J. 171:429-434.
94. Metzger, D.W., Miller, A., and Sercarz, E.E. (1980) Nature
     287:540-542.
95. Smith-Gill, S.J., Wilson, A.C., Potter, M., Prager, E.M.,
     Feldmann, R.J., and Mainhart, C.R. (1982) J. Immunol.
     128:314-322.
96. Kobayashi, T., Fujio, H., Kondo, K., Dohi, Y., Hirayama,
     A., Takagaki, Y., Kosaki, G., and Amano, T. (1982) Molec.
     Immunol. 19:619-630.
97. Ibrahimi, I.M., Eder, J., Prager, E.M., Wilson, A.C., and
     Arnon, R. (1980) Molec. Immunol. 17:37-46.
98. Wilson, I.A., Skehel, J.J., and Wiley, D.C. (1981) Nature
     289:366-373.
99. Varghese, J.N., Laver, W.G., and Colman, P.M. (1983)
     Nature 303:35-40.
100. Laver, W.G., Downie, J.C., and Webster, R.G. (1974)
     Virology 59:230-244.
101. Koprowski, H., Gerhard, W., and Croce, C.M. (1977) Proc.
     Natl. Acad. Sci. USA 74:2985-2988.
102. Yewdell, J.W., and Gerhard, W. (1981) Ann. Rev.
     Microbiol. 35:185-206.
103. Gerhard, W., and Webster, R.G. (1978) J. Exp. Med.
     148:383-392.
104. Webster, R.G., and Laver, W.G. (1980) Virology
     104:139-148.
105. Underwood, P.A. (1982) J. Gen. Virol. 62:153-169.
106. Gerhard, W., Yewdell, J., Frankel, M.E., and Webster, R.
     (1981) Nature 290:713-717.
107. Caton, A.J., Brownlee, G.G., Yewdell, J.W., and Gerhard,
     W. (1982) Cell 31:417-427.
108. Winter, G., Fields, S., and Brownlee, G.G. (1981) Nature
     292:72-75.
109. Krystal, M., Elliott, R.M., Benz, E.W., Jr., Young,
     J.F., and Palese, P. (1982) Proc. Natl. Sci. USA
     79:4800-4804.
110. Breschkin, A.M., Ahern, J., and White, D.O. (1981)
     Virology 113:130-140.
111. Stone, M.R., and Nowinski, R.C. (1980) Virology
     100:370-381.
112. Van Regenmortel, M.H.V., and Hardie, G. (1976)
     Immunochemistry 13:503-507.

113. Friguet, B., Djavadi-Ohaniance, L., Pages, J.,
Bussard, A., and Goldberg, M. (1983) J. Immunol. Meth.
60:351-358.

114. Lubeck, M., and Gerhard, W. (1982) Virology 118:1-7.

115. Colman, P.M., Varghese, J.N., and Laver, W.G. (1983)
Nature 303:41-44.

116. Champness, J.N., Bloomer, A.C., Bricogne, G.,
Butler, P.J.G., and Klug, A. (1976) Nature 259:20-24.

117. Bloomer, A.C., Champness, J.N., Bricogne, G.,
Staden, R., and Klug, A. (1978) Nature 276:362-368.

118. Stubbs, G., Warren, S., and Holmes, K. (1977) Nature
267:216-221.

119. Anderer, F.A., and Ströbel, G. (1972) Eur. J. Immunol.
2:274-277.

120. Anderer, F.A., and Ströbel, G. (1972) Eur. J. Immunol.
2:278-282.

121. Benjamini, E., Young, J.D., Shimizu, M., and Leung, C.Y.
(1964) Biochemistry 3:1115-1120.

122. Young, J.D., Benjamini, E., Stewart, J.M., and Leung,
C.Y. (1967) Biochemistry 6:1455-1460.

123. Young, J.D., Benjamini, E., and Leung, C.Y. (1968)
Biochemistry 7:3113-3119.

124. Benjamini, E., Shimizu, M., Young, J.D., and Leung, C.Y.
(1968) Biochemistry 7:1261-1264.

125. Milton de L., R.C., and Van Regenmortel, M.H.V. (1979)
Molec. Immunol. 16:179-184.

126. Altschuh, D., and Van Regenmortel, M.H.V. (1982) J.
Immunol. Methods 50:99-108.

127. Sengbusch, P., and Wittmann, H.G. (1965) Biochem.
Biophys. Res. Commun. 18:780-787.

128. Loor, F. (1971) Immunology 21:557-564.

129. Hennig, B., and Wittmann, H.G. (1972) In: Principles
and Techniques in Plant Virology. Kado, C.J., and
Agrawal, H.O. (eds.), Van Nostrand-Reinhold, Princeton,
pp. 546-594.

130. Van Regenmortel, M.H.V. (1981) In: Handbook of Plant
Viral Infections and Comparative Diagnosis. Kwistok, E.
(ed.), Elsevier-North Holland, Amsterdam, pp. 541-564.

131. Al Moudallal, Z., Briand, J.P., and Van Regenmortel,
M.H.V. (1982) EMBO J. 1:1005-1010.

132. Briand, J.P., Al Moudallal, Z., and Van Regenmortel,
M.H.V. (1982) J. Virol. Meth. 5:293-300.

133. Wittmann-Liebold, B., and Wittmann, H.G. (1965)
Z. Vererbungsl. 97:305-326.

134. Nozu, Y., Ohno, T., and Okaka, Y. (1970) J. Biochem.
68:39-52.

135. Gallwitz, U. (1973) FEBS Lett. 34:270-272.

136. Wang, A.L., and Knight, C.A. (1967) Virology 31:101-106.

137. Al Moudallal, Z., Altschuh, D., Briand, J.P., and
Van Regenmortel, M.H.V. (1984) J. Immunol. Meth. In
press.

138. Talmage, D.W. (1959) Science 129:1643-1648.
139. Richards, F.F., and Konigsberg, W.H. (1973)
      Immunochemistry 10:545:553.
140. Richards, F.F., Konigsberg, W.H., Rosenstein, R.W., and
      Varga, J.M. (1975) Science 187:130-137.
141. Cameron, D.J., and Erlanger, B.F. (1977) Nature
      268:763-765.
142. Berzofsky, J.A. and Schechter, A.N. (1981) Molec.
      Immunol. 18:751-763.
143. Altschuh, D. (1984) Unpublished results.
144. Lane, D., and Koprowski, H. (1982) Nature 296:200-202.

CHAPTER 4

PROTECTION AGAINST INFECTION BY HERPES SIMPLEX
VIRUS BY USE OF MONOCLONAL ANTIBODIES

ERNEST D. MARQUEZ
College of Medicine
The Pennsylvania State University
Hershey, PA   17033

ABSTRACT

     Herpes simplex viruses types 1 and 2 (HSV-1 and HSV-2)
cause recurrent infections of the mouth, gums, and genital
tract. In addition, these viruses possess oncogenic potential,
and women who suffer from genital infections appear to have a
greater incidence of cervical cancer. Because virus-specific
glycoproteins are involved in immune processes, monoclonal
antibodies were prepared against some of the most common
glycoproteins, termed gA/B, gC, gD, and gE. Although all
monoclonal antibodies did not neutralize HSV-1 or HSV-2 in
vitro, they did protect mice against infection when administered
in a passive immunization procedure. In addition, although
single antibodies offered some protection, combinations of
antibodies offered more complete protection against HSV
infection. Monoclonal antibodies were also used in affinity
chromatography procedures to purify virus antigens that could be
used to actively immunize mice against HSV-2 infection. Lastly,
preliminary efforts to use anti-idiotypic antibodies prepared
against virus-specific monoclonal antibodies as an immunogen are
described.

INTRODUCTION

     Herpes simplex viruses types 1 and 2 (HSV-1 and HSV-2)
cause a variety of clinical conditions that include common
infections of the skin and mucous membranes of the mouth, gums,
and genital tract (1). The incidence of genital infections in
the United States has reached epidemic proportions and is
further complicated by the fact that the infections are
recurrent (2). In addition, HSV-2 and HSV-1 possess oncogenic

potential (3, 4), and women who suffer from genital infections seem to have a greater incidence of cervical cancer (5, 6).

Infection of mammalian cells with HSV-1 results in the appearance of new antigens at the cell surface (7, 8). These antigens induce specific antibodies that can be detected by indirect immunofluorescence (9), antibody-dependent complement-mediated cytolysis (10), and antibody-dependent cytotoxicity (11-13). Virus-specific antigens have been implicated in the destruction of HSV-infected cells by thymus-derived (T-cell) cytotoxic lymphocytes (14, 15), and those antigens involved in immune cytolysis appear to be glycoproteins. Infection of cells by HSV results in incorporation of viral glycoproteins into the infected cell membranes (8). Of the five major glycoproteins (gA, gB, gC, gD, and gE) identified (16, 17), gC, gD, or a mixture of gA and gB have been implicated in antibody-dependent cellular cytotoxicity (18). In this department, Carter et al. (19) have shown that HSV-glycoproteins are involved in T-cell-mediated lysis of HSV-infected cells.

Various investigators have shown that HSV-1 infection produces at least three to four glycoproteins (gA/B, gD, gE) that are biochemically and immunologically related to the glycoproteins specified by HSV-2 (20-23). However, gC of HSV-1 and HSV-2 appears to be unique to each HSV type. HSV-1 gC has been termed type-specific based on observations that antibodies to HSV-1 gC are predominantly type-specific (24-28), heterologous antisera do not react with HSV-1 gC (20), and antisera against gC fail to react with the corresponding HSV-2 glycoprotein (29-31). Although the HSV-2 glycoprotein has been termed "gC" because its electrophoretic migration properties are similar to those of HSV-1 gC (20, 23, 32), the type-specific nature of HSV-2 gC has been substantiated by observations that the biochemical and immunological characteristics of HSV-1 and HSV-2 glycoproteins are quite distinct (20, 32) and that the genes specifying these glycoproteins are not colinear (23).

However, recent evidence indicates that even gC of HSV-1 may not be type-specific because 1) hyperimmune antiserum to HSV-2 can immunoprecipitate gC of HSV-1-infected cells (33) and 2) gC of HSV-1 and gF of HSV-2 share antigenic determinants (34). In spite of these cross-reactions, however, it is apparent that HSV glycoproteins are involved in immune reactions and are the most likely viral products to be used in vaccine production.

PRODUCTION OF MONOCLONAL ANTIBODIES AGAINST HSV GLYCOPROTEINS

Immunization of Animals

BALB/c mouse fibroblasts were infected with HSV-2 (strain 333) at a multiplicity of infection (MOI) of 10 plaque-forming

units (PFU)/cell. When 75 to 85% of the cells displayed
cytopathology, they were fixed in 2% paraformaldehyde for 5 min
at 0°C. After being washed three to five times in
phosphate-buffered saline (PBS), pH 7.4, the cells were
suspended in PBS to a final cell concentration of approximately
10^8 cells/ml. BALB/c mice received 10^7 cells mixed with an
equal volume of Freund's complete adjuvant (0.2 ml total)
intraperitoneally. The animals received a subsequent booster
intraperitoneally consisting of 10^7 cells in Freund's
incomplete adjuvant. After 2 weeks, an intravenous booster
consisting of 10^6 cells in PBS was administered.

Production of Hybridomas

The procedures for production of hybridomas were those
described by Steplewski (personal communication) and reported by
Koprowski, Steplewski, and coworkers (35, 36). Briefly, immune
splenocytes were prepared by mincing the spleen and pressing the
cells through a wire mesh. The spleen cells were washed, the
red blood cells were removed by lysing in 0.85% NH_4Cl, and
both spleen cells and NS-1 myeloma cells were washed three times
in RPMI medium containing 0.01% EDTA. The cells were combined
at a ratio of 10:1 (splenocytes/myeloma cells). For fusion, the
cell pellet was resuspended in 50% polyethylene glycol (Baker
1000) in RPMI medium that was diluted first with RPMI and EDTA
and then with RPMI and serum according to established procedures
(37). The cells were seeded in 96-well plates in the presence
of hypoxanthine, aminopterin, and thymidine (HAT) medium (38),
and incubated at 37°C in an atmosphere of 95% air and 5% added
CO_2. After 7 to 10 days, half to a third of the medium was
replaced with medium containing only hypoxanthine and
thymidine. The supernatants from each of the 96 wells in a
plate were assayed for the presence of anti-HSV-2 MCA.

Radioimmunoassay (RIA) Screening of Monoclonal Antibodies Against HSV-2 Glycoproteins

Glycoproteins were isolated from HSV-2-infected cells using
a modification of procedures described previously (39).
Briefly, infected and noninfected mouse fibroblasts were
extracted with 10 mM carbonate/bicarbonate, pH 9.6, containing
0.06 M β-D-octylglucoside. After centrifugation to remove
cell debris, the glycoproteins were freed from the detergent by
passage through a column of BioRad SM-2 beads. The isolated
glycoproteins (20 to 40 µg) were attached to individual RIA
assay wells (Removawells, Dynatech, Inc.) in the presence of
carbonate/bicarbonate buffer as described (40). Supernatants
from 96-well plates were added to the antigen-containing wells
and incubated at room temperature for 2 hr. After being washed
three times with PBS containing 0.05% Tween 20, ^{125}I-labeled
anti-mouse IgG (IgG fraction, Cappel Laboratories) was added.

After incubation at room temperature for 1 hr, the wells were washed five times with PBS and 0.05% Tween 20, removed individually and assayed for radioactivity with a gamma counter. Cells that showed a positive reaction were subcloned by limiting dilution. Supernatants from the isolated clones were rescreened by RIA and further screened by virus neutralization tests.

Virus Neutralization Assays

Culture fluids from viable clones that were RIA positive were tested in plaque reduction assays for neutralizing activity against HSV-2 as described previously (28). In this test, 100 µl of undiluted supernatant were mixed with identical virus aliquots containing 50 to 100 PFU, plus or minus complement, and incubated at 37°C for 30 min. The mixture was adsorbed to Vero (African green monkey kidney) cell monolayers in 60-mm Petri dishes and grown for 3 days with an overlay of Dulbecco's minimum essential medium (MEM) containing 0.5% methylcellulose. Cultures that produced sufficient antibody to neutralize 50% of the plaques were recloned, retested, and expanded.

Indirect Immunofluorescence Tests

Vero cells were grown in monolayers on coverslips and infected with HSV-1 or HSV-2 (333) at an MOI of approximately 10 PFU/cell. Antibody-containing supernatants were tested against unfixed cells to detect cell surface antigens. Uninfected cells and supernatant from NS-1 cells were used as negative controls. The infected cells were first reacted with normal goat serum to eliminate nonspecific adsorption of IgG to Fc receptors.

Determination of Polypeptide Specificity of Monoclonal Antibodies

The polypeptide specificity of the HSV-2-neutralizing antibodies was determined using procedures described previously (41). Mouse fibroblasts were infected with HSV-2 (MOI = 20) and radiolabeled with [^{35}S]methionine 4 hr after infection. After 12 to 16 hr, the cells were washed twice with PBS and the radiolabeled antigen was extracted with Tris-HCl buffer (pH 7.4) containing 0.15 M NaCl, 1% Nonidet P40, 1 mM phenylmethylsulfonylfluoride (PMSF), and 1 mM L-methionine (EB). After centrifugation to remove insoluble cell debris, the supernatant was assayed for radioactivity and used in immunoprecipitation experiments. Staphylococcal Protein A (available as formalin-fixed staphylococci, Cowan-strain, SA) was precleaned by heating to 90°C for 1 hr in the presence of 2% sodium dodecyl sulfate (SDS) and 5% β-mercaptoethanol and washed several times with EB.

Monoclonal antibody (50 µl) to be tested was allowed to react with 3 to 6 x 10^5 cpm of antigen extract at room temperature for 2 hr. Because some monoclonal antibodies may not bind Protein A, 20 µl of anti-mouse IgG (Cappel Laboratories, 1:100 dilution) was also added and allowed to react for an additional 45 min with constant mixing. Protein A was added (200 µl of a 10% suspension) and allowed to react for 30 min. After 30 min, the mixture was washed five times with EB and the pellet was resuspended in 60 µl of solubilizing buffer (0.125 M Tris buffer, pH 6.8, containing 2% SDS, 5% β-mercaptoethanol, 20% glycerol, and .004% bromphenol blue). The mixture was heated at 100°C for 2 min and separated by polyacrylamide gel electrophoresis in the presence of SDS (SDS-PAGE). Cross-linkage of the gels was accomplished with N-N'-diallyltartardiamide (DATD) instead of bis-acrylamide since resolution of glycoproteins appears to be more efficient (28). The separated components were visualized by autoradiography using X-ray film (Kodak, XAR-5). The precipitated polypeptides were identified based on previously published data (42, 43) (Table 1).

Production of Ascites

BALB/c mice were primed by an intraperitoneal injection of 0.5 ml of pristane (2,6,10,14-tetramethyl-pentadecane) 2 weeks before injection of about 5 x 10^6 cells as described

Table 1--Identification of anti-HSV monoclonal antibodies

MCA designation	Approximate molecular weight of immunoprecipitated polypeptides	Polypeptide designation[*]
7D6 (αgD)	56,000	gD
8C6 (αgC)	125,600	gC
7A6 (αgA/B)	110,400 119,500	gA/B
8A1 (α-gE)	60,000	gE

[*]These polypeptides are assumed to be glycoproteins based on their reactivity by immunofluorescence with unfixed HSV-2-infected cells, coprecipitation with known monoclonal antibody preparations, and neutralization of HSV-2 in vitro. The molecular weights appear to be somewhat smaller than those reported previously (42, 43) and may reflect the greater resolution of glycoproteins using DATD as suggested (28).

previously (42). When tumors developed, the ascites fluids were harvested by intraperitoneal puncture and drainage every other day until the animals died. The fluids were immediately clarified by centrifugation and stored at 5°C.

Determination of Immunoglobulin (Ig) Class

The supernatants from cells grown in culture were collected, and the antibodies were precipitated in the presence of 50% ammonium sulfate. The precipitates were resuspended in PBS to achieve a tenfold concentration and dialyzed against buffer to remove the remaining ammonium sulfate. The Ig class and subclass were determined by double diffusion using 1% agarose gel plates and rabbit antisera to specific mouse Ig (Miles Laboratories). Test samples (19 µl) and typing reagents were added to adjacent wells and allowed to diffuse for 18 to 24 hr in a humidified chamber at room temperature.

Determination of Anti-HSV Titer in Ascites Fluid

The titers of anti-HSV-2 antibodies present in ascites fluids were determined by use of methanol-fixed (-20°C for 5 min) HSV-2-infected cells on coverslips. To determine the degree of cross-reactivity with HSV-1, coverslips containing HSV-1-infected cells were also used. All coverslips were pretreated with normal rabbit serum (1:40 dilution) for 30 min at room temperature. After washing, the cells were subsequently treated with various dilutions of ascites fluid and the presence of bound monoclonal antibody was detected by fluorescein-labeled anti-mouse IgG (1:60, Cappel Laboratories). The highest dilution producing definite fluorescence in infected cells was taken as the antibody titer (Table 2).

Table 2--Characterization of anti-HSV monoclonal antibodies

Antibody HSV-2	Antibody class	Immunofluorescence titers		Neutralization titers[*]	
		HSV-1	HSV-2	HSV-1	HSV-2
Anti-gA/B	IgG1	3125	3125	< 5	< 5
Anti-gC	IgG1	< 5	3125	< 5	< 5
Anti-gD	IgG1	625	3125	< 5	< 5
Anti-gE	IgG1	< 5	3125	< 5	< 5

[*]All neutralization titers were <5 in the presence or absence of complement.

In vitro <u>Neutralization of HSV-1 and HSV-2</u>

To determine the neutralization titers of the various
antibodies in ascites fluids, neutralization tests were carried
out with doubling dilutions of ascites fluids. All ascites
fluids were heat-inactivated by incubation at 56°C for 30 min.
Experiments were then carried out in the presence or absence of
rabbit complement. As before, 100 µl of each ascites
dilution, plus or minus 50 µl of complement, were mixed with
identical aliquots of virus (approximately 200 PFU) in a total
volume of 200 µl and incubated at 37°C for 45 min. The
mixture was then adsorbed to rabbit kidney cells in 60-mm Petri
dishes for 1 hr and overlaid with Dulbecco's MEM containing 0.5%
methylcellulose. The titer was the last dilution at which the
number of plaques was reduced by 50% or more (Table 2).

As can be seen, all antibodies were of the IgG1 class. The
immunofluorescence titers indicated that α-gA/B and α-gD
were type-common in that both HSV-1- and HSV-2-infected cells
were reactive at higher dilutions of MCA. α-gC and α-gE
were type-specific in that only HSV-2-infected cells were
reactive at higher dilutions. All monoclonal antibodies had
neutralization titers of less than 5, which showed a poor
neutralizing ability either with or without complement.

Further experiments were carried out to determine whether
combinations of antibodies could neutralize HSV-2 more
effectively than individual monoclonal antibodies (Table 3).
The <u>in vitro</u> neutralization assays were carried out as described
above except that tripling dilutions of antibodies were used.
In addition, all aliquots of antibodies totaled 100 µl,
whether individually or in combination. The results of these
experiments indicated that a combination of monoclonal
antibodies increased the neutralization potential of the
antibodies <u>in vitro</u>.

Table 3--<u>In vitro</u> neutralization of HSV-2 by monoclonal
antibodies at various dilutions

Monoclonal antibody	Percent plaque reduction			
	1:3	1:9	1:27	1:81
α-gA/B	58	42	10	5
α-gC	60	48	7	10
α-gD	70	50	20	12
α-gC + α-gD	100	100	90	70
α-gA/B + α-gC + α-gD	100	100	98	80

PASSIVE IMMUNIZATION AGAINST HSV INFECTION WITH MONOCLONAL ANTIBODIES

Passive immunization with antiviral antibody prevents ascending neurological illness and death following subcutaneous infection of mice with HSV-1 (44-48). Dix and coworkers (49) used monoclonal antibodies prepared against glycoproteins of HSV-1 to protect mice against acute virus-induced neurological disease after footpad inoculation with either HSV-1 or HSV-2. One exception was that antibody specific for HSV-1 gC failed to protect against HSV-2. Balachandran and coworkers (50) have prepared monoclonal antibodies against gA/B, gC, gD, gE, and the newly described gF of HSV-2. Although these monoclonal antibodies did not neutralize virus _in vitro_, antibodies to individual glycoproteins could protect mice from lethal infection by HSV-2. The results indicated that any of the glycoproteins could serve as antigens for a protective immune response.

We have also carried out similar experiments with monoclonal antibodies prepared against gA/B, gC, gD, and gE of HSV-2. Our experiments applied techniques previously outlined by Dix et al. (49). HSV-2 (333) (about 5 x 10^6 PFU in 50 µl) was inoculated into the hind footpads of BALB/c mice. Twenty-four hours later the mice received 0.2 ml of antibody as undiluted ascites fluid. Combinations of monoclonal antibodies were also given (0.05 ml of each monoclonal antibody in combination). Control mice received either hyperimmune anti-HSV antisera or control ascites fluid. The latter was obtained from mice inoculated with the parent myeloma cell (NS-1) to produce ascites. Experimental and control mice were coded and examined for evidence of neurological disease as evidenced by paraplegia and death. The animals were observed for a 3-week period (Table 4).

These results are in agreement with those recently published by Balachandran et al. (50) in that monoclonal antibodies in ascites form had poor _in vitro_ neutralizing ability but were able to protect animals from ascending neurological disease. In addition, it was also demonstrated that a mixture of monoclonal antibodies seemed to offer more protection than monoclonal antibodies given individually. [Some of the results shown were presented at the Annual Meeting of the American Society for Microbiology in March, 1982 (51)]. These results and those previously published (39, 50) indicate that passive protection against HSV infection is provided by monoclonal antibodies against virus glycoproteins. This could possibly be of benefit in prophylactic treatment of previously uninfected patients who had recently been exposed to HSV-2 or HSV-1.

Table 4--<u>In vivo</u> animal protection studies

Antibody	Virus inoculated	Number of survivors 21 days	% protection
α-gA/B	HSV-2	8/10	80
α-gA/B	HSV-1	5/6	83
α-gC	HSV-2	6/10	60
α-gC	HSV-1	0/6	0
α-gD	HSV-2	6/10	60
α-gD	HSV-1	4/6	66
α-gE	HSV-2	6/10	60
α-gE	HSV-1	1/6	17
α-gA/B + α-gC + α-gD + α-gE	HSV-2	10/10	100
α-gA/B + α-gC + α-gD + α-gE	HSV-1	6/6	100
Anti-HSV-2 antiserum (hyperimmune)	HSV-2	10/10	100
Anti-HSV-1 antiserum (hyperimmune)	HSV-1	6/6	100
Control ascites	HSV-2	0/10	0

PRODUCTION OF A VACCINE AGAINST HSV WITH MONOCLONAL ANTIBODIES

Active Immunization Against HSV

Production of a commercial vaccine against HSV has proven
to be complicated and difficult. To date, a safe and effective
vaccine is not available. Although live attenuated virus
vaccines have proven effective in reducing the incidence of such
diseases as poliomyelitis, rubeola, rubella, mumps, and yellow
fever, the pathogenesis of herpesvirus infections is
considerably different and contributes to the complexity of
vaccine development. For instance, HSV can cause both primary
and recurrent infections, the recurrent infections apparently
resulting from reactivation of the virus, which can exist in a
latent, nonreplicating form. An effective vaccine should
prevent primary infections and either prevent the virus from
entering the latent state or from reactivation.

In addition, HSV has oncogenic potential (3, 4), and it appears that infectious virus is not required for the _in vitro_ transformation of cells (52). It would therefore appear that development of an inactivated vaccine containing subunit viral components without viral nucleic acid would lessen the concern of oncogenicity. Subunit component vaccines, however, require purification processes, are usually poorly immunogenic, and require an adjuvant and/or repeated inoculation to induce an adequate immune response (53, 54). For further discussion of the problems in development of a safe and effective herpesvirus vaccine, the reader is directed to references 55-58.

On the positive side, Price and coworkers (59) have shown that prior immunization with HSV-1 produced substantial protection against development of a latent ganglionic infection. In addition, in humans, it appears that the immune response from a prior HSV infection can influence the manifestations of a subsequent HSV infection. In individuals with pre-existing anti-HSV-1 antibodies, a primary infection with HSV-2 appears to be milder (60, 61). These observations and the data obtained from immunization studies in animals using subunit vaccines (62-65) suggest that the immune response induced by HSV-1 can have a beneficial effect on subsequent HSV infections and that if an HSV vaccine could induce a similar immune response, it might ameliorate clinical manifestations of primary HSV infections.

Production of a Subunit Vaccine With Monoclonal Antibodies

Because the production of a nucleic acid-free vaccine involves the purification of HSV-antigens and because HSV glycoproteins have been implicated in immune mechanisms against HSV infection (14-18), purified HSV glycoproteins should be given primary consideration in vaccine production. Towards this end, immunoaffinity chromatography has proved to be an efficient, reliable method for the purification of glycoproteins from infected cell extracts. Eisenberg et al. (67) have recently reported the purification of gD from HSV-1 and HSV-2 using affinity chromatography columns constructed from type-common monoclonal antibodies, and Arvin et al. (68) have reported purification of gC from HSV-1 using similar techniques. In both instances, the investigators reported that the glycoprotein preparations were free of other viral and host cell proteins and that the major antigenic domains of the glycoproteins were not altered by the purification procedure. In addition, Chan (66) produced two monoclonal antibodies that react with two HSV-1 antigens and were used in immunosorbent columns to purify the viral glycoproteins. Mice immunized with either glycoprotein showed marked resistance to challenge with virulent HSV-1 and, therefore, demonstrated that purified glycoproteins of HSV-1, free of nucleic acid, could be used to protect mice against infection with human HSV-1.

Purification of gA/B, gC, and gE

Briefly, anti-glycoprotein monoclonal antibodies specific
for gA/B, gC, or gE, previously precipitated from mouse ascites
fluid with 40% saturated ammonium sulfate, were combined with
cyanogen bromide-activated Sepharose 4B (Pharmacia Fine
Chemicals, Piscataway, NJ) at a ratio of 20 mg IgG/0.5 g
Sepharose in the presence of coupling buffer (0.2 M $NaHCO_3$, pH
8.5; 0.5 M NaCl; and 0.01 M NaN_3). The reaction was allowed
to proceed overnight at 4°C with constant mixing, and the
mixture was then washed as described previously (68). An
extract (3 to 5 ml) of HSV-2-infected Vero cells, previously
prepared using [^{35}S]methionine-labeled or unlabeled cells and
extracted as described previously (68), was added to
approximately 0.5 ml of antibody-containing Sepharose previously
equilibrated with 10 mM Tris buffer for 6 hr at 4°C. After
mixing for 10 to 12 hr, the immunosorbents were washed ten times
with extraction buffer to remove unreacted antigen. The
purified gC or gE was then eluted with potassium thiocyanate
(3 M KSCN) for 2 hr at 4°C.

To determine the purity of the glycoprotein preparation,
the material eluted with 3 M KSCN was tested by two techniques.
First, the material from the radiolabeled cells was subjected to
electrophoresis in 9.25% SDS-PAGE and visualized by
autoradiography. Second, unlabeled material eluted with 3 M
KSCN was subjected to SDS-PAGE separation, and the glycoproteins
were transferred to nitrocellulose paper by procedures
previously described by Braun et al. (69). A Trans-Blot
electrophoresis unit (Bio-Rad Labs, Richmond, CA) was used for
this purpose. Nitrocellulose sheets, onto which polypeptides
had been transferred, were cut into strips and treated with 3%
gelatin in 0.5 M Tris buffer, pH 7.4, for 1 hr. The strips were
treated with anti-glycoprotein monoclonal antibodies and
hyperimmune anti-HSV-2 antiserum (1:20 dilution). For
visualization, the strips were treated with ^{125}I-labeled
anti-mouse IgG and subjected to autoradiography.

Before immunization experiments with glycoprotein began,
standard procedures for the experiments were established.
First, the minimal lethal dose (MLD) of virus necessary to cause
a lethal infection when inoculated into the hind footpads of
mice was determined. Titers of virus stocks of HSV-1 or HSV-2
were determined just prior to the experiment. BALB/c mice
received increasing concentrations of virus, quantified as PFU,
in the hind footpads. The animals were then observed on a daily
basis for a period of 21 days for the presence of ascending
neurological disease and death. Once the MLD had been
established, all subsequent challenges to immunized animals used
the MLD or multiples of the MLD of virus.

Immunization of Mice With Intact Glycoproteins

Groups of 10 mice were immunized with 50 µg of glycoprotein(s) in Freund's complete adjuvant. These included gA/B, gC, or gE alone or a combination of the three. One week later, the animals were boosted by the identical regimen. Two weeks later the animals were challenged with HSV-2. The animals were observed for 3 weeks for evidence of ascending neurological disease and death (Table 5).

Although these are preliminary experiments, the results agree with those of Chan (66) and others in that protection against infection by HSV can be achieved after immunization with isolated HSV-2 glycoproteins. In addition, a combination of isolated glycoproteins offered more protection than individual preparations. Information from further experiments will include the effects of various adjuvants on efficacy and length of the immunological response. In further experiments, tryptic fragments of glycoproteins containing the antigenic determinant recognized by the specific monoclonal antibody will be tested as immunogens. These experiments are currently underway in this laboratory.

Table 5--Immunization with HSV-2 glycoproteins

Glycoprotein preparation administered	IF titer[*]	Number of survivors after HSV-2 challenge		
		10 days	15 days	21 days
gA/B	1/8	10	8	6
gC	1/16	10	8	5
gE	1/16	10	10	7
gA/B + gC + gE	1/50	10	10	9
Nonimmunized	ND	8	2	1

[*]Indirect immunofluorescence titer was determined with serum obtained by retroorbital puncture just prior to challenge with virus.

IMMUNIZATION AGAINST HSV INFECTION WITH ANTI-IDIOTYPIC ANTIBODIES

Because vaccines are currently unavailable for some very common infectious agents, it is apparent that additional methods of augmenting immunological defenses against disease need to be developed. An ideal method would seem to be one in which the host's immune responses could be specifically stimulated without

injection of the actual infectious agent. This would be of
extreme importance for organisms such as <u>Treponema pallidum</u>, the
agent of syphilis, and <u>Mycobacterium leprae</u>, the agent of
leprosy, which cannot be grown <u>in vitro</u> or for agents such as
the herpesviruses, which can become oncogenic when certain
procedures are used to inactivate the virus (52). One such
method currently being explored involves the use of
anti-idiotypic antibodies as possible immunizing agents.
Antibodies, particularly monoclonal antibodies, that react
specifically with an antigenic determinant possess a particular
idiotype (Id). These antibodies, which are complementary to the
antigenic determinant, will also find their complement in an
anti-idiotypic (anti-Id) antibody. It is then conceivable that
anti-Id antibody could possess an "internal image" of the
antigen. Nisonoff and Lamoyi (70) refer to these as <u>related</u>
<u>epitopes</u> because the term implies similarity but not necessarily
identity between structures in the anti-Id antibody and the
antigenic determinant.

Proof that anti-Id antibodies immunize against an
infectious agent has been presented by Sacks et al. (71). These
investigators immunized mice against African trypanosomiasis
with anti-Id antibodies raised against three protective
monoclonal antibodies, each with a specificity for a different
antigenic determinant. When these anti-Id antibodies were
administered to mice 3 to 4 weeks before challenge with the
trypanosome <u>Trypanosoma rhodesiense</u>, the course of parasitemia
was altered in 19 of 30 animals and the immunity was manifested
as complete protection, reduced parasitemia, or selection
against parasites bearing the specific antigenic determinant.

Urbain et al. (72, 73) described an anti-Id antiserum
directed against anti-tobacco mosaic virus (anti-TMV) antibodies
that induced anti-TMV antibodies when injected into mice. These
investigators concluded that the anti-Id behaved in all respects
as if it possessed one or more antigenic determinants of TMV or
presented at least one epitope related to an epitope on TMV.

Using reovirus as a model, Nepom and coworkers (74)
attempted to determine whether virus receptors on cell surfaces
share idiotypic determinants with antibody to the virus. They
raised a xenogeneic (rabbit) antiserum (anti-Id3) against murine
anti-reovirus type 3 hemagglutinin (anti-HA3) and used it to
define the idiotypic determinant (Id3). Because reovirus binds
to receptors on cells by a specific hemagglutinin (HA),
monoclonal antibodies were produced that reacted specifically
with the HA from reovirus type 3. One of these monoclonal
antibodies (G5) had the same idiotype (Id3) as the murine
anti-HA3 and was used to purify the rabbit anti-Id antibody
raised against anti-HA3. In addition, G5 monoclonal antibody
was injected into syngeneic animals to produce an anti-Id
monoclonal antibody that recognized the same G5 idiotype (Id3)
characterized by the rabbit anti-Id3. By use of these reagents,
the Id3 determinant was shown to be present not only on the G5

monoclonal antibody and the highly purified anti-HA3 antibody, but also on Rl.1 cells, a T-cell-derived cell line that binds reovirus type 3 selectively, and on neuronal cells in primary culture. The Id3 determinant was not present on a) antibody from normal serum, b) antibody specific for HA from reovirus type 1, c) YAC cells that do not bind reovirus, or d) ependymal cells that selectively bind reovirus type 1 but not type 3. Thus, the observation that an anti-Id antibody can detect the presence of an idiotype shared by antihemagglutinin antibodies and by structures on cells containing specific virus receptors suggests a general relationship between disparate receptors that recognize a common determinant. Nepom et al. (74) have suggested that anti-Id antisera generated against idiotypic Ig have potent immunological effects and may interact directly with T-cell receptors or mimic the antigen itself in terms of its ability to prime a nonimmune animal.

The feasibility of inducing antibody production by anti-Id antibody has also been demonstrated by Trenker and Riblet (75). They reported that anti-Id antibodies produced in rabbit could induce mouse lymphocytes bearing the relevant idiotype to produce anti-phosphorylcholine antibody. Optimal conditions under which anti-immunoglobulins can induce resting B lymphocytes to undergo proliferation were described by Braun and Unanue (76). Growth-promoting molecules, from other cells or from 2-mercaptoethanol, were necessary, as was cross-linking of the antibody/antigen. In addition, the anti-Ig molecules and growth-promoting molecules had to be in contact with cells in culture for at least 48 hr and the B-cells had to be from mature mice.

Anti-Id antibodies also affect cell-mediated responses. Anti-Id antibodies raised in guinea pigs were found to have different effects when injected into mice. IgG2 antibodies were suppressive and were able to specifically sensitize suppressor T-cells bearing the relevant idiotype, whereas IgG1 enhanced idiotype expression by sensitizing idiotype-position B and T helper cells (77-80). Mouse anti-Id antibodies can also control idiotypic expression. Small doses of approximately 100 ng of anti-T-15 idiotype could efficiently sensitize T-15 idiotype-bearing T-cells (81), whereas larger doses were found to induce T-15 suppression when injected into neonatal mice (82, 83). In addition, here the class of anti-Id antibody was not thought to determine its regulatory function. Allogeneic monoclonal anti-Id antibodies were also shown to control idiotypic expression (84). As shown previously, low doses of anti-Id monoclonal antibodies led to enhanced expression of corresponding idiotype, while increasing doses induced suppression (84).

In addition, Kennedy et al. (85) have demonstrated that the immune response to hepatitis B surface antigen is enhanced by prior injection of anti-Id antibodies. They found that injection of antibodies to the idiotype before antigenic

stimulation resulted in an increase in the number of cells
secreting immunoglobulin M antibodies to hepatitis B surface
antigen.

The above data strongly suggest that anti-Id antibodies can
be used to stimulate specific immunological responses and could
possibly be used to immunize against infectious agents.

Efforts to determine whether anti-Id antibodies can be used
to enhance the immune response against HSV infection have begun
in our laboratory. Monoclonal antibodies reacting with
antigenic determinants on gC and gE of HSV-2 have been used as
the immunogen in order to produce anti-Id antibodies. The
hybridoma cells, derived from BALB/c mice, were grown in
serum-free media (IVIS System, HANA Media, Inc., Berkeley, CA),
and the supernatants were clarified by centrifugation. The
supernatants containing the monoclonal antibodies were then
added to splenocytes from BALB/c mice in serum-free media in an
in vitro immunization process. After 3 days in culture, the
splenocytes were washed and fused with NS-1 myeloma cells
according to standard procedures. The procedure for screening
the clone supernatants for the presence of anti-Id antibodies
was that described by Morahan (86). This radioimmunoassay
method detects anti-Id antibodies in the supernatant and is
sensitive and specific for anti-Id antibodies but not
anti-allotypic antibodies. The experiments to further
characterize the anti-Id antibodies and determine this potential
for enhancement of immunity against HSV-2 are currently underway.

REFERENCES

1. Kaplan, A.S. (Ed.) (1973) In: The Herpesviruses. Academic
    Press, New York.
2. Douglas, R.G., and Couch, R.B. (1970) J. Immunol. 104:289.
3. Duff, R., and Rapp, F. (1971) J. Virol. 8:469.
4. Duff, R., and Rapp, F. (1973) J. Virol. 12:209.
5. Melnick, J.L., Adam, E., and Rawls, W.E. (1974) Cancer
    34:1376.
6. Plummer, G., and Masterson, J.G. (1971) Am. J. Obstet.
    Gynecol. 111:81.
7. Keller, J.M., Spear, P.G., and Roizman, B. (1970) Proc.
    Natl. Acad. Sci. USA 65:865.
8. Spear, P.G., Keller, J.M., and Roizman, B. (1970) J. Virol.
    5:123.
9. Beth, E., Giraldo, G., and Bernhard, M. (1976) Biomed.
    Express 25:366.
10. Glorioso, J.C., Wilson, L.A., Fenger, T.W., and Smith, J.W.
    (1978) J. Gen. Virol. 40:443.
11. Rager-Zisman, B., and Bloom, B. (1974) Nature 251:542.
12. Siebens, H., Tevethia, S.S., and Babior, B.M. (1979) Blood
    54:88.

13. Oleske, J.M., Ashman, R.B., Kohl, S., Shore, S.L., Starr, S.E., Wood, P., and Nahmias, A.J. (1977) Clin. Exp. Immunol. 27:444.
14. Pfizenmaier, K.H., Jung, H., Starzinski-Powitz, A., Rollinghoff, M., and Wagner, H. (1979) J. Immunol. 119:939.
15. Lawman, M.J.P., Rouse, B.T., Courtney, R.J., and Walker, R.D. (1980) Infect. Immun. 27:133.
16. Spear, P.G. (1976) J. Virol. 17:991.
17. Baucke, R.B., and Spear, P.G. (1979) J. Virol. 32:779.
18. Norrild, B., Shore, S.L., and Nahmias, A.J. (1979) J. Virol. 32:741.
19. Carter, V.C., Schaffer, P.A., and Tevethia, S.S. (1981) J. Immunol. 126:1655.
20. Eberle, R., and Courtney, R.J. (1981) Infect. Immun. 31:1062.
21. Eisenberg, R.J., Ponce de Leon, M., and Cohen, G.H. (1980) J. Virol. 35:428.
22. Para, M.F., Goldstein, L., and Spear, P.G. (1982) J. Virol. 41:137.
23. Ruyechan, W.T., Morse, L.S., Knipe, D.M., and Roizman, B. (1979) J. Virol. 29:677.
24. Vestergaard, B.F., and Grauballe, P.C. (1979) J. Clin. Microb. 10:772.
25. Norrild, B. (1980) Curr. Top. Microbiol. Immunol. 90:67.
26. Powell, K.L., Buchan, A., Sim, C., and Watson, D.H. (1974) Nature (London) 249:360.
27. Pereira, L., Dondero, D.V., Gallo, D., Devlin, B., and Woodie, J.D. (1982) Infect. Immun. 35:363.
28. Pereira, L., Klassen, T., and Baringer, J.R. (1980) Infect. Immun. 29:724.
29. Cohen, G.M., Katze, M., Hydrean-Stern, D., and Eisenberg, R.J. (1978) J. Virol. 27:172.
30. Spear, P. (1975) IARC Sci. Pub. 11:49.
31. Vestergaard, B.F., and Norrild, B. (1979) IARC Sci. Pub. 24:225.
32. Everle, R., and Courtney, R.J. (1982) J. Virol. 41:348.
33. Zweig, M., Heilman, C.J., Jr., Bladen, S.V., Shumutter, S.D., and Hampar, B. (1983) Infect. Immun. 41:482.
34. Zweig, M., Showalter, S.D., Bladen, S.V., Heilman, C.J., Jr., and Hampar, B. (1983) J. Virol. 47:185.
35. Koprowski, H., Steplewski, Z., Herlyn, D., and Herlyn, M. (1978) Proc. Natl. Acad. Sci. USA 75:3405.
36. Koprowski, H., Gerhard, W., and Croce, C.M. (1977) Proc. Natl. Acad. Sci. USA 74:2985.
37. Davidson, R., and Gerhard, P. (1976) Somat. Cell Genet. 2:175.
38. Littlefold, J.W. (1964) Science 145:709.
39. Petri, W.A., and Wagner, R.R. (1979) J. Biol. Chem. 254:4313.
40. Goding, J.W. (1980) J. Immunol. Meth. 39:285.

41. Lampson, L.A. (1980) In: Monoclonal Antibodies-Hybridomas: A New Dimension in Biological Analysis. Kennett, R.H., McKearn, T.J., and Bechtol, K.B. (eds.) Plenum Press, New York, p. 395.

42. Showalter, S.D., Zweig, M., and Hampar, B. (1981) Infect. Immun. 34:684.

43. Balachandran, N., Harnish, D., Killington, R.A., Bacchetti, S., and Rawls, W.E. (1981) J. Virol. 39:438.

44. Davis, W.B., Taylor, J.A., and Oakes, J.E. (1979) J. Infect. Dis. 140:534.

45. McKendall, R.R., Klassen, T., and Baringer, J.R. (1979) Infect. Immun. 23:305.

46. Oakes, J.E., Davis, W.B., Taylor, J.A., and Weppner, W.A. (1980) Infect. Immun. 29:642.

47. Oakes, J.E., and Lausch, R.N. (1979) Infect. Immun. 23:305.

48. Oakes, J.E., and Rosemund-Hornbeak, H. (1978) Infect. Immun. 21:489.

49. Dix, R.D., Pereira, L., and Baringer, J.R. (1981) Infect. Immun. 34:192.

50. Balachandran, N., Bacchetti, G., and Rawls, W.E. (1982) Infect. Immun. 37:1132.

51. Marquez, E.D. (1983) Abs. Ann. Mtg. Amer. Soc. Microb. p. 292.

52. Rapp, F., and Reed, C. (1976) Cancer Res. 36:800.

53. Cappel, R., De Cuyper, F., and Richaert, F. (1980) Arch. Virol. 65:15.

54. Hill, T.J., Field, H.K., and Blyth, W.A. (1975) J. Gen. Virol. 28:341.

55. Hilleman, M. (1976) Cancer Res. 36:857.

56. Wise, T.G., Pavan, P.R., Ennis, F.A. (1977) J. Infect. Dis. 136:706.

57. Rapp, F. (1981) Gyn. Oncol. 12:S341.

58. Allen, W.P., and Rapp. F. (1982) J. Infect Dis. 145:413.

59. Price, R.W., Walz, M.A., Wohlenberg, C., and Notkins, A.L. (1975) Science 188:938.

60. Rawls, W.E., Gardner, H.L., Flanders, R.W., Lowry, S.P., Kantman, R.H., and Melnick, J.L. (1971) Am. J. Obstet. Gynecol. 110:682.

61. Nahmias, A.J., Josey, W.E., Naib, Z.M., Luce, C.F., and Duffey, A. (1970) Am. J. Epidemiol. 91:539.

62. Kutinova, L., Slichtova, V., and Vonka, V. (1982) Dev. Biol. Stand. 52:313.

63. Hilfenhaus, J., Moser, H., Herrmann, A., and Mauler, R. (1982) Dev. Biol. Stand. 52:321.

64. Skinner, G.R.B., Williams, D.R., Buchan, A., Whitney, J., Harding, M., and Bodfish, K. (1978) Med. Microbiol. Immunol. 166:119.

65. Anderson, V.A., and Kilbourne, E.D. (1961) Proc. Soc. Exp. Biol. Med. 107:518.

66. Chan, W.L. (1983) Immunol. 49:343.

67. Eisenberg, R.J., Ponce de Leon, M., Pereira, L., Long, D., and Cohen, G.H. (1982) J. Virol. 41:1099.

68. Arvin, A.M., Koropchak, C.M., Yeager, A.S., and Pereira, L. (1983) Infect. Immun. 40:184.

69. Braun, D.K., Pereira, L., Norrild, B., and Roizman, B. (1983) J. Virol. 46:103.

70. Nisonoff, A., and Lamoyi, E. (1981) Clin. Immunol. Immunopathol. 21:397.

71. Sacks, D.L., Esser, K.M., and Sher, A. (1982) J. Exp. Med. 155:1108.

72. Urbain, J., Wuilmart, C., and Cazenaze, P.A. (1981) Contemp. Top. Mol. Immunol. 8:113.

73. Urbain, J., Cazenaze, P.A., Wikler, M., Franssen, J.D., Mariame, B., and Leo, O. (1980) In: Progress in Immunology, Vol. IV. Fougereau, M., and Dausset, J., (eds.) Academic Press, New York, p. 81.

74. Nepom, J.T., Weiner, J.L., Dichter, M.A., Tardieu, M., Spriggs, D.R., Gramm, C.F., Powers, M.L., Fields, B.N., and Greene, M.I. (1982) J. Exp. Med. 155:155.

75. Trenker, E., and Riblet, R. (1975) J. Exp. Med. 142:1121.

76. Braun, J., and Unanue, E.R. (1980) Immunol. Rev. 52:3.

77. Eichmann, K., and Rawjewski, K. (1975) Eur. J. Immunol. 5:661.

78. Eichmann, K. (1974) Eur. J. Immunol. 4:296.

79. Eichmann, K., (1975) Eur. J. Immunol. 5:511.

80. Rawjewski, K., Hämmerling, G.H., Black, S.H., and Eichmann, K. (1976) In: The Role of Products of the Histocompatibility Gene Complex in Immune Responses. Katz, D.H., and Benacerraf, B., (eds.) Academic Press, New York, p. 445.

81. Julius, M.H., Augustin, A.A., and Cosenza, H. (1977) In: Regulatory Genetics of the Immune System (ICN-UCLA symposia on Molecular and Cellular Biology), Vol. 6. Sercarz, E.E., Harzenberg, E.E., and Fox, C.F., (eds.) Academic Press, New York, p. 179.

82. Augustin, A., and Cosenza, H. (1976) Eur. J. Immunol. 6:496. Balachandran, N., Bacchetti, S., and Rawls, W.E. (1982) Infect. Immun. 37:1132.

83. Cosenza, H., Julius, M.H., and Augustin, A.A. (1977) Immunol. Rev. 34:3.

84. Kohler, H., Richardson, B.C., Rowley, D., and Smyk, S. (1977) J. Immunol. 119:1979.

85. Kennedy, R.C., Adler-Storthz, K., Henkel, R.D., Sanchez, Y., Melnick, J.L., and Dreesman, G.R. (1983) Science 853.

86. Morahan, G. (1983) J. Immunol. Meth. 57:165.

CHAPTER 5

IMMUNOCHEMICAL AND DIAGNOSTIC USES OF MONOCLONAL ANTIBODIES
IN LEPROSY AND GONORRHEA

THOMAS M. BUCHANAN
Immunology Research Laboratory
Seattle Public Health Hospital
Seattle, Washington 98114

ABSTRACT

     Procedures for the generation and characterization of
monoclonal antibodies are described, including methods to
determine the antigen specificity of the monoclonal antibodies
with radioimmunoprecipitation, western blot, autoradiography
following recognition of purified glycolipid antigens or their
hydrolytic products in thin layer chromatography, and a new
procedure employing autoradiography of thinly sliced
polyacrylamide slab gels after their interaction with the
monoclonal antibodies generated and characterized that are
directed at protein, carbohydrate, and glycolipid antigens of
Neisseria gonorrhoeae, and Mycobacterium leprae. Some of these
antibodies have been characterized as to whether the epitope
recognized is species-specific, or with respect to its exact
sugar or amino acid structure. Other monoclonal antibodies
appear to be directed at conformational determinants found only
in the trimeric form of the protein I molecule as it exists
natively in the gonococcal membrane forming an anion-selective
pore through which substances may reach the organism. Several
of these monoclonal antibodies have significant promise for
reagents that may be utilized for culture confirmation on
gonococci, for confirmation of the identity of an organism as M.
leprae, or perhaps for use in "sandwich" immunoassays capable of
rapid and accurate detection of gonococcal or M. leprae
antigen. Such rapid immunoassays may prove superior to culture
methods in certain settings.
     Monoclonal antibodies are proving extremely useful as
high-quality reagents for immunochemical studies and diagnostic
reagents. In this article, some of the uses of monoclonal

antibodies for the study of <u>Neisseria gonorrhoeae</u>, the causative
agent of gonorrhea, and of <u>Mycobacterium leprae</u>, the causative
agent of leprosy, the methodology of producing monoclonal
antibodies used in our laboratory and some shortcuts and
procedures that we have found useful are described. The
manuscript will describe uses of monoclonal antibodies as
molecule-specific markers and as probes of the submolecular
"epitope" or antigenic sites on the molecule. In some instances
the monoclonal antibodies recognize the conformational
determinants of a molecule that are exposed on the surface of
the organism. Diagnostic uses of these monoclonal antibodies
are discussed with respect to their usefulness as reagents for
culture confirmation of gonorrhea and for identification of the
gonococcus or leprosy bacillus by fluorescence, enzyme-linked
immunosorbent assay (ELISA), or peroxidase assays using organism
suspensions or tissue sections containing the organism. The
usefulness of these reagents as epidemiologic markers of strains
with given serotypes, geographic distributions, or clinical
correlates such as the ability to cause dissemination, resist
killing by human serum, or produce penicillinase is discussed.
Finally, the potential usefulness or these reagents in
"sandwich" assays to permit rapid definitive identification of
the etiologic agent, in some instances perhaps even replacing
the culture, is considered.

PRODUCTION OF MONOCLONAL ANTIBODIES

<u>Antigen Purification and Immunization Protocols</u>

     Researchers at the Immunology Research Laboratory, Pacific
Medical Center, have previously published procedures for the
purification of pili (1, 2) and protein I (3) antigens of <u>N.</u>
<u>gonorrhoeae</u>, and of the phenolic glycolipid of <u>M. leprae</u> (4).
We have immunized with whole gonococci or leprosy bacilli as
well as with purified or partially purified antigen preparations
and have consistently obtained the best results with partially
purified or purified preparations. In general, we immunize
10-week-old BALB/c mice by weekly intraperitoneal injection with
20 to 50 µg of carbohydrate or glycolipid antigens, or 50 to
200 µg for protein antigens. Responses to pure carbohydrate
or glycolipid antigens were generally poor, but this same amount
of antigen complexed to methylated BSA in the case of glycolipid
(4), or noncovalently bound to proteins of the homologous
organism in the case of arabinomannan of <u>M. leprae</u> (5, 6) or
lipopolysaccharide of the gonococcus, produced excellent immune
responses in the mice. Mice were screened for their antibody
response to the immunogen, and those selected with the best
responses were given their final immunizations intravenously 3
days before the fusion.

Screening for Antibody Responses

We generally screen for antibody production by the immunized mice by ELISA with microtiter plates coated with the appropriate purified immunogen (3, 4, 6-8). To screen for antibody in the culture fluid supernatant of fused cells, either ELISA (8, 9) or a solid-phase radioassay employing ^{125}I-labeled protein A (10) was used. The microtiter plates have been coated with sodium carbonate buffer (0.05 M, pH 9.5), Tris buffer (0.05 M, pH 8.0), or acetate (0.2 M, pH 4.5), in concentrations of 2 to 20 µg/ml of antigen, at 37°C to 60°C, for 3 to 5 hours, depending upon the antigen. All assays of both serum antibody responses and the antibody levels in culture supernate were done in duplicate and variations of more than 15 to 20% in the duplicate values were considered unreliable and were repeated. Most of the coated plates containing purified antigens can be stored in an antigenic state at 4°C for up to 1 year. The ELISA methodology has the advantage of not requiring radioactive reagents and can detect IgM and IgG subclasses of mouse antibodies not recognized by protein A. A precise assay for detecting antibody to the antigen of interest is the single most important ingredient to obtaining high yields of monoclonal antibodies to the antigen of interest in our experience.

Fusions

The BALB/c myeloma cell line NSI/1 was kindly provided by R. C. Nowinski with the permission of C. Milstein (Medical Research Council, Cambridge). The fusion protocol, cell culture methods and adaptations thereof have been previously reported (8-10). The 2 mice with the strongest antibody response were anesthetized with ether, killed by cervical dislocation, and their spleens were harvested. A single cell suspension of splenocytes in RPMI 1640 (GIBCO Laboratories, Grand Island, New York) supplemented with penicillin and streptomycin but without fetal calf serum (FCS) was prepared by extensive mincing of the tissue and passage through a sterile nylon mesh. The cells were washed three times with RPMI without FCS, being pelleted by centrifugation at 160 X g for 5 min after each wash. Viability was assessed by trypan blue exclusion, and the cells were counted with a hemocytometer. NSI/1 myeloma cells were maintained in log phase growth for 3 days before fusion. On the day of fusion, they were harvested by pelleting at 160 X g for 5 min, washed once with RPMI 1640 without FCS, then resuspended and counted. The splenocytes (10^8) were combined with the NSI/1 myeloma cells in a ratio of 4:1 in a single 50-ml round-bottomed glass centrifuge tube and pelleted by centrifugation at 200 X g for 5 min. The supernatant was aspirated, and 1.0 ml of 40% polyethylene glycol 1450 (Kodak Laboratories, Rochester, NY), pH 7.8, was added with gentle dispersion of the pellet. The cell mixture was pelleted at

350 X g for 10 min, resuspended and gently washed with 10 ml of
RPMI 1640, and recentrifuged at 160 X g for 5 min to remove the
PEG. The supernate was aspirated, and the pellet gently taken
up into RPMI 1640 supplemented with 15% fetal calf serum (FCS,
GIBCO or HyClone), penicillin (100 U/ml), streptomycin (100
µg/ml), 1 mM sodium pyruvate, and 2 mM L-glutamine ("complete
media") plus 0.1 mM hypoxanthine, 0.4 mM aminopterin, and 0.015
mM thymidine (HAT medium). The cells were dispersed into
96-well microtiter plates at a density of 5 x 10^5 cells/well
(200 µl volume) and incubated at 37°C in 6.4% CO_2 and 100%
humidity. A thymocyte feeder-cell layer was used in some
instances and not used in other fusions. After the cells
reached 50% confluency in the majority of wells (generally after
8 to 11 days and occasionally as late as 13 to 14 days, with HAT
medium replaced between days 5 to 8), they were screened for
antibody production. Each strongly positive well W as diluted
with HAT media and replated onto half a microtiter plate (48
wells) at a density of 5 cells per well ("minicloning"). These
cells were screened after growing to 50% confluency in the
majority of wells (after 8 to 14 days of growth); 4 positive
wells per miniclone were diluted in complete media without
hypoxanthine, aminopterin, or thymidine, and placed into 96-well
sterile microtiter plates at a density of 1 cell per well.
Wells containing discrete, single outgrown clones (visually
assessed) were coalesced into a single plate that was again
tested for antibody activity. Strongly positive clones were
selected and amplified by growth in 24-well plates in complete
media to a density of 10^7; clones to be frozen in liquid N_2
were grown to a density of 10^8. BALB/c mice primed with
pristane (2, 6, 10, 14--tetramethyl pentadecane, Aldrich
Chemical Co., Milwaukee WI)--were injected intraperitoneally at
a concentration of 5 to 10 x 10^6 for production of antibody
enriched malignant ascites.

CHARACTERIZATION OF MONOCLONAL ANTIBODIES AND DETERMINATION OF
THEIR SPECIFICITY

The immunoglobulin class and subclass of the monoclonal
antibodies produced were determined by immunodiffusion
precipitation with antibodies specific for given murine
immunoglobulin species (Miles Laboratories, Elkhart, IN).
Discrete monoclonal bands were detected using electrophoresis on
cellulose acetate (Microzone System, Beckman Instruments,
Fullerton, CA).
Immunologic specificity of the monoclonal antibodies was
determined by a) radioimmunoprecipitation (10), b) western
blotting (9, 11), c) reaction of ^{125}I-labeled purified
monoclonal antibody with purified glycolipid or its degradation
hydrolytic products separated on thin-layer chromatography (8),
or d) the procedure referred to as gel-immunoradioassay (GIRA)

(12). In the GIRA procedure, discontinuous polyacrylamide gel
electrophoresis was performed using a 3-mm-thick 5%
polyacrylamide stacking gel and 10% polyacrylamide separating
gel in a Biorad (Richmond, CA) gel apparatus. Samples were
prepared in a buffer containing sodium dodecyl sulfate and
applied to the gel at 25 mA constant current through the
stacking gel and 50 mA constant current through the separating
gel. The buffer front was allowed to migrate only 4 cm into the
separating gel, then the gel was removed and the 4 cm x 4 cm
section of the separating gel containing the samples was cut
from the larger slab. This section was mounted onto a brass
cryostat specimen holder using a filter paper spacer between the
gel and holder (Whatman #1 paper), and O.C.T. embedding compound
(Tissue-Tek II^{TM}, Lab-Tek Products division of Miles
Laboratories). The specimen was frozen by immersion of the
lower portion of the brass holder (but not the gel itself) in
liquid nitrogen. The specimen was then placed in a
microtome-cryostat (Damon/IEC, Needham Heights, MA) adjusted to
cut 50-μm-thick slices and precooled to -30°C. After the gel
was allowed to reach thermal equilibrium in the cryostat
overnight, it was sliced into about 40 thin sections. These
were stored in precooled 80% ethanol at -20°C until needed. The
assay itself was performed as follows: the slices were
rehydrated in PBS for 10 min, then placed in 1.5-ml
microcentrifuge tubes containing 500 μl of PBS with 0.2%
Triton X-100 and 0.5% BSA (reaction buffer) and the various
monoclonal ascites diluted 1:25. A soft artists' #5 paint brush
was used to pick up and manipulate the thin, rehydrated gel
slices. The antibody and gel mixtures were inverted four to
five times, then placed at 4°C overnight. After this
incubation, the slices were washed twice in separate plastic
Petri dishes containing 30 to 40 ml of PBS with 0.5% Triton
X-100 (wash buffer) for 30 min per wash. The slices were then
placed in new 1.5-ml microcentrifuge tubes containing 500-μl
reaction buffer and approximately 10^6 cpm of ^{125}I-labeled
sheep anti-mouse immunoglobulin. This labeled antibody was
prepared by purifying the IgG fraction from sheep anti-mouse
immunoglobulin sera (Cappel Laboratories) using a staphylococcal
protein A column (Pharmacia Fine Chemicals, Piscataway, NJ),
then iodinating the IgG fraction with ^{125}I as described by
Greenwood et al. (5). The gel-radiolabel mixtures were inverted
four to five times, then incubated for 1 h at room temperature.
The gels were then placed into a Petri dish and washed
extensively (eight to nine washes, each consisting of 30 to 40
ml of wash buffer and lasting 30 min). After the final wash,
the gels were mounted onto Gel Bond film (Marine Colloids
Division, Bioproducts, Rockland, ME), allowed to slowly air dry
at room temperature, and autoradiographed. Our best results for
characterization of antigen specificity were with the GIRA
procedure (approximately 70% of monoclonal antibodies directed
at protein antigens can be characterized using GIRA).

RESULTS

Using the above methodologies, our laboratory has produced
large numbers of IgG1, IgG2, IgG3, and IgM monoclonal antibodies
to the protein I of N. gonorrhoeae, the principal outer membrane
protein of the gonococcus (10). Monoclonal antibodies have also
been produced to the proteins II and III of the gonococcus, as
well as to the lipopolysaccharide. Immunoglobulin class IgM
monoclonal antibodies have been produced to the phenolic
glycolipid (8) and to the arabinomannan polysaccharide of M.
leprae (5). Monoclonal antibodies of IgG and IgM classes have
been produced to the 68,000- and 14,000-Da proteins of the
leprosy bacillus (9). We have demonstrated (using mild
hydrolytic conditions that remove a single sugar at a time from
the terminal trisaccharide of the phenolic glycolipid molecule)
that the monoclonal antibodies with greatest specificity for the
leprosy bacillus that are directed at this molecule are reactive
with only the intact molecule. Removal of the terminal
3,6-di-O-methyl-glucose single sugar moiety produced almost
total loss of recognition of the phenolic glycolipid by the
monoclonals, indicating that the immunodominant antigenic site
was the terminal sugar and/or its linkage to the adjacent
penultimate sugar. Other monoclonal antibodies to the pili of
gonococci have been identified as directed at the antigenically
common portion of the pilus subunit, or even at epitopes within
specific small peptides within the subunit primary amino acid
structure. Still other monoclonal antibodies directed at
protein I that are serotype specific appear to be directed at
confirmational determinants found only when the molecule is in
its trimeric or native form as it is found in the gonococcal
membrane. When the protein I is dissociated into the monomeric
subunit, these monoclonals recognize the monomer only weakly or
not at all compared to the trimer.

Some monoclonals directed at protein I recognize serotypes
of gonococci that make up as few as 10% of gonococcal strains,
to as many as 50%, of the strains present in a randomized
collection of organisms obtained from nine U.S. cities.
Similarly, the monoclonals directed at the arabinomannan antigen
of M. leprae recognizes all mycobacteria (5), whereas two
monoclonals directed at the 68,000-Da subunit mass protein of M.
leprae recognize only M. leprae among 22 species of mycobacteria
tested (9). The monoclonals directed at the phenolic glycolipid
of M. leprae recognized primarily the leprosy bacillus (4[+]),
and weakly Mycobacterium terrae and Mycobacterium
nonchromogenicum, two non-pathogens, among 25 species of
mycobacteria tested (8). Several monoclonal antibodies may be
combined to provide species-specific reagents that can be useful
for culture confirmation of a given species. For example, such
products are currently commercially available for the
culture-confirmation of N. gonorrhoeae by coagglutination. The
monoclonal antibodies also make excellent reagents for use in

direct or indirect fluorescence tests, providing a strong
fluorescent signal and less background reactivity than
polyclonal antisera. Monoclonal antibodies to the protein I of
gonococci or to the phenolic glycolipid of the leprosy bacillus
each produce strong ring-like fluorescence around their
respective organisms, indicating the surface exposure of these
molecules. These reagents may prove useful for the detection of
the organism, or of its antigen, in tissue specimens.

Monoclonal reagents can serve as epidemiological aids that
correlate with specific disease syndromes, with strain
variations in different geographic locations, with nutritional
requirements or "auxotypes" of the organism, with the ability to
resist killing by normal human serum, or with penicillin
susceptibility or the production of penicillinase. They may
also be useful as quality reagents to recognize molecules that
are important for the pathogenesis of disease or for human
immunity to that disease. For example, we found that women who
had experienced one episode of gonococcal pelvic inflammatory
disease and then were reinfected with gonococci became infected
with organisms bearing a different serotype of protein I if
their reinfection was associated with pelvic inflammatory
disease, but not if their reinfection was confined to the
mucosal surface (13). This implied that immunity to recurrent
PID with gonococci of the same protein I serotype developed
following their first episode of PID, and that host defense
mechanisms promoted by the protein I immunization were effective
only against invasive forms of the disease. This can be
explained on the basis that protein I stimulates bactericidal
antibody, and that bactericidal antibody is effective during
PID, when serum can reach the highly inflamed infection site,
but not in localized infection where complement is present at
only 5 to 10% of serum levels. It has been determined that the
protein I serotype 1 gonococci are more frequently resistant to
killing by normal human serum (3, 14, 15), are more likely to
cause disseminated gonococcal infection (16), and are more
frequently associated with the production of penicillinase by
gonococci in the Philippines than with other serotypes. Using
monoclonal antibody to phenolic glycolipid of <u>M</u>. <u>leprae</u> in a
quantitative immunoassay, we have been able to measure nanogram
quantities of the molecule and to determine precisely the amount
of this molecule in purified heat-killed organisms used for
human vaccination and in lepromin and other skin test reagents.
This information has provided a new understanding and explains
why many of the previously observed responses to these reagents
may have varied as directly correlated with their different
quantities of phenolic glycolipid.

The high quality of monoclonal antibodies offers the
promise that combinations of these antibodies may be used in
"sandwich" immunoassays to detect small quantities of antigen in
human tissues or swab samples. If sufficiently sensitive, these
immunoassay tests could be performed within 1 h and might

replace culture techniques. Also, since they recognize only
antigen and not live organisms, such tests might make it
possible to obtain samples under conditions where culturing was
impossible, and test the samples several days later when the
organisms were dead but the antigen was still intact. Initial
evaluations of these possibilities indicate that sensitivities
in the range of 10^3 to 10^4 may be achieved, and that buffer
conditions can be developed under which most protein, glycolipid
and carbohydrate antigens will remain stable for periods of 1
week or longer. Thus these monoclonal reagents are extremely
promising for new rapid diagnostic tests that will be extremely
accurate and cost-effective.

REFERENCES

1. Hermodson, M.A., Chen, K.C.S., and Buchanan, T.M. (1978)
    Biochemistry 17:442-445.
2. Pearce, W.A., and Buchanan, T.M. (1978) J. Clin. Invest.
    61:931-943.
3. Buchanan, T.M., and Hildebrandt, J.F. (1981) Infect. Immun.
    32:985-994.
4. Young, D.B., and Buchanan. T.M. (1983) Science
    221:1057-1059.
5. Miller, R.A., and Buchanan, T.M.: Production and
    characterization of a murine monoclonal antibody
    recognizing a shared mycobacterial polysaccharide.
    Unpublished data.
6. Miller, R.A., Dissanayake, S., Buchanan, T.M. (1983)
    Am. J. Trop. Med. Hygiene 32:555-564.
7. Buchanan, T.M. (1978) J. Infect. Dis. 138:319-325.
8. Young, D.B., Khanolkar, S.R., Barg, L.L., Buchanan, T.M.
    (1984) Generation and characterization of monoclonal
    antibodies to the phenolic glycolipid of Mycobacterium
    leprae. Infect. Immun. (in press).
9. Gillis, T.P., and Buchanan, T.M. (1982) Infect. Immun.
    37:172-178.
10. Tam, M.R., Buchanan, T.M., Sandstrom, E.G., Holmes, K.K.,
    Knapp, J.S., Siadak, A.W., and Nowinski, R.C. (1982)
    Infect. Immun. 36:1042-1053.
11. Towbin, H., Staehelin, T., and Gordon, J. (1979)
    Proc. Natl. Acad. Sci. USA 76:4350-4354.
12. Poolman, J.T., Hopman, C.T.P., Zanen, H.C. (1983)
    Infect. Immun. 40:398-406.
13. Buchanan, T.M., Eschenbach, D.A., Knapp, J.S., and Holmes,
    K.K. (1980) Am. J. Obstet. Gynecol. 138:978-980.
14. Cannon, J.G., Buchanan, T.M., and Sparling, P.F. (1983)
    Infect. Immun. 40:816-819.
15. Sandstrom, E.G., Knapp, J.S., and Buchanan, T.M. (1982)
    Infect. Immun. 35:229-239.

16. Hildebrandt, J.F., and Buchanan, T.M. (1978) In:
     Immunobiology of Neisseria gonorrhoeae. Brooks G.F., et
     al. (eds.) American Society for Microbiology, Washington,
     DC, p. 138.

CHAPTER 6

ANTIGENIC ANALYSIS OF AFRICAN TRYPANOSOMES WITH MONOCLONAL
ANTIBODIES

TERRY W. PEARSON, MICHAEL W. CLARKE, NICOLE M. PARISH,
JENNIFER P. RICHARDSON, LESLIE A. MITCHELL, and LINDA E. SAYA
Department of Biochemistry and Microbiology, University of
Victoria, Victoria, British Columbia, Canada V8W 2Y2

ABSTRACT

Monoclonal antibodies have been used over the past few years to
study antigenic sites on the major variable surface
glycoproteins (VSGs) of bloodstream and metacyclic (salivary
gland) forms of African trypanosomes. These studies have been
directed towards understanding the antigenic polymorphisms of
VSGs and the extent of the antigenic repertoire of the
metacyclics. Recently, monoclonal antibodies specific for other
(non-VSG) antigens of trypanosomes have been derived. Some of
these are useful as species-specific markers and others for
analysis of non-VSG membrane antigens on bloodstream and
procyclic (midgut) forms of trypanosomes. Information from
these monoclonal antibodies and from recent amino acid and DNA
sequence studies on VSGs suggests new possibilities for
vaccination strategies despite the supposed impasse established
by the phenomenon of antigenic variation.

INTRODUCTION

     African trypanosomes are infamous for their ability to
undergo antigenic variation, a process by which they, as a
population, avoid elimination from the host. In this process,
individual variants within the population bear antigenically
different major variable surface glycoproteins (VSG) and thus
are not "seen" by the specific antibodies induced by the
predominant VSG produced by the majority of parasites in a
parasitemic wave. Thus the antigenically different trypanosomes
are selected by host antibodies and establish the antigenic
phenotype of the next parasitemic wave. It is this process that

gives rise to the successive waves of parasites characteristic of trypanosome infections (1).

The African trypanosomes are extremely successful parasites (partly because of antigenic variation), and they have a tremendous impact on the health and economic well-being of many millions of people in Africa, either directly by causing sleeping sickness or indirectly by infecting their domestic livestock. Clearly, control of the trypanosome, at least in humans, is desirable and has been selected as one of the goals of the World Health Organization (2).

Many believe that immunological control of the African trypanosome is unlikely, if not impossible, because of the difficulties imposed by antigenic variation. However, recent information gained in part using monoclonal antibodies suggests new strategies for immunological control of these parasites and has kindled renewed enthusiasm in the study of their immunological behavior.

THE TRYPANOSOME LIFE-CYCLE

Most African trypanosomes are cyclically transmitted with the tsetse fly as an intermediate host and vector. In the fly, uncoated midgut forms (procyclics) migrate forward, ultimately lodging in the salivary glands where they differentiate into the metacyclic stage. Here the parasites acquire a surface coat, the VSG, which (at least in the bloodstream form (3) and probably in the metacyclic form) covers the entire surface of the parasite. Metacyclic trypanosomes are passed into the mammalian host when the tsetse fly takes a blood meal. Once in the host, the metacyclic trypanosomes differentiate into long slender bloodstream forms and then into short stumpy forms, which are infective for the fly. In the fly midgut, the bloodstream forms differentiate into uncoated procyclic forms, completing the cycle.

The bloodstream forms of the African trypanosomes are the most well studied, primarily because these are the stages of the parasite associated most strongly with antigenic variation and because, with most species at least, large quantities of material can be obtained (4). The use of these stages allows the purification of milligram quantities of VSG. Metacyclic trypanosomes are much more difficult to obtain in quantities normally considered sufficient for biochemical analysis and are much less studied than the bloodstream stages. Uncoated culture forms that are established and maintained <u>in vitro</u> (5) are thought to be identical to procyclic midgut forms and can be grown in large quantities for biochemical and antigenic analysis.

The fact that monoclonal antibodies can be used to dissect the antigens from mixtures even when only small quantities of antigenic material are available (6) has allowed their use with

the antigenically complex trypanosomes at all stages in the
life-cycle.

MONOCLONAL ANTIBODIES TO VARIABLE SURFACE GLYCOPROTEINS OF
BLOODSTREAM TRYPANOSOMES

The first monoclonal antibodies to African trypanosomes
were produced between 1977 and 1979, with purified VSGs from T.
brucei as immunizing antigens (6, 7). These antibodies were
used primarily to investigate the epitopes on VSGs and their
relationship to the trypanosome surface. Of 11 monoclonal
reagents, all were specific for the immunizing VSG, for none of
them cross-reacted with any of eight other antigenically
different VSGs. Only two of the antibodies bound to the surface
of living trypanosomes as tested by indirect immunofluorescence,
whereas all 11 bound to acetone-fixed parasites of the
appropriate antigenic type. This finding was the first
indication that the antigenic variability of the VSGs extends
through parts of the molecules other than those exposed to the
external environment. Several of the monoclonal antibodies were
shown to bind epitopes on the protein portions of the molecules
as they precipitated nonglycosylated molecules translated from
reticulocyte lysate systems in vitro. The monoclonal antibodies
could be used to make immunoadsorbents that allow purification
of VSGs from antigen mixtures (8).
     Several other groups have described monoclonal antibodies
specific for VSGs from T. rhodesiense (9, 10). Again, both
exposed and non-exposed antigenic sites were detected by the
monoclonal probes. In no instance was any cross-reactive
anti-VSG monoclonal antibody found. To our knowledge, none of
several hundred different monoclonal antibodies tested over the
past 4 or 5 years detect cross-reacting epitopes on
antigenically different VSGs.
     About 1 year ago we initiated a project with Drs. Tony
Barbet and Travis McGuire to study the distribution of epitopes
on trypanosome VSGs. To do this we chose two weakly
cross-reacting T. brucei VSGs, WaTat 1.1 and WaTat 1.12 (11).
Of 27 monoclonal antibodies specific for these purified
antigens, 11 react with surface-exposed epitopes on living
trypanosomes (Table 1). Of these, five react with both VSGs,
two react with VSG 1.1 only, and four react with VSG 1.12 only.
Of further interest is the reactivity of the monoclonal
antibodies in different assays. Some of the antibodies react
with VSGs in both liquid-phase and solid-phase assays, some only
when the VSG is in liquid-phase, and some only when the VSG is
bound to a solid-phase. These results imply that antigen
conformation is extremely important for maintenance of many of
the VSG antigenic sites as would be expected, and stress the
importance of using intact trypanosomes for identification of
biologically important antigenic sites. Peptide mapping studies

Table 1--Monoclonal antibodies specific for WaTat 1.1 and 1.12 variable surface glycoproteins (VSG) tested for binding in various assays

Immunogen	Hybridoma	Surface[a]		Solid phase[b]		Liquid phase[c]	
		1.1	1.12	1.1	1.12	1.1	1.12
1. Trypanosomes 1.1	TRYP 1E1	+	+	++	++	++	++
2. Trypanosomes 1.1	TRYP 2B1	+	+	+	++	++	++
3. Trypanosomes 1.1	TRYP 4C1	+	+	-	-	++	+
4. Trypanosomes 1.1	TRYP 22A1	+	-	-	-	++	-
5. VSG 1.1	WAT 8A1	-	-	++	++	+	+
6. VSG 1.1	WAT 20A1	-	-	++	++	+	+
7. VSG 1.1	WAT 24A1	+	+	-	-	++	+
8. VSG 1.1	J2/71.15.27	-	-	++	++	++	+
9. VSG 1.1	J2/116.32.18	+	-	+	-	-	-
10. VSG 1.1	J2/165.26.18	-	-	++	++	-	-
11. VSG 1.1	J2/166.32.5	ND	ND	+	-	-	-
12. VSG 1.1	J2/174.25.26	-	-	++	++	+	-
13. VSG 1.1	J2/242.14.26	-	-	++	++	++	+
14. VSG 1.1	J2/391.28.12	ND	ND	+	-	-	-
15. VSG 1.1	J2/426.29.35	-	-	++	++	+	-

Table 1--Continued

Immunogen	Hybridoma	Surface[a]		Solid phase[b]		Liquid phase[c]	
		1.1	1.12	1.1	1.12	1.1	1.12
16. VSG 1.12	248.33.35	−	−	+	++	+	++
17. VSG 1.12	258.32.8	−	+	+	++	+	++
18. VSG 1.12	261.34.33	−	+	+	++	+	++
19. VSG 1.12	284.31.1	−	−	++	++	−	−
20. VSG 1.12	396.32.11	−	+	+	++	+	++
21. VSG 1.12	399.13.6	−	−	+	+	−	−
22. VSG 1.12	437.30.35	−	+	+	+	−	−
23. VSG 1.12	438.28.35	−	−	++	++	−	−
24. VSG 1.12	448.26.5	−	−	++	++	−	++
25. VSG 1.12	448.28.10	−	−	+	+	−	+
26. VSG 1.12	449.32.5	+	+	+	++	−	+

ND = Not done.
[a]Tested by indirect immunofluorescence on living trypanosomes.
[b]Hybridoma culture supernatants were tested in solid-phase radioimmunometric assays (12).
[c]Hybridoma culture supernatants were tested in liquid-phase radioimmunoassay with ^{125}I-labeled VSGs as antigen (7, 8).

and ordering of peptides by amino acid sequencing are underway
so that localization of epitopes on the protein primary
structure can be achieved by use of monoclonal reagents as
probes for individual peptides.

Preliminary studies with monoclonal antibodies that bind
to surface-exposed epitopes on living trypanosomes WaTat 1.1
and 1.12 indicate that these epitopes are dependent on the
three-dimensional structure of the VSG, for none of the
antibodies bind to cyanogen bromide cleavage fragments
produced from reduced and alkylated molecules. Thus it may be
necessary to produce monoclonal reagents with the peptides
themselves as immunizing antigens, or at least to use
fragments obtained from non-reduced VSG, in order to localize
the surface-exposed antigenic sites. Whether or not regions
on the VSG that are normally immunosilent can be made to
induce an immune response when isolated from the intact VSG,
as has been done with other antigens (13, 14), remains to be
seen. Information recently obtained from amino acid and DNA
sequencing data (see below) indicates that there is a strong
possibility that the overall three-dimensional structure of
the trypanosome VSGs is conserved, thus allowing optimism that
common immunosilent regions could exist on many different
VSG. If such regions exist on VSG in surface-exposed portions
of the molecules, then new immunization strategies become a
distinct possibility.

That the overall three-dimensional structure of
trypanosome VSG may be conserved is suggested by data obtained
from some recent sequencing and VSG purification studies.
Comparison of amino terminal amino acid sequences of eight
antigenically different VSGs of sequential parasitemic peaks
and of published amino terminal sequences of VSGs from various
trypanosome species shows a remarkable degree of sequence
homology among the VSG (Olafson, R.W., Clarke, M., Kielland,
S., Pearson, T.W., Barbet, A.F., and McGuire, T.C., manuscript
submitted). These data, along with previously published
C-terminal homologies found by DNA sequencing (15), suggest
that the VSGs may have more of their primary structures
conserved than was previously thought to be the case. In
addition, reverse-phase high-performance liquid
chromatographic (HPLC) analyses of VSGs (Clarke, M., Olafson,
R.W., and Pearson, T.W., manuscript submitted) showed that the
overall polarity of the different VSG molecules is conserved.
This fact has allowed both analytical and preparative HPLC
purification of trypanosome VSGs from lysed trypanosomes.

MONOCLONAL ANTIBODIES TO VARIABLE SURFACE GLYCOPROTEINS OF
METACYCLIC TRYPANOSOMES

Metacyclic trypanosomes are much more difficult to work
with than bloodstream forms simply because it is difficult to

obtain them in large quantities and because of the need for
fly-rearing facilities. Nevertheless, substantial progress
has been made in the study of the VSG. Early reports
suggested, that on passage in the fly, the VSG phenotype of
bloodstream populations reverted to a single antigenic type in
the metacyclic population (16, 17). Subsequent studies with
more completely defined antisera (18, 19) showed that the
metacyclic populations in the single tsetse fly are
antigenically heterogeneous but at least exhibit a more
limited antigenic repertoire than bloodstream trypanosomes.

Monoclonal antibodies produced by immunization with whole
living metacyclic trypanosomes (20, 21) have confirmed the
findings obtained with antisera that metacyclic populations
are antigenically heterogeneous with respect to surface
epitopes. In addition, the monoclonal reagents have allowed
relatively accurate determinations of the number of different
antigenic phenotypes present in a single fly and the antigenic
relationship between metacyclic trypanosomes from the same or
different geographical areas. That the parasites in the first
parasitemic wave of bloodstream trypanosomes share surface
antigenic sites with the initiating metacyclic population has
been clearly demonstrated (20, 21). This observation suggests
the possibility of using large quantities of bloodstream forms
as a source of VSG that at least cross-reacts with metacyclic
surface antigens and could be used for immunization against
metacyclic populations. Protection of mice against infection
with metacyclics has in fact been attained after immunization
with bloodstream forms taken from the first parasitemic peak
(20). Biochemical characterization of any of the metacyclic
VSGs has not been performed, and their precise relationship to
those VSG appearing on bloodstream forms is unknown. In
addition, the extent of the metacyclic VSG repertoire has not
been determined even for a given serodeme. The likelihood of
successful vaccination against trypanosomes therefore cannot
be predicted at this time.

MONOCLONAL ANTIBODIES TO ANTIGENS OTHER THAN VARIABLE SURFACE
GLYCOPROTEINS

Although most investigators have been interested in the
antigenic analysis of VSG, others have (inadvertently or
otherwise) derived monoclonal antibodies to other trypanosome
antigens (9). None of these monoclonals (or their specific
antigens) have been characterized at all except that they have
been found not to be variant-specific.

We have, however, recently identified an antigen specific
for all T. congolense parasites by use of two different
monoclonal antibodies derived in different laboratories
(Parish, N., Cleever, H., Morrison, W.I., and Pearson, T.W.,
in preparation). This antigen is found on both bloodstream

and procyclic forms of several $\underline{T}$. $\underline{congolense}$ isolates tested in both solid-phase radioimmunometric and indirect immunofluorescence assays. The antigen is a protein or glycoprotein and has an SDS-molecular weight of 31,000. The epitopes bound by the monoclonal antibodies are not exposed on the surface of either the bloodstream or procyclic forms although they appear to be membrane associated. The antigen thus appears to be a species-specific marker for $\underline{T}$. $\underline{congolense}$ and may have use as a reagent for epidemiological studies. A search for homologous proteins in the $\underline{T}$. $\underline{brucei}$ species and associated subspecies is underway using two-dimensional gel analysis.

Monoclonal antibodies specific for nonvariable membrane antigens of other species of trypanosomes have also been derived by using uncoated procyclic culture forms of $\underline{T.b.}$ $\underline{rhodesiense}$ (Richardson, J., Jenni, L., and Pearson, T.W., in preparation). Fourteen different monoclonal reagents specific for surface epitopes of the $\underline{T.b.}$ $\underline{rhodesiense}$ were selected using indirect immunofluorescence on living procyclics. All of these antibodies bound to living procyclics of other $\underline{T.b.}$ $\underline{rhodesiense}$ clones and to procyclics of $\underline{T.b.}$ $\underline{brucei}$ and $\underline{T.b.}$ $\underline{gambiense}$ species. None bound to $\underline{T}$. $\underline{congolense}$ procyclics. Several of the antibodies tested so far bound to acetone-fixed bloodstream forms of $\underline{T.b.}$ $\underline{rhodesiense}$ in immunofluorescence assays and appeared to bind along the plasma membrane. Again, these monoclonal antibodies show species specificity and may have epidemiological utility. Preliminary studies in collaboration with Dr. Leo Jenni (Basel) have shown that the monoclonal reagents bind to procyclic trypanosomes taken from the tsetse fly midgut.

Of perhaps more importance is the potential value of these membrane antigens for use in immunization strategies. For example, it may be possible to decrease transmission rates by influencing the differentiation of procyclics in the tsetse fly. Alternatively, the possibility that immune responses directed against invariant membrane antigens could be selectively amplified is worth exploring, especially if done in conjunction with immunization against variable or nonvariable epitopes on bloodstream or metacyclic variable antigens.

ACKNOWLEDGMENT

Some of the work presented here received financial support from the UNDP/World Bank/WHO Special Programme for Research and Training in Tropical Diseases and from the Natural Sciences and Engineering Research Council of Canada. We thank Jennifer Duggan for typing this manuscript.

REFERENCES

1. Vickerman, K. (1978) Nature 273:613-617.
2. UNDP/World Bank/WHO (1977) Special Programme for Research
    and Training in Tropical Diseases. WHO Publications,
    Geneva.
3. Cross, G.A.M. (1975) Parasitology 71:393-417.
4. Turner, M.J. (1982) Adv. Parasitol. 21:69-153.
5. Brun, R., and Schönenberger, M. (1979) Acta Tropica
    36:289-292.
6. Pearson, T.W., and Anderson, N.L. (1980) Anal. Biochem.
    101:377-382.
7. Pearson, T.W., Pinder, M., Roelants, G.E., Kar, S.K.,
    Lundin, L.B., Mayor-Withey, K.S., and Hewett, R.S. (1980)
8. Pearson, T.W., Kar, S.K., McGuire, T.C., and Lundin, L.B.
    (1981) J. Immunol. 126:823-828.
9. Campbell, G., Griswold, S., Cain, G., Giorgi, J.V., and
    Warner, N.L. (1981) In: Monoclonal Antibodies and T-cell
    Hybridomas. Hämmerling, G.J., Hämmerling, U., and
    Kearney, J.F. (eds.), Elsevier/North-Holland, Amsterdam,
    pp. 323-328.
10. Lyon, J.A., Pratt, J.M., Travis, R.W., Doctor, B.P., and
    Olenick, J.G. (1981) J. Immunol. 126:134-137.
11. Barbet, A.F., Davis, W.C., and McGuire, T.C. (1982)
    Nature 300:453-456.
12. Tsu, T.T., and Herzenberg, L.A. (1980) In: Selected
    Methods in Cellular Immunology. Mishell, B.B., and
    Shiigi, S.H. (eds.), W.H. Freeman and Co., San Francisco,
    pp. 373-397.
13. Sutcliffe, J.G., Shinnick, T.M., Green, N., and Lerner,
    R.A. (1983) Science 219:660-666.
14. Bittle, J.L., Houghten, R.A., Alexander, H., Shinnick,
    T.M., Sutcliffe, J.G., Lerner, R.A., Rowlands, D.J., and
    Brown, F. (1982) Nature 298:30-33.
15. Matthyssens, G., Michiels, F., Hamers, R., Pays, E., and
    Steinert, M. (1981) Nature 293:230-233.
16. Gray, A.J. (1975) Gen. Microbiol. 41:195-199.
17. Jenni, L. (1977) Acta Tropica 34:35-41.
18. LeRay, D., Barry, J., and Vickerman, K. (1978) Nature
    273:300-302.
19. Barry, J., Hajduk, S., Vickerman, K., and LeRay, D.
    (1979) Trans. R. Soc. Trop. Med. Hyg. 73:205-208.
20. Esser, K., Schoenbechler, M.J., and Gingrich, J.B. (1982)
    J. Immunol. 129:1715-1718.
21. Nantulya, V.M., Musoke, A.J., Moloo, S.K., and Ngaira,
    J.M. (1983) Acta Tropica 40:19-24.

CHAPTER 7

THE IDENTIFICATION AND ANALYSIS OF MAJOR
FUNCTIONAL POPULATIONS OF DIFFERENTIATED CELLS

W.C. DAVIS, L.E. PERRYMAN, and T.C. MCGUIRE
Department of Veterinary Microbiology and Pathology
College of Veterinary Medicine
Washington State University, Pullman, WA   99164-7040

The advent of monoclonal antibody technology (1-6) and the
recent technology for the direct transfer of genomic DNA (7-9)
has revolutionized the approaches available for the analysis of
the immune response in outbred species and the elucidation of
the genetic basis of disease susceptibility. Through cell
fusion, it is now possible to capture and immortalize B
lymphocytes producing antibody of desired isotype and
specificity. This breakthrough has provided a means of using
the full potential of antibodies as a bioanalytical tool and
conducting investigations in outbred species with a precision
previously possible only with inbred strains of mice. Through
transfection, the transfer and insertion of genomic DNA into the
chromosomes of replicating eukaryotic cells, it is now possible
to isolate and identify genes coding for specific gene products
(8, 9) and to establish cell lines in the species of choice.
The latter discovery has provided a means of producing cell
lines for use in genetic typing and cell lines of lymphoid cells
that have differentiated to perform specific immunologic
functions; it may also provide an additional method for the
development of cell lines producing monoclonal antibodies (9).
     In this report, we describe the techniques we have evolved
to exploit monoclonal antibody technology in studying the immune
response in domestic animals and the strategies currently being
used to prepare monoclonal antibodies to major histocompatibility
gene products and leukocyte differentiation antigens; we also
present a preliminary description of a set of monoclonal
antibodies being prepared for use in studying the immune
responses of domestic animals.

PREPARATION AND CHARACTERIZATION OF CROSS-REACTIVE MONOCLONAL
ANTIBODIES

To elucidate the mechanisms of immunity in domestic
species and apply the information obtained to control disease,
it will be essential to identify the cell subpopulations that
participate in a given response and define the constellation
of membrane molecules (differentiation antigens) regulating
the interactions of cells. Where multiple species must be
studied, this presents an intriguing challenge with two major
components. At this juncture, there is a need to a) define
species differences in the immune system that influence host
response to disease and b) elucidate the genetic factors that
affect the way an animal responds to an infectious agent.
Because primary emphasis has been placed on the analysis of
the immune response in humans and mice, few well-characterized
immune reagents exist as yet that meet the needs of
investigators interested in food animal research. Before
significant progress can be made, an extensive set of immune
reagents are needed for the identification and
characterization of leukocyte differentiation antigens and
gene products of the major histocompatibility gene complex
(MHC) in the species of interest. Monoclonal antibody
technology affords an expedient way to achieve this objective.
There are two approaches available for the production of
monoclonal antibodies to lymphocyte membrane antigens. The
first is to develop antibodies to relevant antigens on a
species by species basis. Although this has proven to be an
effective approach in the study of the immune system in humans
and mice (7-22), it may not be the method of choice for food
animal research. The reason is that researchers using
antibodies to species-specific antigens must rigorously
establish that the antigens expressed on the cell membrane of
a given cell type in each species are indeed homologous
counterparts, and that they carry out the same regulatory
functions. In studies with humans, a number of monoclonal
antibodies that appear to recognize the same differentiation
antigens have been developed, as have monoclonal antibodies
that appear to detect molecules with similar patterns of
expression on leukocytes; however, some are indeed different
(10, 17, 18, 20). Without rigorous proof of identity, one is
left with the question of whether investigators using
different sets of monoclonal antibodies are studying the same
cell populations and/or differentiation antigens. The finding
that a number of antigens are differentially expressed on
subpopulations of one or more cell types (15-21, 23, 24)
emphasizes that this is not a trivial problem and that it may
well complicate correct interpretation of data, especially
when interspecies comparisons of immune function are being
made.

The second approach is to produce cross-reactive antibodies useful in two or more species. When an animal is immunized with cells from another species, two types of antibodies are formed: antibodies that recognize antigenic epitopes on molecules that are essentially species specific and antibodies that recognize antigenic epitopes expressed on molecules in one or more species. The majority of these antibodies recognize molecules with a common phylogenetic origin. Such antibodies (both xenoantisera and monoclonal antibodies) have been used to detail the relation of proteins, most notably the immunoglobulins (25-27) and products of the MHC (28-41). The success with cross-reactive antibodies has indicated that it should be possible to use monoclonal antibody technology to develop a library of monoclonal antibodies reactive with homologous gene products derived from two or more species. The development of such a set of antibodies would have a number of advantages. It would a) permit the direct demonstration that homologous gene products have indeed been identified, that they are expressed in an identical way on the same cell subpopulations and function in an identical manner; b) permit inter- and intraspecies comparison of molecules identified by monoclonal antibodies that recognize species-restricted antigenic epitopes; c) facilitate the elucidation of the relation of families of molecules coded for by genes with a common phylogenetic origin; d) provide a means of generating monoclonal antibodies to polymorphic antigenic determinants coded for by each allelic set of genes; and, importantly, e) minimize the time and redundancy of effort required to prepare monoclonal antibodies for each species of interest.

PRODUCTION OF CROSS-REACTIVE MONOCLONAL ANTIBODIES

The strategy for developing a library of cross-reactive antibodies has included the following steps: 1) apply tissue culture methodology that optimizes the outgrowth of cell hybrids and permits the primary analysis and preservation of large numbers of antibody-producing hybridomas; 2) refine methodology for optimizing the production of hybridomas producing cross-reactive monoclonal antibodies; 3) develop and/or modify assays to facilitate characterization of the patterns of reactivity of cross-reactive monoclonal antibodies and determine their apparent specificity; 4) determine the biochemical and functional properties of the cell membrane molecules recognized by the monoclonal antibodies; 5) determine the phylogenetic relationship of the cell membrane molecules recognized by the monoclonal antibodies; and 6) define the chromosomal locus of the genes coding for the molecules and the number of allelic variants present in each species.

The procedure we have found most useful in the production of large numbers of hybrids has been a two-phase approach (5). One of the major difficulties encountered in screening of cultures for hybrids producing an antibody of interest has been the short interval between the time of assay and the time of cell transfer to a second culture vessel. This short interval is necessary to avoid overgrowth and cell death. A second difficulty has been the management of large numbers of cultures while determining which cell lines should be maintained. However, with a two-step procedure, it is possible to work with 400 to 600 primary cultures with reasonable effort (Figure 1). In phase I of production, fused cells are dispersed in 96-well plates and cultured for 10 to 14 days. Supernatants are collected and tested for antibody activity. Cell lines producing the antibodies of interest are then transferred to 24-well plates and maintained in static culture for 2 weeks. By removing excess cells every 2 to 3 days, the cell line is not allowed to expand beyond one well during long-term culture. At the end of 2 weeks, a duplicate plate is prepared and maintained without thinning for a week. Supernatants are then collected and tested for antibody activity. Selected cell lines are expanded into 6-well plates and then preserved in liquid nitrogen.

Once an aliquot of cells has been preserved, the remaining cells are allowed to overgrow and die, thus making a final antibody-rich supernatant for further analysis. In phase I, all cell lines of interest can be maintained and antibody-producing cell lines can be assessed for cloning.

In phase II, each antibody is systematically analyzed to select the cell lines to clone for the establishment of permanent cell lines. A hybrid cell line is plated at high cell dilution to yield 50 to 100 single colonies in four 96-well plates. The clones are tested for antibody activity. Positive clones are taken and carried through the same steps as in phase I. The four to six clones are selected and preserved to establish the cell line. In phase II, the selection cloning of the cell lines of immediate interest can be performed and the stability of antibody production determined.

The factors found critical for consistent production of hybrids are use of 2-mercaptoethanol (2-ME) and lipid A (Ribi Immuno-Research, Hamilton, MT and List Biological Laboratories, Inc., Campbell, CA) in the culture medium and the use of thymocytes as feeder cells. Comparative studies have shown that there is an additive effect when 2-ME is used in conjunction with feeder cells. Though the mechanism has not been elucidated, the addition of 1 µg lipid A per well at the time the fused cells are dispensed into 96-well plates increases the yield of hybrids by 20 to 40%. Other factors, such as the myeloma cell line used as a fusion partner, source of plastic ware, purity of the water used to make medium, and

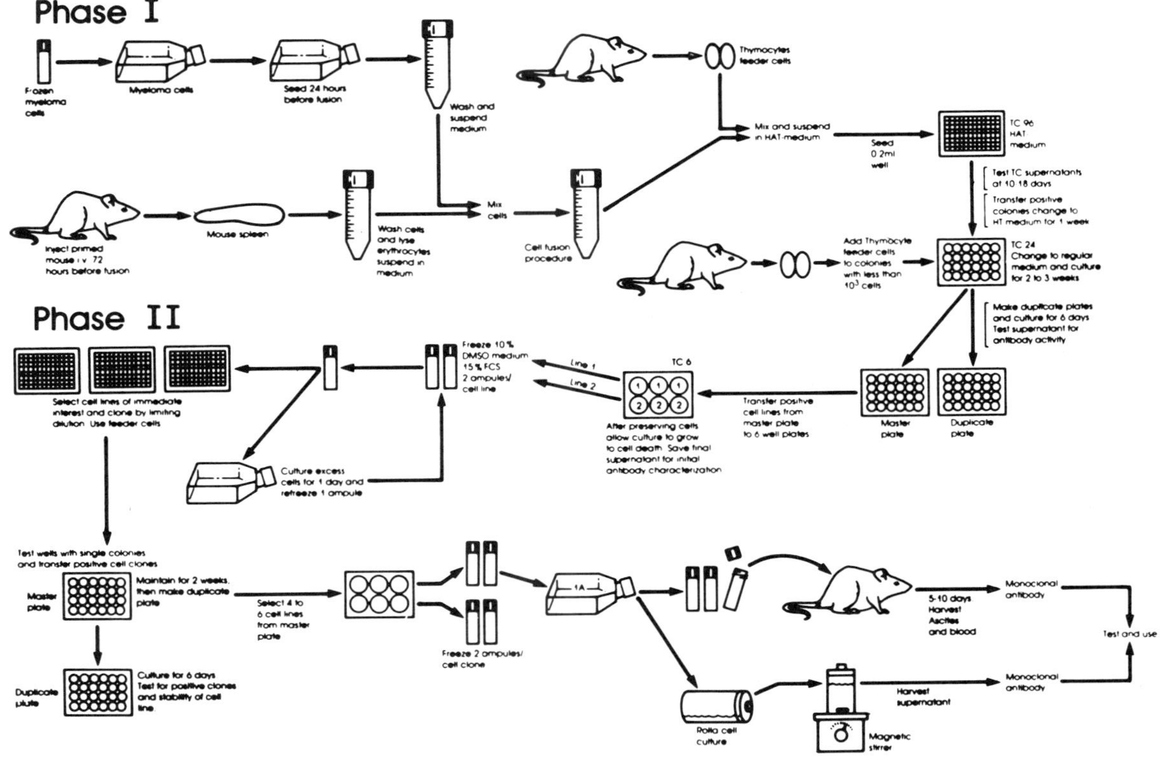

Figure 1--Flow diagram of a two-phase procedure for producing and analyzing large numbers of antibody-producing hybridomas. See text for details.

variation in fetal bovine serum lots are important but only
secondary to the culture conditions mentioned.

HYBRIDS SECRETING CROSS-REACTIVE ANTIBODIES

When mice are immunized with xenogeneic cells, they usually
produce antibodies to antigenic epitopes that are species
restricted and antibodies to epitopes expressed on cells from
other species. If an animal has been exposed to cells from a
single species, however, the predominant clonotypes present in
the system produce antibody to species-restricted epitopes. We
explored different strategies of detection of monoclonal
antibodies as well as strategies of immunization that might
increase the number of clonotypes producing cross-reactive
antibody. We found that hyperimmunization with leukocytes from
multiple species, given sequentially or as a mixture, alter the
composition of the population of B-cell clonotypes and the
specificity of the antibodies being produced. The results
obtained thus far indicate that the potential for producing
hybrids synthesizing antibodies that are broadly cross-reactive
increases as more species are used as cell donors.
We found that identification of cross-reactive antibodies
could be facilitated by designing an immunization protocol to
exclude cells from one of the species of interest. This enabled
us to use cells from that species in the primary screening for
cross-reactive antibody activity. In our studies, we usually
tested for antibody activity with cells from one of the species
used in the immunization protocol as well as a species excluded
from the immunization protocol.

METHODS OF ASSAY

Several methods of assay have been used to select hybrids:
the complement-mediated microcytotoxicity assay (CT) (42), the
enzyme-linked immunosorbent assay (ELISA) (43-45), and flow
microfluorimetry (FMF) (46). In the initial phases of the
program, CT and ELISA were sufficient for selecting monoclonal
antibodies that had potential for use in genetic typing and
analysis of leukocyte subsets. However, FMF proved essential
for selecting and sorting monoclonal antibodies for functional
analysis and comparison of cross-species reactivity. In the
studies described here, we used the Becton Dickinson FACS™
research analyzer, a new instrument similar to the laser-based
fluorescence-activated cell sorters (FACS) currently used in
many laboratories. The principal differences are the mode of
fluorescence excitation, volume sensing, and transmission of
cells through a sensing orifice. The FACS analyzer rapidly
measures individual cells simultaneously for fluorescence
(autofluorescence and fluorescence imparted by fluorochrome-

coupled antibodies bound to a cell membrane), color of
fluorescence (using a mercury-arc illuminator), and for size (as
measured by electronic volume). Up to 1000 cells a second can
be analyzed, and the data are displayed in two modes: a) a
histogram profile based on relative cell number and relative
change in fluorescence intensity of unlabeled and
fluorochrome-antibody tagged cells and b) a dot plot profile
based on cell size and intensity of fluorescence. By comparing
data displayed in these modes, we could estimate the size of the
cell population expressing a given membrane antigen, the
staining intensity (dim or bright), and the size of the cells
expressing the antigen, and could construct a profile of the
antigens differentially expressed on cell subsets.

SELECTION AND CHARACTERIZATION OF MONOCLONAL ANTIBODIES

     Because we are interested in the identification of potential
genetic markers as well as differentiation molecules, our initial
endeavors have focused on the selection and cloning of cell lines
producing antibody to any and all antigenically distinct leukocyte membrane molecules. The CT assay was used as the primary
assay to select and establish cell lines in our initial studies.
Later, both CT and the ELISA were used to select cell lines for
maintenance in static culture and analysis by FMF to distinguish
cell lines producing antibody to cell subpopulations. The
following is a description of the monoclonal antibodies that
have been produced from cloned cell lines that react with
bovine, ovine, caprine, equine, and human leukocytes. At this
juncture 120 cell lines have been selected from a library of 700
lines carried through static culture. These were derived from
approximately 12,000 hybrids produced from 27 fusions. The data
on 94 of the monoclonal antibodies are summarized in Table 1 and
Figures 2-6. (The following codes denote the acronyms for the
monoclonal antibodies and immunization schemes: B, E, or G =
immunized with bovine, equine, or caprine HF-PBL, respectively;
CA = CBA (H-2^k) mice immunized with B10.A (H-2^a) mouse lymph
node cells; HT = immunized with equine thymocytes; TH = immunized
with bovine, equine, caprine, and rat thymocytes; PT = immunized
with porcine thymocytes; PIg = immunized with porcine and bovine
IgM; PORC, SPAM, and CAPP = immunized with bovine and porcine
platelets; RH = immunized with rat lymph node cells and equine
fribrocytes; ANA = immunized with bovine erythrocytes infected
with Anaplasma marginale; H = immunized with horse, rabbit, rat,
dog, and goat HF-PBL; HELP and SAG = immunized with BoLA class I
antigen isolated from a membrane lysate using monoclonal
antibody H58A bound to protein A-staphylococcus.)

Table 1--A library of cross-reactive monoclonal antibodies for use in biological, biomedical, and agricultural research (analyzed with a Becton Dickinson FACS research analyzer)

Monoclonal antibody no.	Ig isotype	Species pattern of cross reactivity (% cells positive for antigen)						Apparent specificity
		Cow	Sheep	Goat	Pig	Horse	Human	
PNA	Lectin	50-86	54	57	69	60-84	31	Species variable
H4	IgG3	15-25	15-25	10-28	15-35	15-25	15-30	MHC-class II
Leu2b	IgG2a	-	-	-	-	15-20	28	T cytotoxic/ suppressor cell
Leu8	IgG2a	15-25	-	-	-	15-20	68	T cell subset
H1A	IgG2a	85+	-	-	90	87	95	-
H6A	IgG2a	85+	98	50	50-90(P)	97	97	-
H11A	IgG2a	90	87	65	62(P)	98	95	MHC-class I
H12A	IgM	-	-	-	-	-	95(P)	-
H17A	IgG2ab	85(P)	-	-	90(P)	98	90(P)	MHC-class I
H18A	IgM	8-10	8-11	8-11	3-14	10	15-19	Monocyte
H20A	IgG1	60(P)	-	-	60	74	60	-
H21A	IgM	90(P)	38-90(P)	16(P)	34-80(P)	97	90	MHC-class I
H22A	IgG2a	-	-	-	88	78	90	MHC-class I
H25A	IgM	-	-	-	-	-	40-90(P)	-
H26A	IgG1	-	-	40-57(P)	-	78-90	72	-
H26H	IgM	19-36(P)	-	-	-	64	16-30(P)	-
H34A	IgG2a	26-36	27	14-30(P)	-	34-55	20	MHC-class II
H42A*	IgG2a	20-28	20-29	11-20	28	32-59	19	MHC-class II
H58A*	IgG2a	95	94	85	83(P)	99	95	MHC-class I
H64A	IgG2a	-	-	61(P)	-	97	?(P)	MHC-class I
E4C	IgM	8-30(P)	5	3-16	5-10	10-40	5-10(P)	T subset

Table 1--(Continued)

Monoclonal antibody no.	Ig isotype	Species pattern of cross reactivity (% cells positive for antigen)						Apparent specificity
		Cow	Sheep	Goat	Pig	Horse	Human	
E18A	IgG2a	26	–	–	–	80	–	–
E22E	IgM	–	–	–	–	95	80(P)	–
E23A	IgM	15(P)	–	–	–	36	–	Subpop
E40A	IgM	15-30(P)	10	12-23	11	19-37	10-20(P)	T subset
E47B	IgM	10-30	9	8-20(P)	–	40-56	9-23(P)	Subpop
E48C	IgM	40(P)	–	?(P)	–	41	38(P)	Subpop
E50A	IgM	40(P)	?(P)	32	–	44	40(P)	Subpop
HT6	IgG3	–	–	10-22	–	63	–	Subpop
HT8	IgM	–	–	–	–	67	–	Subpop
HT14E	IgG1	–	–	–	–	22	–	Subpop
HT16A	IgG1	–	–	–	–	33	–	Subpop
HT23A	IgG1	–	–	–	–	46	–	Pan T
HT53A	IgM	13-25(P)	30	–	–	32	–	Subpop
B1A	?	15-30	–	–	–	–	–	T subset
B5C	IgG2b	95(P)	94	89	–	98	90(P)	MHC-class I
B16A	IgM	30	11(P)	16-21	19	20-30	9-18	Monocyte and granulocyte
B18A	IgG3	30	–	–	12-20	20(P)	–	Monocyte and granulocyte
B24A	IgM	40	–	–	–	–	–	Pan T
B26A	IgM	50-60	–	36	–	–	+(P)	Pan T
B29A	IgG2a	50-60(P)	–	–	30(P)	22	30	T subset
G1A	IgM	–	64	80	–	–	–	–
G2A	IgM	30(P)	10	80	–	–	–	Subpop
SAG 4A	IgG2a	85(P)	–	60-80	–	–	–	MHC-class I
SAG 7A	IgG2a	85(P)	(P)	–	–	–	–	MHC-class I

Table 1--(Continued)

Monoclonal antibody no.	Ig isotype	Species pattern of cross reactivity (% cells positive for antigen)						Apparent specificity
		Cow	Sheep	Goat	Pig	Horse	Human	
SAG 8A	IgG2a	85(P)	35(P)	72(P)	–	–	–	MHC-class I
HELP 1	IgM	85(P)	–	–	–	–	–	MHC-class I
HELP 2A	IgM IgG1	85(P)	–	–	50-80(P)	–	–	MHC-class I
HELP 2B	IgM	85(P)	22(P)	–	50-80(P)	–	–	MHC-class I
CA 48P-A	IgM	85(P)	84(P)	50-94(P)	70-80(P)	98(P)	98(P)	MHC-class I
CA 48P-A$_1$	IgG2a	85(P)	84(P)	50-94(P)	70-80(P)	98(P)	98(P)	MHC-class I
CA 4C-A	IgM	85(P)	21(P)	–	30-80(P)	–	–	MHC-class I
TH 1A	IgG2a	20-42	–	–	–	–	–	Subpop
TH 2A	IgG2a	15-33	–	–	20-30(P)	–	–	Subpop
TH 4B	IgM	20-30	12-20	20-30	16-37	37	21	MHC-class II
TH 11E	IgG2a	90	25(P)	20(P)	+(P)	–	–	–
TH 12A	IgG2a	20-45	9-25	–	–	–	–	Subpop
TH 14B	IgG2a	20-30	30-33	34	–	20-41	20	MHC-class II
TH 16A	IgG2a	20-35	24-28	34	32	P?	20(P)	MHC-class II
TH 17A	IgM	86	76(P)	80	30(P)	–	90(P)	–
TH 18A	IgG3	90	–	–	80(P)	–	?	T and B cells
TH 21A*	IgG2b	20-30	23-34	20-32	23-32	33-55	19	MHC-class II
TH 22A	IgG1	15-30	28-37	20	24	9	20-30(P)	MHC-class II
TH 31B	IgG1	30-46	–	–	–	–	–	Subpop
TH 57A	IgG2b	31	–	–	19	12	–	Subpop
TH 61A	IgM	20	10	7	57	25	22	Subpop
TH 62A	IgM	10	–	–	–	–	–	Subpop
TH 71A	IgM	10	–	28	9	27	46	Subpop
TH 81A*	IgG2a	30+	22-30	15-20	30	–	20-30(P)	MHC-class II
TH 82D	IgG1	30+	23-30	10-15	–	–	–	T subset
TH 87A	IgM	20+	–	–	–	–	–	Subpop

Table 1--(Continued)

Monoclonal antibody no.	Ig isotype	Species pattern of cross reactivity (% cells positive for antigen)						Apparent specificity
		Cow	Sheep	Goat	Pig	Horse	Human	
TH 90A	IgM	12-30	?	20	-	23	10	Subpop
TH 92A	IgM	14-30	?	41	34	-	38	Subpop
TH 97A	IgG2a	5-10	?	6	-	-	5	Subpop
PT9A	IgG2a	20-37	15-27	15-20	35	-	32	MHN-class II
PT15A	IgG2a	-	-	-	57	-	-	Pan T + monocyte
PT25D	IgG3	42	24-79	63	25-35	30-60	44	-
PT35A	IgG3	40+	57-70	68	25-38	21-46	40	-
PT36A	IgG1	?	-	-	32	-	-	T subset
PT37A	IgG3	?	-	-	37-45	-	-	Pan T + monocyte
PT40A	IgG3	40+	71	64	21-36	21-36	36	-
PT80B	?		-	-	75	-	-	Pan T + monocyte
PT81B	IgG2b	-	-	-	40-50	-	-	T subset
PT85B	IgG2a	90	80(P)	80	85	90	90(P)	MHC-class I
PT90A	IgG2a	-	-	-	20-30	-	-	T subset
PT91A	IgG3	-	-	-	45	-	-	Pan T + monocyte
PIg31B	IgG1	15	12-15	12-15	10-15	-	-	Light chain
PIg45A	IgG2b	15	12-21	10-17	10-17	12-17	-	IgM
PIg47B	IgG1	15	10-20	10-17	10-17	12-18	-	IgM
RH 1A	IgG3	21-30	23	8-18	-	20-30	32	Subpop
RH 12A	IgG2a	85(P)	-	-	-	-	-	MHC-class I
RH 16A	IgG2a	85(P)	-	-	-	-	-	MHC-class I
PORC 9A	IgG2a	17	38	40-45	19-30	-	-	MHC-class II

Table 1--(Continued)

Monoclonal antibody no.	Ig isotype	Species pattern of cross reactivity (% cells positive for antigen)						Apparent specificity
		Cow	Sheep	Goat	Pig	Horse	Human	
PORC 24A	IgG2a	–	–	–	91	–	–	MHC–class I
SPAM 11A	IgM	8–13	5–11	5–8	7–11	–	–	Monocyte?
CAPP2A	IgGl	19–40	34	30–50	–	–	6	Monocyte?
MPGE17A	IgGl	49(P)	–	–	–	–	–	Subpop
ANA8A	IgGl	–	–	–	–	–	–	Bovine RBC
ANA13A	IgM	41	–	–	–	–	–	Bovine RBC + subpop
ANA18A	IgM	31	–	15	20	–	–	Bovine RBC + subpop

Summary of data obtained from FACS analysis of patterns of cross reactivity of monoclonal antibody with HF–PBL derived from different species. The percentage of cells labeled is based on a statistical analysis of histogram profiles of labeled cells. The apparent specificity is indicated where sufficient data have been obtained. (Where data are limited or the histogram is uninformative the designation has been left blank.) Where information is limited but the histogram reveals a distinct population of cells is labeled, a temporary designation of subpopulation is indicated. PNA = peanut agglutinin, P = polymorphic antigen.

*Reacts with mouse leukocytes.

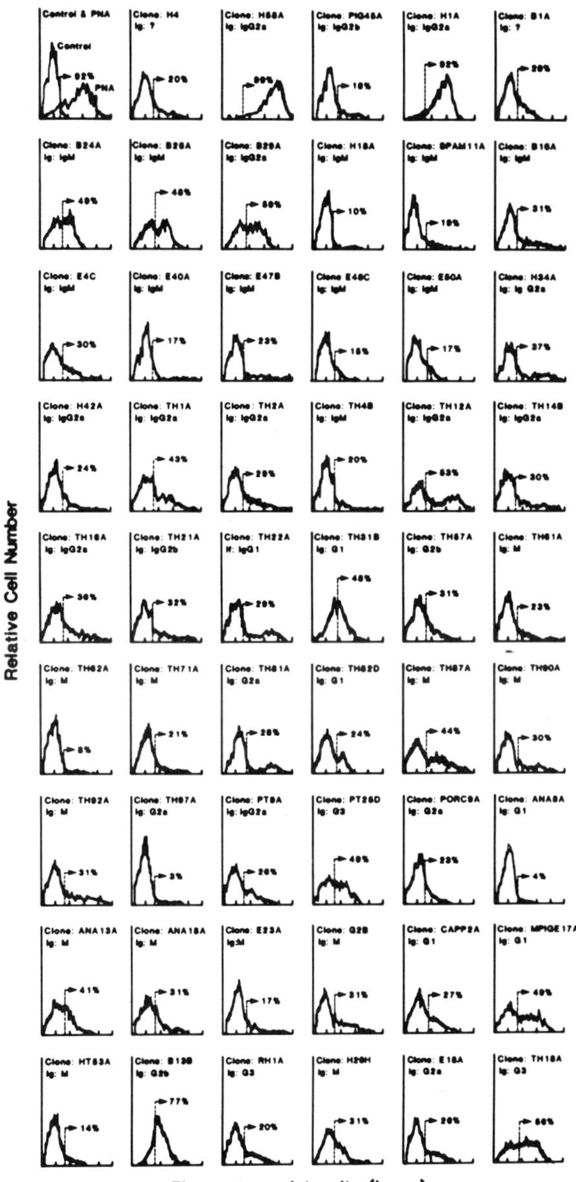

Figure 2--Comparison of histogram profiles of bovine HF-PBL labeled with monoclonal antibody specific for cell membrane antigens with broad or narrow species distribution.

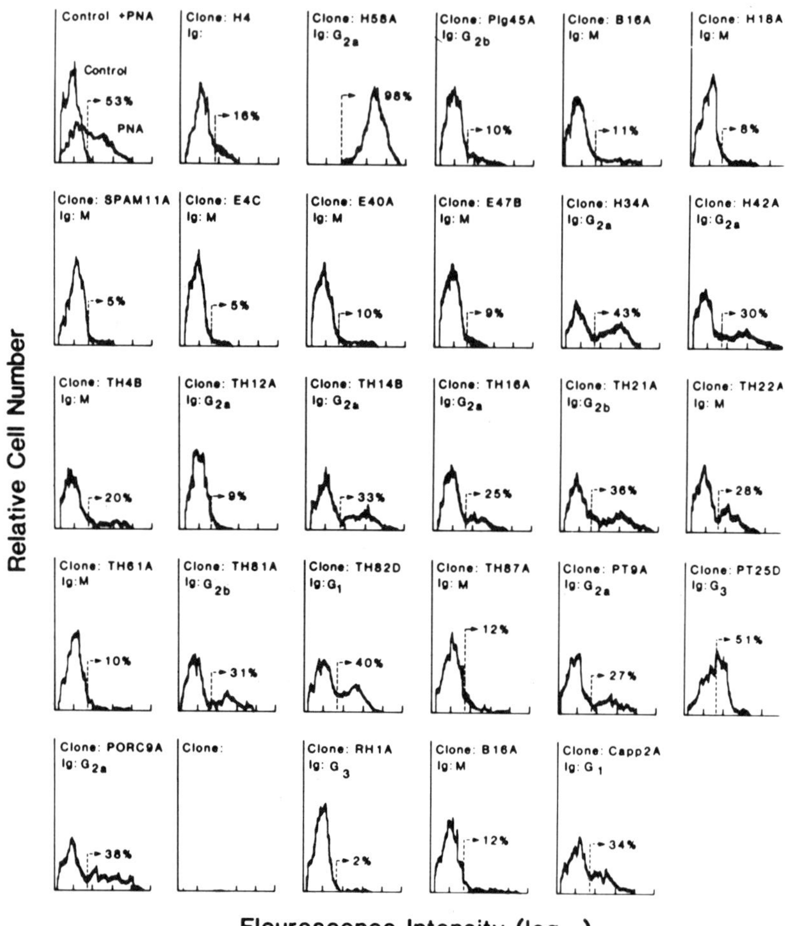

Figure 3—Comparison of histogram profiles of ovine HF-PBL labeled with monoclonal antibody specific for cell membrane antigens with broad or narrow species distribution.

MONOCLONAL ANTIBODIES REACTIVE WITH PRODUCTS OF THE MAJOR HISTOCOMPATIBILITY GENE COMPLEX

To prepare a useful set of monoclonal antibodies cross-reactive with MHC gene products in multiple species, we first identified antibodies that clearly recognize known MHC antigens in humans and mice. We then used these antibodies as standards in conjunction with flow cytometry and other assays to identify monoclonal antibodies to monomorphic and/or polymorphic MHC antigens expressed on cells in one or more species. Using the

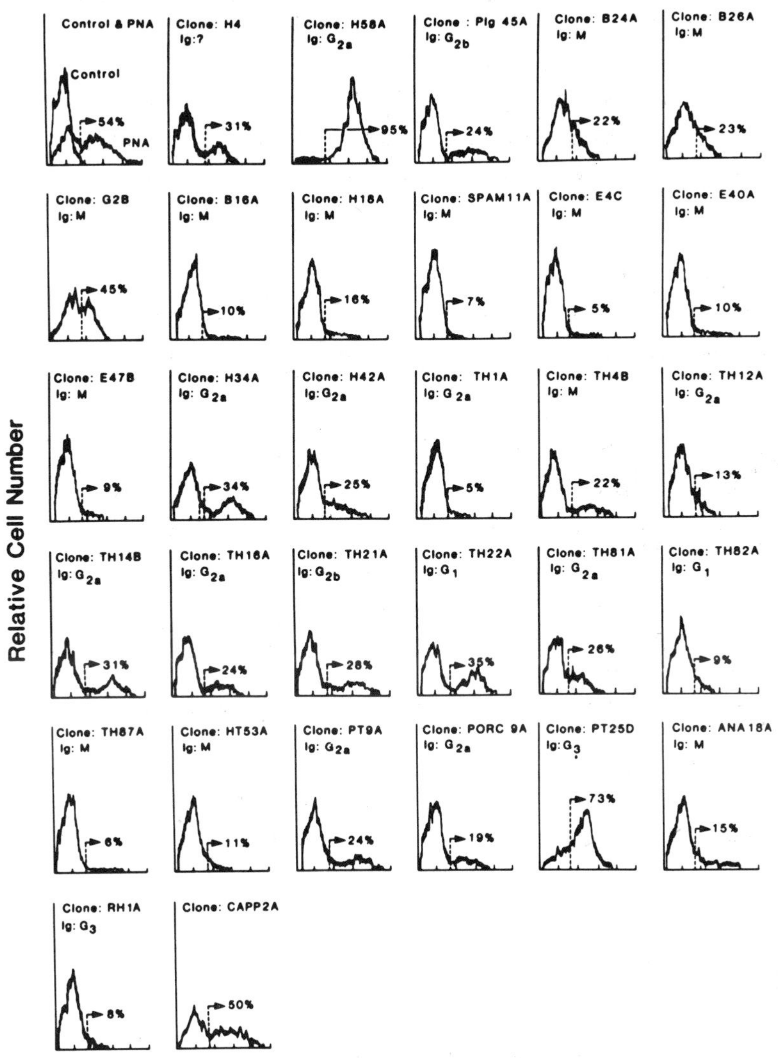

Figure 4--Comparison of histogram profiles of caprine HF-PBL labeled with monoclonal antibody specific for cell membrane antigens with broad or narrow species distribution.

CT assay, we screened a series of 100 monoclonal antibodies (reactive with products of the H-2 gene complex) against bovine leukocytes and identified four antibodies (CA4C, CA13A, CA48P,

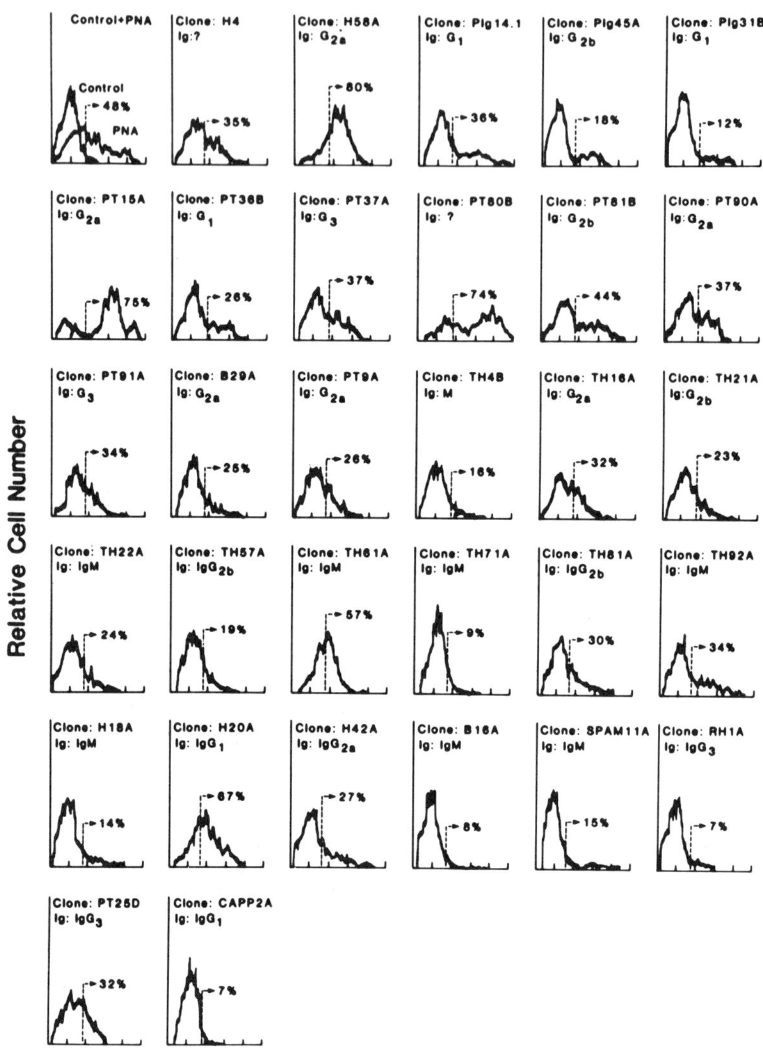

Figure 5--Comparison of histogram profiles of porcine HF-PBL labeled with monoclonal antibody specific for cell membrane antigens with broad or narrow species distribution.

and CA51C). Two of the hybridomas producing the antibodies, CA4C and CA48P, were cloned independently at the Abteilung Immunogenetik Max-Planck-Institut für Biologie in Tübingen, Germany (MPI) and at Washington State University (WSU).

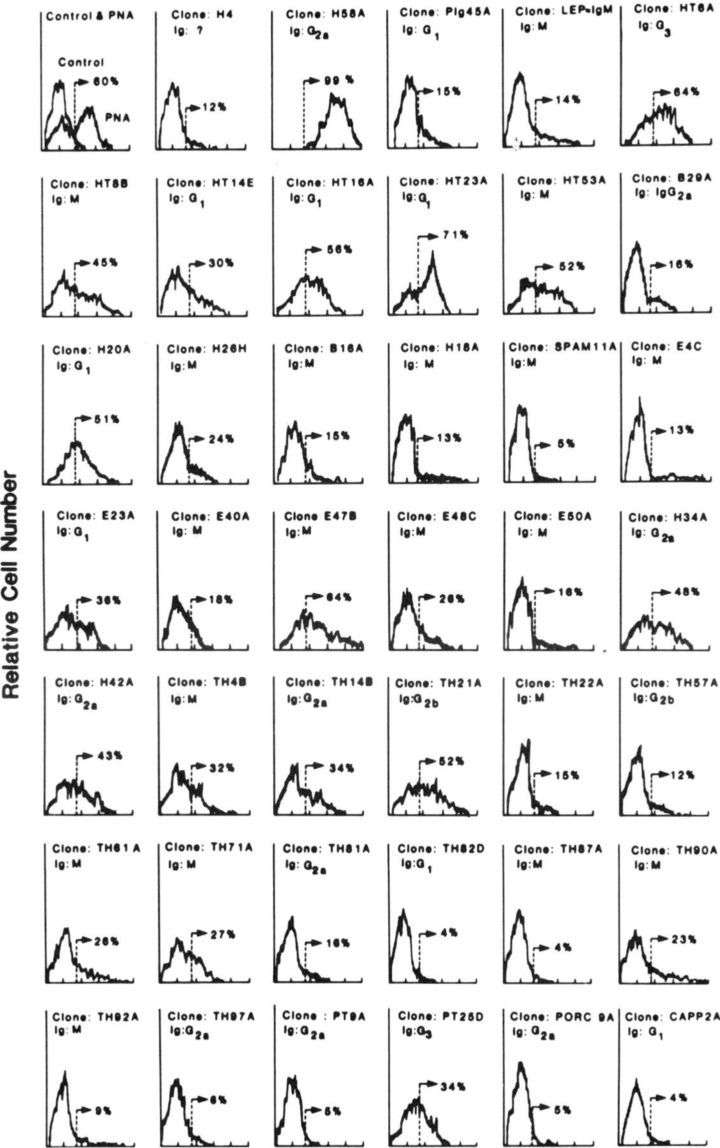

Figure 6--Comparison of histogram profiles of caprine HF-PBL labeled with monoclonal antibody specific for cell membrane antigens with broad or narrow species distribution.

Analysis of the subclones TI.4C and TI.48P from MPI, and CA4C-A
and CA48PA (and CA48P-A$_1$ a subclone producing a different Ig
isotype) from WSU, revealed that the antibodies were specific
for H-2D class I antigens (designated m84 and m89, respectively)
(35, 47). Subsequent analysis revealed that CA4C-A detects a
polymorphic antigen in cattle and pigs and CA48P-A, a
polymorphic antigen in cattle, goats, sheep, pigs, horses, and
humans (Table 1). In further studies, we screened monoclonal
antibodies from 70 cloned cell lines, derived from different
cell fusions, against a panel of 29 inbred strains of mice
differing only at the H-2 complex (unpublished data). Four
antibodies were identified that reacted with one or more strains
of mice: H58A, an antibody that reacts with class I antigens
($H-2K^{f,p,pv}$, $H2D^{wl6}$), and H42A ($H-2^{w3}$), TH21A, and TH81A
($H-2^{b,f,r,s,u,v,w4,w6,w7,wl7,w23}$), antibodies that react with
class II antigens (Mandic and Klein, unpublished data). As
noted in Table 1, the antibodies were found to be broadly
cross-reactive.

To prepare for use of these antibodies as standards in
comparative studies, the antibodies were examined by FMF for
patterns of labeling of Hypaque-Ficoll-purified human peripheral
blood leukocytes (HF-PBL). The labeling patterns were compared
with those of two commercially available monoclonal antibodies
specific for HLA-DR human class II gene products, HLA-DR
(donated by Becton Dickinson Monoclonal Center, Mountain View,
CA) (19) and H4 (donated by P. Terasaki, University of
California, Los Angeles, CA) (48). The analysis revealed that
monoclonal antibodies to class I and class II antigens yield
distinctive dot plot profiles that can be used to identify
additional anti-MHC monoclonal antibodies (6). Regardless of
whether a given antibody stained cells brightly (strong shift in
position of stained cells relative to unstained cells) or dimly
(small shift in stained cells), the patterns remained consistent
(typical patterns of staining profiles are illustrated in
Figure 7). Further analysis revealed that HLA-DR reacts with
horse HF-PBL as well as with human cells and that H4 reacts with
all species tested. When H4 and H42A were used to screen for
additional monoclonal antibodies (with apparent specificity for
class II antigens) using human and bovine cells, dot plot
profiles were found to differ between species but to yield
consistent intraspecies labeling profiles (6).

Nine additional antibodies were identified that react with
HF-PBL in multiple species (H34A, H42A, TH4B, TH12A, TH14A,
TH16A, TH21A, TH22A, and TH81A). Except for TH12A, the
monoclonal antibodies were found to be broadly cross-reactive.
Of special interest, H34A was found to consistently label more
cells of ruminants and horses than of humans, i.e., approximately 10% more than noted with HLA-DR, H4, and the other WSU
monoclonal antibodies. Studies with purified T and B cells in
the cow revealed that TH4B, TH12A, TH14B, TH16A, TH21A, and
TH22A react with B cells (48 and unpublished data).

Biochemical studies with pig cells, though preliminary, have indicted that H42A, TH16A, and TH21A precipitate pig class II antigens (Lunney et al., unpublished data). Comparative analysis of labeling with CA48P-A and H58A revealed that similar dot plot profiles are obtained with minimal intraspecies and interspecies differences.

Thirteen additional monoclonal antibodies with apparent specificity for class I antigens were identified using H58A as a standard for comparison. These include B5C, RH12A, RH16A, H11A, H17A, H21A, PT85B, SAG4A, SAG7A, SAG8A, HELP1A, HELP2A, and PORC24A.

Genetic studies with B5C, RH12A, RH16A, H17A, H21A, SAG4A, SAG7A, SAG8A, HELP1A, and HELP2A (and HELP2B, an isotype variant) have yielded preliminary data consistent with the premise that they recognize MHC class I antigens, with RH12A and RH16A showing a strong association with the inheritance of certain BoLA haplotype antigens (Bernoco, unpublished data). Sufficient data are not yet available to assign haplotype specificity.

MONOCLONAL ANTIBODIES THAT RECOGNIZE B CELLS

Because it has been difficult to identify B-cell-specific monoclonal antibodies, our initial efforts to obtain a B marker have focused on identification of cross-reactive monoclonal antibodies specific for surface-bound IgM (sIgM). For these studies, a monoclonal antibody (DAS6) specific for bovine IgM was obtained from S. Srikumaran and R.A. Goldsby (Amherst College, MA) and one (USDA-#5C9.C12.1) specific for pig IgM from P. Paul (USDA, Ames, IA). A monoclonal antibody specific for horse IgM (LEP-IgM) was prepared by us (49). One hundred monoclonal antibodies known to be reactive with pig immunoglobulin (Kelley and Davis, unpublished data) by ELISA were screened for reactivity against semipurified bovine immunoglobulins (Ig). Monoclonal antibodies positive by ELISA for bovine Ig were then screened for reactivity with bovine (and other species) HF-PBL by FMF. Three antibodies were identified; PIg31B, PIg45A, and PIg47A. Comparative analysis with DAS6 on bovine cells, LEP-IgM on equine cells, and USDA-#5C9,C12.1 on pig cells, revealed patterns of labeling consistent with the reagents PIg45A and PIg47A being specific for sIgM and PIg31B being specific for a light-chain antigen. (PIg31B reacts with all classes of pig Ig by ELISA.) Further studies by FMF revealed that PIg45A and PIg47A react with goat, sheep, and horse cells and that PIg31B reacts with goat, sheep, and pig cells (Table 1, Figures 2-6).

An additional antibody (PIg14.1) was identified that reacted only with pig cells (Figure 5), but it is not yet established whether this monoclonal antibody detects a specific Ig isotype.

MONOCLONAL ANTIBODIES THAT RECOGNIZE ANTIGENS PREDOMINANTLY ON T CELLS

Several approaches were employed to identify monoclonal antibodies with apparent specificity for T cells: comparative analysis with the FACS analyzer; analysis by a cytotoxicity assay using purified T and B cells; and dual fluorescence labeling of purified T and B cells and HF-PBL.

To define antibodies that could be used directly or indirectly as standards in FMF analysis, monoclonal antibodies of known specificity for T cells and other cell types were obtained from Becton Dickinson (Leu 2b, Leu 3a, Leu 3b, Leu 4, Leu 5, Leu 7, Leu 8, Leu 10, and Leu Ml; gifts from the BD Monoclonal Center, Mountain View, CA) and screened for patterns of reactivity with human, cow, goat, sheep, pig, and horse HF-PBL. In addition, peanut agglutinin (PNA), a plant lectin found useful as a T-cell marker in mice (50, 51) and a monocyte marker in humans (53, 54) was evaluated. The studies revealed that two of the BD reagents (Leu 2b and Leu 8) reacted with horse cells and one (Leu 8) reacted with bovine cells (Table 1). The studies also revealed that PNA yields entirely different patterns of labeling of HF-PBL derived from different species (6). PNA labeled predominantly monocytes in humans, T lymphocytes and possibly some monocytes in horses, and mixed populations in ruminants and pigs (Table 1 and Figures 2-6).

Comparative analysis of the staining profiles of BD and WSU monoclonal antibodies on human cells revealed only one similar pattern of labeling (Leu Ml and H18A). One antibody (B28A) was found to label a small population of lymphocytes in humans, pigs, and horses and a large population in cattle.

Analysis of our monoclonal antibodies revealed a set of 10 with apparent specificity for antigens present predominantly on T cells (BlA, B24A, B26A, B29A, TH82D, TH87A, MPGE17A, HT23A, E4C, and E40A). As noted in Table 1, the patterns of cross reactivity of individual antibodies varied. The data on specificity are still limited. However, the findings obtained thus far are consistent with the premise that the antibodies detect T-cell antigens. Dual fluorescence labeling studies and FMF analysis of bovine leukocytes have shown that BlA, B24A, B26A, B29A, and TH82D label T cells and not B cells. Extensive field studies with B26A (in conjunction with TH21A, which is specific for a class II MHC antigen) have indicated it is a pan-T antibody that reacts with cells from many breeds of cattle (48). Comparative analysis of labeling patterns by FMF indicate that BlA and TH82D detect subpopulations of T cells; that B24A detects a population of cells comparable to or identical with the population of cells detected by B26A; that B29A detects a population of cells consistently larger than detected with B26A; and that TH87A and MPGE17A detect the majority of lymphocytes as well as a population of large cells (indicating that the antigen detected may not be restricted to T cells).

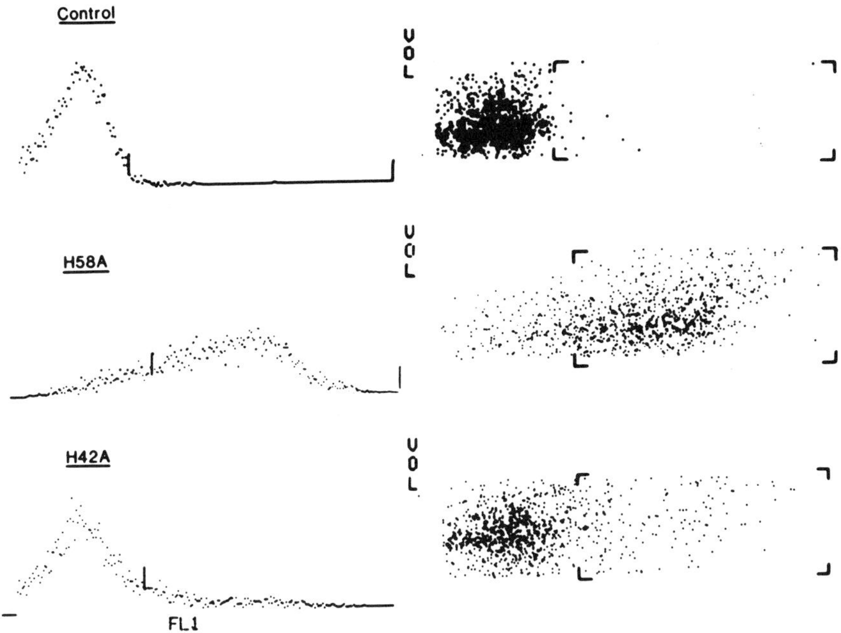

Figure 7--Comparison of histogram and dot plot profiles of
bovine HF-PBL labeled with H58A, a monoclonal antibody specific
for MHC class I antigens, and H42A, a monoclonal antibody
specific for MHC class II molecules. The cell preparations were
indirect labeled according to techniques described in the Becton
Dickinson Monoclonal Antibody Methods Manual. HF-PBL were
processed in 96-well microtiter plates--5 x 10^5 cells per well
in 50 µl of monoclonal antibody for 30 min at 4°C, washed
three times at 4°C, then reacted with affinity-purified
fluoresceinated goat anti-mouse immunoglobulin (TAGO, Inc.,
Burlingame, CA). The cells were washed three times, then
resuspended in a 1.5% solution of paraformaldehyde in buffered
saline. The cells were kept refrigerated until examined with
the Becton Dickinson FACS analyzer.

Analysis of HT23A on equine cells by FMF indicated it is a
pan-T antibody with species-restricted activity.
Analysis of E4C and E40A has indicated that, in many
species, these antibodies detect small populations of HF-PBL in
multiple species that are composed of both large and small
cells. Evidence for these antibodies reacting with T cells is
supported by findings obtained from immunofluorescent staining
of thymus tissue. Studies in the horse have shown that E4C
labels a small population of cells in the medulla and that E40A
labels a large population of cells in the cortex.

MONOCLONAL ANTIBODIES REACTIVE WITH T CELLS AND MONOCYTES

Comparative studies revealed four monoclonal antibodies that detect an antigen present on T cells and monocytes (PT15A, PT37A, PT80B, and PT91A) in the pig. Dot plot profiles of fluorochrome-labeled cells suggest that PT15A and PT80B detect the same antigen or an antigen with a similar cell distribution. The staining profiles obtained with PT37A and PT91A are similar, but distinctly different from those obtained with PT15A and PT80B.

MONOCLONAL ANTIBODIES REACTIVE WITH T AND B LYMPHOCYTES

One monoclonal antibody (TH18A) that reacts with both T and B cells has been identified by Davis and Lalor (unpublished data). Comparative analysis by dual fluorescence and FMF on bovine HF-PBL indicated that only small lymphocytes are labeled.

MONOCLONAL ANTIBODIES REACTIVE WITH MONOCYTES AND MONOCYTES/GRANULOCYTES

Seven antibodies were identified that exhibited predominant labeling of monocytes in one or more species: H18A, B16A, B18A, E48C, E50A, RH1A, and CAPP2A. H18A is one of the most interesting monoclonal antibodies. When reacted with human HF-PBL, it labels monocytes intensely with the dot plot display of data corresponding closely to those obtained with PNA and Leu M1 (6). With cells from other species, however, only a small population of cells are detected.

B16A and B18A labeled populations of bovine HF-PBL of similar size and exhibited similar staining profiles by histogram and dot plot analysis. B16A, however, detected an antigen with a broader species distribution. In addition, comparison of staining profiles of bovine cell preparations enriched for granulocytes and eosinophils revealed that B16A labels most of the cells (90 to 95%) whereas B18A labels only 75 to 85%). As with H18A, the patterns of labeling in different species varied. As noted in Table 1, B16A usually labels 30% of bovine HF-PBL. In other species the antigen detected was usually present on a small population of cells (possibly reflecting a different cell composition of HF-PBL).

E48C and E50A yielded similar staining profiles when reacted with cells from cattle, horses, and humans as analyzed by FMF. The percentage of cells labeled appeared to vary proportionately with the content of monocytes present in the HF-PBL. With human cells, where monocytes and lymphocytes are readily distinguished on dot plot display, all monocytes were labeled, as was a population of cells in the size range of lymphocytes. Based on intensity of staining, the expression of antigen detected with E48C and E50A varied more than the antigen detected with H18A (compare histogram profiles in Figures 2 and 6). Intra- and interspecies variation in antigen expression indicated that E48C and E50A detect different antigenic epitopes and possibly different molecules.

RH1A detected an antigen that is present in multiple species. Dot plot analysis revealed a complex distribution of labeled cells, suggesting variable expression of antigen on multiple cell types, but with the antigen predominantly expressed on monocytes.

CAPP2A detected an antigen present in multiple species but as with B16A and H18A, the antigen was present in different concentrations in HF-PBL from different species. In cattle, CAPP2A appeared to detect an antigen on a subpopulation of monocytes and some, but not all, granulocytes.

MONOCLONAL ANTIBODIES THAT RECOGNIZE ANTIGENS DIFFERENTIALLY EXPRESSED ON LEUKOCYTES FROM SPECIES TO SPECIES

Seven monoclonal antibodies (SPAM11A, TH61A, TH62A, TH71A, TH90A, TH92A, and TH97A) were identified that recognize antigens differentially expressed on HF-PBL from different species. Comparative analysis of dot plot displays of labeled cells has not yielded information, however, that suggests the antigens are present on a predominant cell type. The antibodies have been grouped here merely for discussion. SPAM11A detects an antigen expressed on a subpopulation of HF-PBL that is usually small in ruminants and pigs. Occasionally, the number of cells detected by FMF reaches 20% in cows. TH61A, TH62A, TH71A, TH90A, and TH92A exhibit a wide range of reactivity in different species. Analysis thus far indicates that the variation is primarily attributable to variations in the labeling of large cells. In some species, only dim labeling occurs and the shift in stained cell profiles is nominal; in others, staining is intense. Thus far, TH97A has been found to label only a small population of bovine HF-PBL. Preliminary data suggest the antigen is present on a large population of bovine thymocytes.

MONOCLONAL ANTIBODIES REACTIVE WITH BOVINE ERYTHROCYTES

Three antibodies (ANA8A, ANA13A, and ANA18A) were identified that exhibit interesting patterns of reactivity with bovine erythrocytes and leukocytes. The hybrids producing these antibodies were selected from a group of 15 hybrids because the monoclonal antibodies they produced agglutinated bovine erythrocytes. When the antibodies were tested against HF-PBL, two of the monoclonal antibodies (ANA13A and ANA18A) were found to react with subpopulations of leukocytes. Only ANA8A was found specific for erythrocytes. ANA13A and ANA18A reacted with

different sized populations of HF-PBL, with the population
recognized by ANA18A including some granulocytes.

MONOCLONAL ANTIBODIES REACTIVE WITH MIXED POPULATION OF BLOOD
LEUKOCYTES

     Twenty-nine monoclonal antibodies selected for further
analysis yielded patterns of labeling by histogram and dot plot
analysis that have not as yet suggested a predominant
specificity for known cell types (H20A, H25A, H26A, H26H, B13B,
E18A, E23A, E47B, HT6A, HT8B, HT14E, HT16A, HT53A, G1A, G2A,
TH1A, TH2A, TH11E, TH17A, TH31A, TH57A, PT9A, PT25D, PT35A,
PT40A, PORC9A, PT36B, PT18B, and PT90A) (Table 1 and Figures
2-6). Most of the hybrids producing these monoclonal antibodies
were selected for cloning and further analysis based on their
patterns of killing of HF-PBL in the CT assay. Eight of the
antibodies yielded similar histogram profiles of stained cells
when reacted with bovine HF-PBL (E18A, E23A, HT53A, TH2A, TH57A,
PT9A, H26H, and PORC9A). However, the percentage of cells
labeled with each antibody appeared to vary when patterns of
reactivity with HF-PBL were compared for animals of the same and
different species. HT6A, HT8B, HT14E, HT16A, PT36A, PT81B, and
PT90A yielded varied patterns also. Here, however, reactivity
was species restricted. B13B, TH17A, and TH11E appeared to
stain the majority of HF-PBL taken from cattle, but in
granulocyte-enriched preparations, TH17A and TH11E reacted with
different sized populations of cells. TH31B reacted with both
HF-PBL and granulocyte preparations with no major difference in
the staining profile (the pattern of labeling obtained with both
types of cell preparations is illustrated in Figure 2). The
antigen detected appeared to be in low concentration, for the
shift in fluorescence of labeled cells was nominal. PT25D,
PT35A, and PT40A yielded identical histogram profiles on FMF
analysis and appeared to detect the same antigen. Though the
profiles were similar to the one obtained with TH31B, the
antibodies appeared to distinguish at least two populations of
cells that expressed different concentrations of antigen on
their cell membranes. H20A yielded a more complex pattern of
labeling. The antigen detected is differentially expressed on
multiple populations of monocytes, granulocytes, and
lymphocytes. The majority of lymphocytes appeared to express
the antigen but at a gradation of very low (dim fluorescence) to
high (bright fluorescence) with no distinct division between
stained and unstained populations of cells. Comparative studies
showed that TH31B, PT25D, PT35A, PT40A, and H20A yield
consistent patterns of labeling with HF-PBL from all species
tested. E47B, TH1A, TH11E, and TH17A yielded contrasting
patterns of labeling by FMF analysis on bovine cells. The
monoclonal antibodies also exhibited different patterns of
species cross reactivity (Table 1).

Studies are in progress to rigorously establish the specificity of the set of monoclonal antibodies described here and also to determine the molecular weight and function of the membrane molecules detected.

DISCUSSION

It is evident, from the studies completed thus far, that the development of a set of cross-reactive monoclonal antibodies specific for homologous gene products is possible and moreover, that the availability of such antibodies will facilitate the elucidation of the immune system in food animals as well as other species. The strategies for optimizing production of hybrids producing cross-reactive antibodies have proven effective and the availability of FMF instrumentation has provided a way to rapidly distinguish monoclonal antibodies reactive with unique leukocyte subpopulations. In addition, FMF has provided a way to analyze complex cell populations and conduct extensive comparative studies to detail the patterns of distribution of cross-reactive monoclonal antibodies on leukocytes in different species. To date, we have succeeded in identifying sets of cross-reactive monoclonal antibodies specific for gene products coded for by class I and class II genes of the MHC. We have also succeeded in producing a large set of cross-reactive monoclonal antibodies that react with gene products differentially expressed on leukocyte cell membranes. Though it is obvious that more work needs to be done to fully characterize many of the monoclonal antibodies, some are immediately useful, e.g., those that detect polymorphic antigenic determinants and those that detect defined membrane molecules. The antibodies to monomorphic class I and class II antigens can be used to dissect the role of MHC in the regulation of immunity to antigens and pathogens of interest. The sets of antibodies are sufficiently large to permit intraspecies and interspecies comparisons and the demonstration of any dynamic changes in antigen expression on cells both in vitro and in vivo. The antibodies with apparent specificity for T cells and monocytes can be used to begin detailing T-cell function in relation to MHC restriction phenomena (37, 38, 57, 58). The monoclonal antibodies to monomorphic determinants (MHC and non-MHC) can also be used in two different ways to facilitate the generation of specific types of monoclonal antibodies. For example, the anti-MHC and non-MHC monoclonal antibodies can be used to purify gene products for use in the generation of monoclonal antibodies to antigens that define each allelic gene product present in a given species (31). In addition, monoclonal antibodies specific for a given differentiation antigen expressed only in a single species can be used to purify that antigen. Once purified, the defined antigen can be used to generate monoclonal antibodies to

different epitopes on the same molecule that are expressed on
the homologous molecule in other species. The strategy has
already proven useful in preliminary experiments designed to
make monoclonal antibodies to polymorphic class I BoLA antigens
(i.e., SAG4A, SAG7A, SAG8A, HELP1A, and HELP2A were developed
with antigen isolated with monoclonal antibody H58A).

The studies completed suggest that it would also be useful
to screen existing monoclonal antibody reagents from other
sources for cross reactivity (59-62), especially well-defined
monoclonal antibodies that are commercially available and being
used extensively in the study of immunity in humans. Our
initial screening of the Becton Dickinson series has revealed
three antibodies (HLA-DR, Leu 2b, and Leu 8) that react with
equine cells and one (Leu 8) that reacts with bovine cells as
well. Although it appears that only limited cross reactivity
will be found, each additional cross-reacting monoclonal
antibody will be useful.

The results obtained thus far pose two questions that must
be considered when cross-reactive antibodies are used. First,
are the antigens detected by cross reactivity expressed in
different concentrations in different species? If so, does this
reflect differences in cell function? Second, are the antigens
detected closely related or comprised of sets of phylogenetically related molecules (63-65) that share common epitopes but
mediate different functions? There are no complete answers yet,
but studies of the MHC provide evidence that species differences
in antigen expression occur (61) and that epitopes can be
expressed on more than one phylogenetically related molecule
(36, 37). Extensive studies with cross-reactive monoclonal
antibodies generated against class II A and E antigens in the
mouse have permitted the identification of the E homologue in
the pig (62) and humans (42, 63) (and possibly the A homologue).
Comparative analysis of antigen distribution has shown that, in
contrast to the situation in mice and humans, class II antigen
in pigs is highly expressed on circulating T cells (66-68). In
mice and humans, the homologous antigen is expressed in low
concentration or is absent on resting cells. Stimulation by
antigens is necessary for antigen expression. Thus, the question exists as to whether such species variation in antigen
expression reflects altered or augmented function or merely a
difference unrelated to function. Analysis of distribution of
class I MHC antigens in the mouse has recently revealed the
expression of a common epitope expressed on K and Qa 1 molecules
(37). This finding indicates that similar situations might
exist in other species and, in addition, suggests that
cross-reactive antibodies, though specific in a given context,
might detect the same and/or different molecules in
cross-species comparisons (39, 61). The finding that some of
the cross-reactive monoclonal antibodies found in our studies
appear to detect antigens expressed in different concentration
shows that this is a possibility and must be taken into account.

On a final note, the studies with cross-reactive monoclonal antibodies have revealed a quite different problem with important implications: cross reactivity does not always connote close homology (69, 70). Recent studies have shown that a number of monoclonal antibodies yield unusual cross reactions; e.g., anti-Thy-1 with vimentin (69), anti-Thy-1 (different monoclonal antibody) with V_k TEPC 15 idiotype determinant (71), anti-κ light chain with β_2 microglobulin, anti-tropomyosin with vimentin (72), anti-large T of SV40-transformed cells with normal growth-related nuclear protein (73). These studies show that the detection of a useful antibody against an antigen in a given species may not be useful in another species. Also, the studies suggest that the results obtained in functional studies can include specific and nonspecific components.

In summary, we have described methods for producing monoclonal antibodies that can be used for analysis of the immune response in food animals, strategies for developing useful cross-reactive monoclonal antibodies for direct species comparison of cell function, and presented preliminary findings on a series of monoclonal antibodies currently being analyzed. Though further characterization is required, we are now in a position to address important questions concerning the genetic basis for disease susceptibility.

ACKNOWLEDGMENTS

The studies reported here were supported in part by funds from the Department of Veterinary Microbiology and Pathology, College of Veterinary Medicine, Washington State University; Carnation Research Farms; The National Pork Producers Council; USDA-ARS--Animal Disease Research Unit; USDA-ARS--Hemoparasite Diseases Research Unit cooperative agreement 58-9AHZ-2-663, USDA-SEA 82-CRSR-2-2045, 83-CRSR-2-2281, 83-CRSE-2-2253 and 83-CRSR-2-2194; Formula Funding--Public Law 95-113; and by Agricultural Research Center Project 3073.

The data presented here were presented in part at the International Conference on the Application of Ruminant Immunology to the Control of Bovine Diseases held in September 1983 at the International Laboratory for Research on Animal Diseases, Nairobi, Kenya.

The flow diagram and description for the preparation and assay of monoclonal antibodies were prepared, in part, for a workshop on Priorities in Biotechnology Research for International Development (July 1982, Washington, D.C.) sponsored by the Board on Science and Technology for International Development, the National Academy of Sciences/National Research Council, and the U.S. Agency for International Development.

REFERENCES

1. Köhler, G., and Milstein, C. (1975) Nature 256:495.
2. Oi, V.T., and Herzenberg, L.A. (1980) In: Selected methods
    in cellular immunology, Mishell, B.B., and Shiigi, S.M.
    (eds.) W.H. Freeman and Co., San Francisco, pp. 351-372.
3. R.H. Kennett, T.J. McKearn, and K.B. Bechtol (eds.) (1981)
    Monoclonal antibodies, hybridomas: A new dimension in
    biological analysis. Plenum Press, New York.
4. G.J. Hämmerling, and J.F. Kearney (eds.) (1981) Monoclonal
    Antibodies and T-cell Hybridomas: Perspectives and
    Technical Advances. Elsevier/North-Holland, Amsterdam,
    pp. 587.
5. Davis, W.C., McGuire, T.C., and Perryman, L.E. (1983)
    Periodicum Biologorum 85 (in press).
6. Davis, W.C., Perryman, L.E., and McGuire, T.C. (1984)
    In: Application of Ruminant Immunology to the Control of
    Bovine Diseases. Morrison, I., Black, S., and Ole Moi Yoi,
    O., (eds.) in press.
7. Rigby, P.W. (1983) J. Gen. Virol. 64:255.
8. Jonak, Z.L., Braman, V., and Kennett, R.H. (1983) Second
    Annual Congress for Hybridoma Research Proceedings.
    (In press.)
9. Kavathas, P., and Herzenberg, L.A. (1983) Proc. Nat. Acad.
    Sci. USA 80:524.
10. Reinherz, E.L., and Schlossman, S.F. (1981) Immunol.
    Today 2:69.
11. Bhan, A.K., Nadler, L.M., Stashenko, P., McCluskey, R.T.,
    and Schlossman, S.F. (1981) J. Exp. Med. 154:737.
12. Goldstein, G. (1982) T cells and immunoregulation: A
    contemporary overview. J. Clin. Immunol. 2:5S-7S (Suppl.).
13. Kung, P.C., Talle, M-A., DeMaria, M., Ziminski, N.,
    Look, R., Lifter, J., and Goldstein, G. (1981) Int. J.
    Immunopharmacol. 3:175.
14. Lampson, L.A., and Lenz, R. (1980) J. Immunol. 152:393.
15. Gatenby, P.A., Kansas, G.S., Chen, Y.X., Evans, R.L., and
    Engleman, E.G. (1982) J. Immunol. 129:1997.
16. Hanjan, S.N.S., Kearney, J.F., and Cooper, M.D. (1982)
    Clin. Immunol. Immunopathol. 23:172.
17. Engleman, E.G., Warnke, R., Fox, R.I., and Levy, R. (1981)
    Proc. Natl. Acad. Sci. USA 78:1791.
18. Ledbetter, J.A., Evans, R.L., Lipinski, M.,
    Cunningham-Rundles, C., Good, R.A., and Herzenberg, L.A.
    (1981) J. Exp. Med. 153:310.
19. Engleman, E.G., Benike, C.J., Glickman, E., and Evans, R.L.
    (1981) J. Exp. Med. 154:193.
20. Howard, F.D., Ledbetter, J.A., Wong, J., Bieber, C.P.,
    Stinson, E.B., and Herzenberg, L.A. (1981) J. Immunol.
    126:2117.
21. Abo, T., and Balch, C.M. (1981) J. Immunol. 127:1024.
22. Haynes, B.F. (1981) Immunol. Rev. 57:127.

23. Pinder, M., Pearson, T.W., and Roelants, G.E. (1980) Vet. Immunol. Immunopathol. 1:303

24. Bluestone, J.A., and Hodes, R.J. (1983) Immunol. Today 4:256.

25. Neoh, S.H., Jahoda, D.M., Rowe, D.S., and Voller, A. (1973) Immunochemistry 10:805.

26. Balch, C.M., Dagg, M.K., and Cooper, M.D. (1976) J. Immunol. 117:447.

27. Vaerman, J.P., Kobayashi, K., and Heremans, J., Sr. (1974) Adv. Exp. Med. Biol. 45:251.

28. Iha, T.H., Gerbrandt, G., Bodmer, W.F., McGary, D., and Stone, W.H. (1973) Tissue Antigens 3:291.

29. Kvist, S., Ostberg, L., and Peterson, P.A. (1978) Scand. J. Immunol. 7:265.

30. Kvist, S., Klareskog, L., and Peterson, P.A. (1978) Scand. J. Immunol. 7:447.

31. Stallcup, K.C., Springer, T.A., and Mescher, M.F. (1981) J. Immunol. 127:923.

32. Teillaud, J.L., Crevat, D., Chardon, P., Kalil, J., Goujet-Zalc, C., Mahouy, G., Vaiman, M., Fellous, M., and Pious, D. (1982) Immunogenetics 15:377.

33. Brodsky, F.M. and Parham, P. (1982) Immunogenetics 15:151.

34. Sachs, D.H. (1983) Transplant. Proc. 15:40.

35. Figueroa, F., Davis, W.C., and Klein, J. (1981) Immunogenetics 14:177.

36. Ivanyi, P. (1981) In: Current Trends in Histocompatibility: Immunological profiles, Vol. 1. Reisfeld, R.A., and Ferrone. S. (eds.), Plenum Press, New York, pp. 133-181.

37. Klein, J., Figueroa, F., and Nagy, Z.A. (1983) Ann. Rev. Immunol. 1:119.

38. Hood, L., Steinmetz, M., and Malissen, B. (1983) Ann. Rev. Immunol. 1:529.

39. Soloski, M.J., Uhr, J.W., Flaherty, L., and Vitetta. (1981) J. Exp. Med. 153:1080.

40. Hoang-Xuan, M., Leveziel, H., Zilber, M-T., Parodi, A-L., and Levy, D. (1982) Immunogenetics 15:207.

41. Pierres, M., Mercier, P., Madsen, M., Mawas, C., and Kristensen, T. (1982) Tissue Antigens 19:289.

42. Stocker, J.W., and Bernoco, D. (1979) Technique of HLA typing by complement-dependent lympholysis. In: Immunological Methods. Lefkovitz, I., and Pernis, B. (eds.), Academic Press, New York, pp. 217-226.

43. Douillard, J.Y., Hoffman, T., and Herberman, R.B. (1980) J. Immunol. Meth 39:309.

44. Sutter, L., Bruggen, J., and Sorg, C. (1980) J. Immunol. Meth. 39:407.

45. Daniels, D.J. and Gielkins, L.J. (1980) J. Immunol. Meth. 30:325.

46. Herzenberg, L.A. (1978) In: Knapp, W., et al. (eds.). Immunofluorescence and Related Staining Techniques. Elsevier/North-Holland, Amsterdam, pp. 99-109.
47. Klein, J., Figueroa, F., and David, C.S. (1983) Immunogenetics 17:533.
48. Lewin, H.A., and Bernoco, D. (1983) Fed. Proc. 42:1230.
49. McGuire, T.C., Perryman, L.E., and Davis, W.C. (1983) Am. J. Vet. Res. 44:1284.
50. London, J., Berrih, S., and Bach, J.F. (1978) J. Immunol. 121:438.
51. London, J., and Horton, M.A. (1980) J. Immunol. 124:1803.
52. O'Keefe, D., and Ashman, L. (1982) Clin. Exp. Immunol. 48:329.
53. Banks, K.L., and Greenlee, A. (1981) Am. J. Vet. Res. 42:1651.
54. Banks, K.L. (1982) Am. J. Vet. Res. 43:314.
55. Stone, W.H. (1981) In: Butler, J.E. (ed.). The Ruminant Immune System. Plenum Press, New York, pp. 433-450.
56. Davis, W.C., McGuire, T.C., Anderson, L.W., Banks, K.L., Seifert, S.D., and Johnson, M.I. (1982) Proceedings of the 7th National Anaplasmosis Conference, 1981. pp. 285-305.
57. Lawman, M.J.P. (1982) J. Am. Vet. Med. Assoc. 181:1022.
58. Antczak, D.F. (1982) J. Am. Vet. Med. Assoc. 181:1030.
59. Ault, K.A., and Springer, T.A. (1981) J. Immunol. 126:359.
60. McMaster, W.R., and Williams, A.F. (1979) Immunol. Rev. 47:117.
61. Lunney, J.K., Osborne, B.A., Sharrow, S.O., Devaux, C., Pierres, M., and Sachs, D.H. (1983) J. Immunol. 130:2786.
62. Pierres, M., Rebouah, J.P., and Kourilsky, F.M. (1981) J. Immunol. 126:2424.
63. Stout, R.D., Yutoku, M., Grossberg, A., Pressman, D., and Herzenberg, S.A. (1975) J. Immunol. 115:509.
64. Ades, E.W., Zwerner, R.K., Acton, R.T., and Balch, C.M. (1980) J. Exp. Med. 151:400.
65. Dalchau, R., and Fabre, J.W. (1979) J. Exp. Med. 149:576.
66. Charron, D.J., Engleman, E.G., Benike, C.J., and McDevitt, H.O. (1980) J. Exp. Med. 152:127s.
67. Hercend. T., Ritz, J., Schlossman, S.F., and Reinherz, E.L. (1981) Human Immunol. 3:247.
68. Mann, D.L., and Sharrow, S.O. (1980) J. Immunol. 125:1889.
69. Lennox, E.S. (1983) Transplant. Proc. 15:45.
70. Jensenius, J.C., and Williams, A.F. (1982) Nature 300:586.
71. Pillemer, E. and Weissman, I.L. (1981) J. Exp. Med. 153:1068.
72. Blose, S.H., Matsumara, and Lin, J.J.C. (1981) Cold Spring Harbor Symp. Quant. Biol. 46:455.
73. Lane, D.P., and Hoeffler, W.K. (1980) Nature 288:167. Natl. Acad. Sci. USA 80:524.

CHAPTER 8

APPLICATION OF MONOCLONAL ANTIBODIES TO THE STUDY OF IMMUNE
RESPONSES

KEIKO OZATO
Laboratory of Developmental and Molecular Immunity
National Institute of Child Health and Human Development
National Institutes of Health, Bethesda, MD   20205

Immune responses have two remarkable characteristics.
First, recognition of specific antigens takes place on a clonal
basis. Each lymphocyte clone is predetermined for its
specificity and recognizes only a single antigenic determinant.
Thus, changes in immune responses mean changes in clonal
populations. Second, the products of the major
histocompatibility complex (MHC) are prerequisites for immune
recognition of foreign antigens. A foreign antigen and its
recognition by lymphocytes are not analogous to a simple lock
and key relationship. Rather, T-cell recognition of antigens is
possible only in conjunction with own MHC (self). This MHC
restriction is extensive and profound and is found in various
T-effector responses as well as antibody (B cell) responses
through the activity of helper and suppressor T-cells.
     Because of these very characteristics, studies of immune
responses benefited greatly from the advent of hybridoma
technology. This technology obviated a long-term problem of
immunology, i.e., the necessity to deal with heterogeneous
mixtures of lymphocytes bearing heterogeneous specificities. In
recent years, the isolation and amplification of numerous
lymphocyte clones exhibiting exquisite specificities have been
achieved by the application of hybridoma technology.
Significantly, it extended from antibody-producing B-cell
hybridomas to T-cell hybridomas that showed suppressor or helper
functions in MHC-restricted manner (1, 2). In addition,
cytotoxic T-cell hybridomas have been reported (3). By yet
unknown mechanisms, T cells seem to preferentially fuse and
establish stable hybridomas if the fusion partner is of T-cell
origin (1-3). The same preference holds true for B cells, that
is, antibody-producing hybridomas are generated exclusively by
using myeloma or equivalent tumor cells. These cells provided

clonal products characteristic for the function of each clone in
quantities that allow dissection of specificity at the molecular
level.

STUDIES OF MAJOR HISTOCOMPATIBILITY COMPLEX PRODUCTS BY THE USE
OF MONOCLONAL ANTIBODIES

    The MHC encompasses a chromosome segment that includes
genes encoding class I and class II antigens. They are
expressed on the surface of lymphocytes and other cells and are
signified by their unusually high polymorphism. Notably, more
than 50 alleles are known in mice, and each molecule has
multiple amino acid substitutions. Because of the polymorphism
and because there are multiple genes encoding similar molecules,
these antigens present complex reactivities with conventional
alloantisera raised in MHC-disparate animals. Genes in the MHC
also have been shown to control immune responses (4) and to be
responsible for H-2-restricted T-cell lysis (5). Further,
strong association with some diseases has been noted (6). These
features of the MHC are shared among different mammalian species.
    In the Immunology Branch at NIH, we have attempted to sort
out this complex nature of MHC by studying monoclonal antibodies
reacting with its products. A series of monoclonal antibodies
were obtained through immunizations of mice with cells of
different haplotypes and subsequent fusion with nonproducer
SP2/0 cells. Testing of these antibodies with a panel of
strains enabled us to map their reactivities to different
regions of MHC (7, 8). Further, the expected class I and class
II molecules were immunoprecipitated by these antibodies. In
most cases the monoclonal antibodies showed reactivity to
multiple antigens with heterogenous reactivity patterns. That
is, the majority of antibodies were cross-reactive and differed
from each other in terms of reactivity modes. However, some
antibodies reacted only with one antigen. These cross-reactive
and unique specificities have been classified into two groups,
public and private. Monoclonal antibodies were found to detect
fine specificities that previously were difficult to dissect.
Thus, even antibodies showing identical reactivities to
wild-type strains were often different with respect to their
reactions to mutant strains carrying discrete changes in MHC
(9). For example, monoclonal antibodies 34.2.12 and 34.5.8
appeared identical in reactions to a panel of wild-type strains,
specifically being assigned to a private antigen of H-2D^d.
However, they reacted differently with B10.D2.H-2^{dm1}, a mutant
strain that appears to have arisen by intragenic recombination
in the H-2D region (Sears, unpublished data) (Figure 1). A
similar fine specificity difference was found with antibodies
reacting with I-A^b antigens; strain bml2 has a mutation in the
1-A region (10). Most, if not all, of these complex reactivity
patterns revealed by monoclonal antibodies represented new

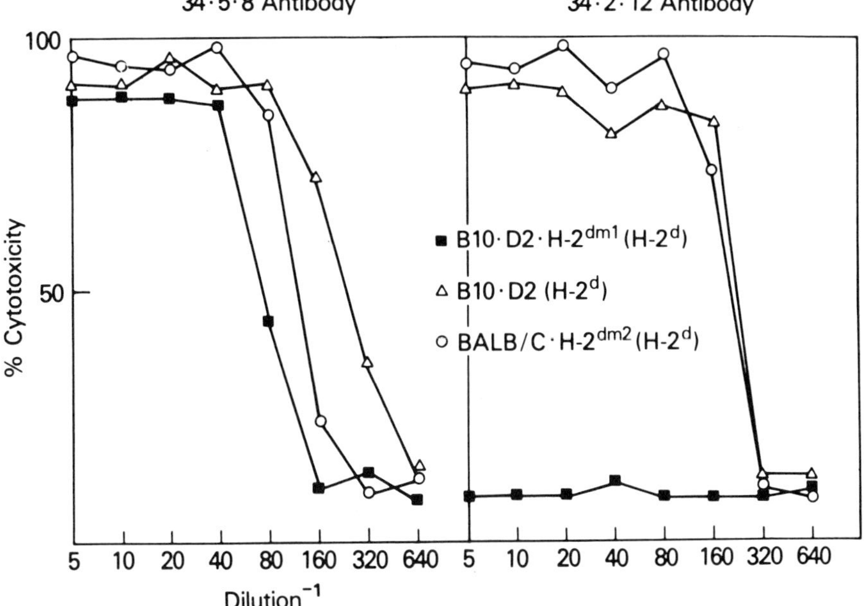

Figure 1--Cytotoxicity of monoclonal anti-H-2D^d antibodies against B10.D2 H-2^{dm1} spleen cells. Monoclonal antibodies in ascites were diluted as indicated. Cytotoxicity was measured by ^{51}Cr-release from spleen cells by two-stage cytotoxicity assay (7) using rabbit complement. The values represent mean of duplicates. These two antibodies showed reactivity patterns identical to those of other strains (8).

specificities (11) that conventional allosera had failed to detect. An important point is that a monoclonal antibody truly defines a discrete single antigenic entity rather than heterogenous sites. Thus, monoclonal antibodies obviate elements of uncertainty associated with conventional H-2 serology.

These specificities undoubtedly reflect the structural complexity of MHC antigens that results from high polymorphism. Consistent with this interpretation, it has been shown most extensively in mouse class I molecules that there are multiple amino acid substitutions throughout the molecules, although a higher degree of polymorphism is found in the first two domains of the molecule. This has been found by amino acid and by DNA sequence analyses (12, 13).

Because a monoclonal antibody defines a single binding entity, it also enabled us to undertake detailed studies on kinetics of antibody binding in order to estimate the number of H-2 molecules per cell (Dower, Ozato, Segal, unpublished data). With radiolabeled anti-H-2 antibodies, we found complex modes of equilibrium binding and dissociation kinetics attributable to the coexistence of monovalent and bivalent binding of the antibodies to antigens localized in the plane of the lymphocyte membrane. Based on these kinetic studies, we have estimated the average number of H-2K^k and D^d molecules on a B10.A splenic lymphocyte to be 5 x 10^4 and 1 x 10^5, respectively.

In addition, monoclonal antibodies contributed to determining the molecules responsible for MHC restriction. An example has been shown in a test of cytolytic activity of T effectors against target cells coupled with a hapten, restricted to H-2^d haplotype (14). In this system, T killer cells were generated by _in vitro_ stimulation of lymphocytes with stimulators pretreated with a small hapten, _N_-iodoacetyl-_N_-ethylenediamine (AED). The stimulators were syngeneic, i.e., identical to responding cells in the MHC haplotype. The killer cells lysed AED-hapten-coupled target cells only if the MHC haplotypes were matched. The cytolysis was completely inhibited by monoclonal antibodies reacting with H-2D^d or H-2L^d antigen (Table 1). Antibodies that reacted with H-2K^k antigen that did not serve as restricting antigen in these effector cells showed no inhibition, even though the antigens were present on the target lymphocytes. Antibodies reacting with H-2K^d antigen present on effectors but not on target fibroblasts had no effect. Thus the components responsible for H-2 restriction are indeed class I molecules, not other gene products encoded by nearby class I genes.

The role of I-A molecules in T-cell recognition in an antibody response presumably under MHC restriction has been demonstrated by Hodes et al. (15). The antibody response was inhibited by an anti-IA antibody, and it was specific for the haplotypes responsible for the antigen presentation. This specificity could be verified by the use of radiation-induced bone marrow chimeras in which donor lymphocytes mature in MHC-mismatched host animals and mount immune responses against various antigens.

THE USE OF MONOCLONAL ANTI-H-2 ANTIBODIES FOR THE STUDY OF MOUSE H-2 GENE STRUCTURE

In recent years, much information on the genetic structure of MHC has been obtained through advances in molecular genetics. Currently it appears that more than 20 class I genes exist (in BALB/c mice as many as 36 genes) in the classic MHC and the adjacent Qa, Tla regions (13). Thus, class I genes belong to an extensive multigene family. Because of this multigene nature, the identification of each gene has been difficult. It has been achieved by DNA-mediated gene transfer in which cloned class I genes were transferred into mouse L cells using the herpes Tk gene as a selectable marker. The expression of the transferred genes has been observed by

Table 1--Monoclonal antibody inhibition of BALB/c anti-AED-self lysis (14)

Monoclonal antibody pretreatment	Percent lysis against AED-hapten-treated cells*	
	T111 (L^d)	T483 (D^d)
Medium alone	12.8	14.8
Anti-L^d (30-5-7)	1.6	19.6
Anti-D^d (34-5-8)	12.6	3.6
Anit-K^k + D^k (3-83)	12.1	10.5

*T111 and T483 are transformed mouse fibroblasts that express products of exogenously transferred L^d and D^d genes, respectively. These target cells were labeled with ^{51}Cr, then incubated with monoclonal antibodies in ascites for 20 min at 4°C. Cytotoxic effector cells generated in vitro by stimulation with syngeneic spleen cells coupled with hapten AED were then added. They were incubated at 37°C for 5.5 h. Lysis was measured at an effector target ratio of 80:1. No lysis was found when target cells were not treated with AED.

monoclonal antibodies with predefined specificities. This enables functional class I genes to be unequivocally identified (16, 17), a significant contribution of monoclonal antibody technology. In addition, an entirely new approach has emerged from technologies in molecular genetics, i.e., altering the structure of class I genes at the DNA level, thereby permitting examination of the structural basis for immunological polymorphism and functions. Employing previously cloned H-2D^d and H-2L^d genes, we constructed hybrid genes in which the DNA sequences encoding the C2 domains have been exchanged. The C2 domain is the part of class I molecule closest to the membrane. Testing many monoclonal antibodies reactive with these hybrid gene products revealed that a majority of serological specific- ities are mapped to the N and/or the C1 domain. However, we detected serologically polymorphic sites in the C2 domain as well (18) (Figure 2). Further, alloreactive T cells and hapten- specific H-2-restricted cytotoxic T cells appear to predominantly recognize polymorphism present on the first two domains (19).

More specific modification such as single nucleotide substitution has become possible. In our laboratory, oligonucleotide-directed site-specific mutagenesis (20) has been applied to a class I gene to delineate the role of carbohydrates and S-S linkage in the H-2 gene. More extensive application of this approach will allow us to identify a particular amino acid substitution responsible for a particular site for monoclonal antibodies recognition and T-cell functions.

Mapping of Serological Determinants to the Domains of Mouse Class I Antigens as Detected by Hybrid Gene Products

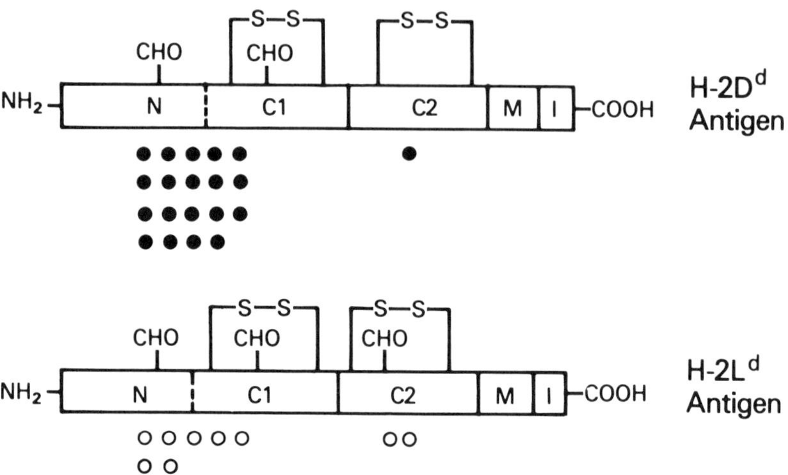

Figure 2--Mapping of serological determinants to the domains of mouse class I antigens as detected by hybrid gene products. Mapping was carried out with radioimmunoassays using ^{125}I-labeled goat (or sheep) anti-mouse Ig on the products of hybrid genes in which the C2 domains were exchanged between L^d and D^d genes.

CONCLUSIONS AND PERSPECTIVE

Hybridoma technology allowed immunologists to isolate and amplify single lymphocyte clones that bear specific receptors and retain functions. Intensive research in molecular genetics of the clonal products obtained from the hybridomas has now begun to elucidate the structure of the products.

MHC products that play an essential role in T-cell recognition have been defined by studying monoclonal antibodies reactive with MHC. These antibodies allow us to identify each MHC molecule and each determinant on a molecule unequivocally.

Recent work in molecular genetics provided conclusive information on the genetic structure of MHC. It further extended our ability to modify MHC genes in vitro and to transfer such genes into a new environment. This provides the potential to delineate the structure-function relationship of

MHC genes and to elucidate the evolutionary mechanisms that
brought about the high polymorphism and the tight association
with T-cell recognition.

ACKNOWLEDGMENTS

I am grateful to all of my collaborators for their
enthusiastic participation in the studies described here. I
especially acknowledge Dr. D. H. Sachs with whom I initiated
these studies under his warm encouragement. The expert
secretarial assistance of Mrs. K. Miceli is gratefully
acknowledged.

REFERENCES

1.   Taniguchi, M., and Miller, J.F.A.P. (1978) J. Exp. Med.
     148:373.
2.   Kappler, J., Skidmore, B., White, J. and Marrack, P. (1981)
     J. Exp. Med. 153:1198.
3.   Kaufman, Y., Berke, G., and Eshhar, Z. (1981) Proc. Natl.
     Acad. Sci. USA 78:2502.
4.   McDevitt, H.O., Deak, B.D., Schreffler, D.C., Klein, J.,
     Stimplfling, J.H., and Snell, G.D. (1972) J. Exp. Med.
     135:1259.
5.   Zinkernagel, R.M., and Doherty, P.C. (1974) Nature 248:701.
6.   Svejgaard, A., and Ryder, L.P. (1976) Lancet ii:547.
7.   Ozato, K., Epstein, S.L., Henkart, P., Hansen, T.H., and
     Sachs, D.H. (1981) Transplant. Proc. 13:958.
8.   Ozato, K., Mayer, N.M., and Sachs, D.H. (1982)
     Transplantation 34:113.
9.   Hansen, T.H., Ozato, K., and Sachs, D.H. (1983) Adv.
     Immunol. (in press).
10. Hansen, T.H., Walsh, W.D., Ozato, K., Arn, J.S., and Sachs,
     D.H. (1981) J. Immunol. 127:2228.
11. Klein, J., Figueroa, F., and David, C.H. (1983)
     Immunogenetics 17:553.
12. Kimball, E.S., and Coligan, J.E. (1983) Contemp. Top.
     Immunobiol. (In press.)
13. Hood, L., Steinmetz, M., and Malissen, B. (1983) Ann. Rev.
     of Immunol. (In press.)
14. Levy, R.B., Richardson, J.C., Margulies, D.H., Evans, G.A.,
     Seidman, J.G., and Ozato, K. (1983) J. Immunol. 130:2514.
15. Hodes, R.J., Hathcock, K.S., and Singer, A. (1980) J. Exp.
     Med. 152:1779.
16. Evans, G.A., Margulies, D.H., Camerini-Otero, R.E.,
     Ozato, K., and Seidman, J.G. (1982) Proc. Natl. Acad. Sci.
     USA 79:1994.

17. Goodenow, R.S., McMillan, M., Nicolson, M., Taylor
     Sher, B., Eakle, K., Davidson, N., and Hood, L. (1982)
     Nature 300:231.
18. Evans, G.A., Margulies, D.H., Shykind, B., Seidman, J.G.,
     and Ozato, K. (1982) Nature 300:755.
19. Ozato, K., Evans, G., Shykind, B., Margulies, D., and
     Seidman, J. (1983) Proc. Natl. Acad. Sci. USA 80:2040.
20. Wallace, R.B., Schold, M., Johnson, M.J., Dembek, P., and
     Itakura, K. (1981) Nucleic Acid Res. 9:3647.

CHAPTER 9

SURFACE CHANGES OCCURRING IN MYOBLASTS PRIOR TO FUSION

JUDITH V. SCHOLLMEYER
Roman L. Hruska Meat Animal Research Center
Agricultural Research Service, U.S. Department of Agriculture
P.O. Box 166, Clay Center, NE   68933

ABSTRACT

The effects of a calcium chelator, EGTA, and a calcium
ionophore, A23187, on fusion of a cloned muscle cell line, L_6,
are reported. The results confirm that EGTA essentially blocks
all myoblast fusion because the lateral alignment of presumptive
myoblasts cannot occur in the absence of extracellular calcium.
A23187, however, promotes the precocious fusion of myoblasts,
apparently by facilitating Ca^{2+} transport into myoblasts. The
calcium-activated protease, CAF, appears to relocate in response
to the Ca^{2+} flux, changing from a random, dispersed
distribution in proliferative myoblasts to a predominantly
peripheral distribution in prefusion myoblasts. Coincident with
the CAF relocation is an altered distribution of a surface
glycoprotein, fibronectin. Extracellular fibronectin is seen in
abundance in proliferating myoblasts but is essentially absent
from the surface of fusing myoblasts. We conclude that CAF when
activated by Ca^{2+} influx may act to promote the release of
fibronectin from the cell surface, thus providing a mechanism by
which the membrane of the fusing myoblast may be rearranged to
accommodate fusion.

INTRODUCTION

Myogenesis in cultures derived from embryonic skeletal
muscle has often been described as a succession of distinct
phases (1-4): a) <u>prefusion myogenesis</u>--proliferation of the
myogenic cells (myoblasts); b) <u>recognition-alignment</u>--withdrawal
of myoblasts from the proliferative rounds, cessation of DNA
synthesis, recognition of other myoblasts, and lateral alignment
of myoblasts; c) <u>fusion</u>--formation of multinucleated myotubes

from the fusion of mononucleated myoblasts; and d) <u>postfusion
myogenesis</u>--differentiation of functional muscle fibers.
Although it is very likely that such a sequence occurs <u>in vivo</u>,
the identification of distinct phases <u>in vitro</u> becomes difficult
because muscle cultures are not synchronized, and therefore all
the phases may be present at once in the same culture dish.

Some specific information regarding the evolutionary
differences in the muscle progenitor cells (myoblasts) has been
noted as these cells progress through differentiation. First,
the <u>proliferative myoblasts</u>, those involved in active mitotic
activity, are clearly involved in DNA synthesis but display few
other features at this stage that would distinguish them from
fibroblasts. The second myogenic cell to evolve is the
<u>presumptive myoblast</u>. This cell derives from the proliferative
myoblast but differs in that it can no longer undergo mitosis
and does not synthesize DNA. Presumptive myoblasts are also
cells that have been identified as being "committed" to fusion
(3) and therefore are the same cells that participate in the
recognition or clustering of myoblasts at the same stage of
differentiation. Ultimately, these spindle-shaped cells will
begin to stack longitudinally in preparation for fusion (5).
Numerous biochemical (6-8) and immunological (9) analyses have
revealed consistent alterations in the myoblast cell surface
prior to fusion. Primary among these is a loss of a surface
glycoprotein, fibronectin, from the surface of fusing myoblasts
(6). In addition, consistent with the rationale that the cell
membrane of a fusing myoblast is a region in considerable flux,
Pauw and David (8) have shown that various membrane proteins are
not consistently accessible to lactoperoxidase-catalyzed
iodination during alignment and fusion. Agents that alter
membrane structure or cell-cell attachment have been shown to
alter or inhibit myoblast fusion. Because an increase in
membrane fluidity appears to precede fusion (10, 11),
perturbations that alter fluidity also alter the recognition and
fusion processes (12, 13). Reductions in extracellular calcium
levels (14, 15) also inhibit the recognition-alignment and
therefore, the fusion process.

Precise knowledge of how the evolving process of muscle
differentiation might be controlled is not yet available;
however, substantial documentation has been presented that
implicates extracellular calcium as a primary requirement for
the initial recognition-alignment phases of myogenesis (14, 16,
17) and also suggests that a net influx of calcium into
fusion-competent myoblasts is a requisite event for myoblast
membrane fusion (18).

In this article, evidence is presented linking the activity
of a calcium-activated neutral protease (CAF) with the
Ca^{2+}-mediated membrane alterations accompanying myoblast
fusion. These data represent the first evidence that suggests a
mechanism for the translation of the Ca^{2+} influx event into
effective action within the cell.

MATERIALS AND METHODS

Materials

Supplies for tissue culture other than fetal bovine serum were obtained from Gibco Laboratories (Grand Island Biological Co., Grand Island, NY). Fetal bovine serum was purchased from HyClone Laboratories, Logan, UT. Glass coverslips were obtained from Bellco Co., Vineland, NY.

Cell Culture

L_6 myoblasts were subcloned from Yaffe's L_6 line (generously supplied by D. Schubert, Salk Institute) and were selected for fusing (Y_1) or non-fusing (Y_6) properties. Cells that were to be used for the immunofluorescence studies were grown on coverslips placed in six-well Costar plates (35 mm^2/well).

Determination of $^{45}Ca^{2+}$ Influx

Cultures to be assayed were incubated in Dulbecco's minimum essential medium (DMEM) plus 10% fetal bovine serum and approximately 33 μCi/ml $^{45}CaCl_2$ (20.7 mCi/mg; New England Nuclear) for 1 h at 37°C in a 5% CO_2 atmosphere. Medium was removed and the plates were rinsed twice with phosphate-buffered saline (PBS). Then, 1.5 ml of $LaCl_3$-saline (5.4 mM KCl, 137 mM NaCl, 1 mM $LaCl_3$, 6.1 mM glucose) was added to each 35-mm^2 well, and the dishes were rocked gently on a Buchler platform rocker for 5 min at 30°C. Dishes were rinsed twice with $LaCl_3$-saline buffer at room temperature over a period of 30 min. Cells were lysed by the addition of 0.2 ml of 0.1 mM Tris and 0.1% SDS (pH 7.4). The lysate was added to a glass scintillation vial, and 1.0 ml tissue solubilizer (NCS; Amersham/Searle) was subsequently added. The vials were capped and stored in the dark overnight. Samples were counted after the addition of 10 ml of scintillant (Omnifluor-toluene; New England Nuclear). All time-points represent the average of triplicate samples. The net $^{45}Ca^{2+}$ influx was normalized to the number of nuclei as determined in parallel cultures.

Fibronectin Production and Addition

Cell-surface fibronectin (FN) was prepared from roller bottle cultures of chick embryo fibroblasts as described previously (19) and stored in buffer (pH 11) at -60°C. Antibodies against chicken cell-surface fibronectin eluted from preparative SDS gels were prepared in rabbits and were further affinity-purified by use of fibronectin dimer coupled to agarose as described previously (20).

Treatment with EGTA and A23187

Ethylene glycol-bis(β-aminoethyl ether)N,N,N',N'-tetraacetic acid (EGTA; Sigma Chemical Co.) was prepared as a 50 mM stock solution (pH 7.0). Because different batches of serum have slightly variable Ca^{2+} concentrations, the optimal EGTA concentrations varied somewhat and were titrated at each change of serum. An EGTA level (usually 1.83 mM) was chosen that produced maximal inhibition of fusion with minimal effect on cell growth and adhesion to the substratum. The Ca^{2+} ionophore A23187 (Calbiochem) was prepared as a 20 mM stock solution in 100% dimethylsulfoxide (DMSO; Sigma Chemical Co.). This stock was diluted, with vigorous mixing on a vortex mixer, 1:2000 with DMEM to prepare a 10 μM working solution just before use. The final concentration of A23187 used was 1 μM.

CAF and Fibronectin Antibodies/Immunofluorescence

Monoclonal antibodies to porcine skeletal CAF were prepared in the following manner. BALB/c mice were injected intraperitoneally with 250 μg CAF mixed 1:1 with incomplete Freund's adjuvant (Difco Laboratories, Detroit, MI). One week after the second injection, each mouse received daily intravenous of 25 μg CAF for 4 days.

For the cell fusion, on day 5, the mice were killed, and spleen cells were harvested and washed once in serum-free DMEM; 10^8 lymphocytes were mixed with 10^7 mouse myeloma cells (NS1). The combined cells were rinsed one additional time in serum-free DMEM, the medium was removed, and 1 ml of 40% polyethylene glycol (PEG-1000, J. T. Baker Co.) in DMEM containing 15% dimethylsulfoxide (DMSO) was added as the mixture was agitated for 1 min. One milliliter of 40% PEG-1000 in DMEM without DMSO was added for an additional minute. The cells were then diluted dropwise with 20 ml of serum-free DMEM, centrifuged at 500 X g for 10 minutes, and resuspended in DMEM with HAT components (10^{-2} M hypoxanthine, 5 X 10^{-3} M aminopterin, 1.6 X 10^{-3} thymidine, and 3 X 10^{-4} M glycine). Hybridomas were screened by the enzyme-linked immunoassay (ELISA) for CAF-specific antibody. The solid-phase antigen for the ELISA was the same antigen used for injection. Specific antigens were identified by western immunoblot analysis (21). All hybridomas examined elicited antibody specific for the 80-kDa subunit of CAF. Polyclonal antibodies directed against chick embryo fibronectin were prepared as described previously (22). Coverslips from the respective treatments were rinsed with calcium- and magnesium-free PBS, fixed for 10 min at room temperature in 3.5% formaldehyde (Touismis) in 0.1 M phosphate buffer (pH 7.2), made permeable by acetone treatment (22), and rinsed again with the phosphate buffer. Monoclonal anti-CAF and rabbit anti-chick embryo fibronectin were utilized as the primary antibodies. Rhodamine-labeled goat anti-mouse IgA, IgG

and IgM (Cappel Laboratories) diluted 1:50 were used as second
antibodies to the monoclonal primary antibody, anti-CAF.
Rhodamine-labeled goat anti-rabbit IgG (Cappel Laboratories)
diluted 1:50 was the secondary antibody to anti-fibronectin.
The preparations were incubated as described previously (22). A
Zeiss photomicroscope III equipped with epifluorescence optics
and a 63X phase Planachromat oil immersion objective was used to
examine and photograph all preparations.

Measurement of Fusion

Muscle cells grown in monolayer were rinsed in PBS, treated
with trypsin/EDTA, and incubated at 37°C. The trypsin was then
inactivated with serum or soybean trypsin inhibitor, and cells
were fixed in Lugol's iodine. Samples were taken, and 200 to
300 nuclei were counted as being single or multiple. Time
courses showed that trypsinization was virtually complete at 5
min, with little change over the next 3 min.

Inhibitor Addition

An inhibitor that specifically inhibits CAF activity was
prepared from porcine skeletal muscle according to the procedure
detailed in Cottin et al. (23). The inhibitor was purified to
homogeneity by the use of an affinity column comprised of
monospecific anti-inhibitor linked to CM-Sephadex (Pharmacia).
Utilizing a caseinolytic assay described by Cottin et al. (23),
The purified inhibitor was shown to be capable of binding to and
completely inhibiting the CAF-Ca^{2+} complex.

RESULTS

Influence of EGTA on Cell Alignment and Fusion

Addition of EGTA at 24 h after plating appears to alter the
capacity of L_6 myoblasts to align laterally. Control cultures
demonstrate the characteristic alignment of myoblasts at 48 hrs
(Figure 1a). Figure 1b displays the random orientation of
myoblasts after 48 h exposure to 1.8 mM EGTA. As has been shown
for primary chick myoblasts, the addition of EGTA to L_6
myoblasts prior to fusion results in a complete blockage of
fusion by preventing alignment of myoblasts.

Influence of the Ionophore A23187 on Fusion

Our findings demonstrate that the addition of A23187 about
47 h after plating results in an accelerated rate of fusion, so
that fusion is essentially complete in 72 h (Figure 1c).
Control cultures did not display this degree of fusion until
120 h. These observations are in substantial agreement with

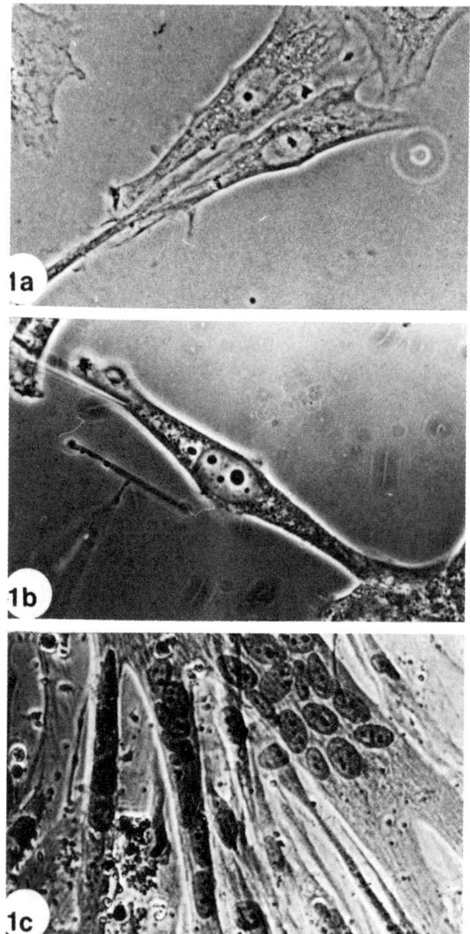

Figure 1--Phase micrographs of L_6 myoblasts grown normally without the addition of EGTA (a); with the addition of 1.8 mM EGTA (b); and grown in the presence of 1 mM A23187 (c). Cells grown normally (a) display lateral alignment after 24 hrs, while those exposed to EGTA (b) remain randomly oriented. A23187 (c) clearly promotes precocious fusion in cultures treated after 24 h.

those published earlier utilizing primary cultures (18).
However, the results from our studies would indicate that,
unlike the primary cultures, L_6 myoblasts cannot be induced to
achieve the degree of fusion, with or without the addition of
ionophore, that is seen with the primary chick myoblast cultures.

Distribution of CAF and FN after EGTA and A23187 Treatment

In myoblasts treated with EGTA, CAF displays a random
intracellular distribution (Figures 2c and d), whereas in
untreated cultures (48 h) CAF is located primarily in the
peripheral regions of the fusing myoblasts (Figures 2a and b).
Cells treated with A23187 displaying precocious fusion, show a
predominantly peripheral distribution of CAF (Figures 2e and
f). Fibronectin localization patterns are quite distinct from
those of CAF, in that EGTA-arrested cells display a substantial
extracellular array of fibronectin (Figure 3b), whereas
untreated cultures (48 h) exhibit the continual loss of
fibronectin from the fusing myoblast surfaces (Figure 3a).
A23187-treated cells that are already demonstrating fusion
(Figure 1c) possess much reduced amounts of surface-associated
fibronectin (Figure 3c).

Effect of Exogenous Inhibitor on Fibronectin Accumulation

Myoblasts treated with A23187 displaying lateral alignment
have been treated with purified CAF inhibitor (3 mg/ml) and
subsequently processed for anti-fibronectin immunofluorescence
(Figures 4a and b). Substantial amounts of fibronectin are
identified at the intensities of the aligned myoblasts (Figure
4b). Myoblasts treated with A23187 but not with inhibitor show
no such accumulation of fibronectin (Figures 4c and d).

Uptake of ^{45}Ca Using A23187

The amount of ^{45}Ca^{2+} uptake was monitored and
correlated with the proportion of fusing cells demonstrated in
cultures treated with A23187 (Table 1). Shortly before the
initiation of fusion, the net ^{45}Ca^{2+} influx began to
increase, and by 60 h, when fusion was 30% complete, the net
calcium influx began to increase. After an additional 20 h
there was a dramatic increase; this latter increase corresponds
to the expected time of appearance of sarcoplasmic reticulum in
these cultures.

DISCUSSION

These results obtained with EGTA or A23187 essentially
confirm the work of others (14, 18) in that the addition of EGTA
precludes the alignment of myoblasts (Figure 1b), thus

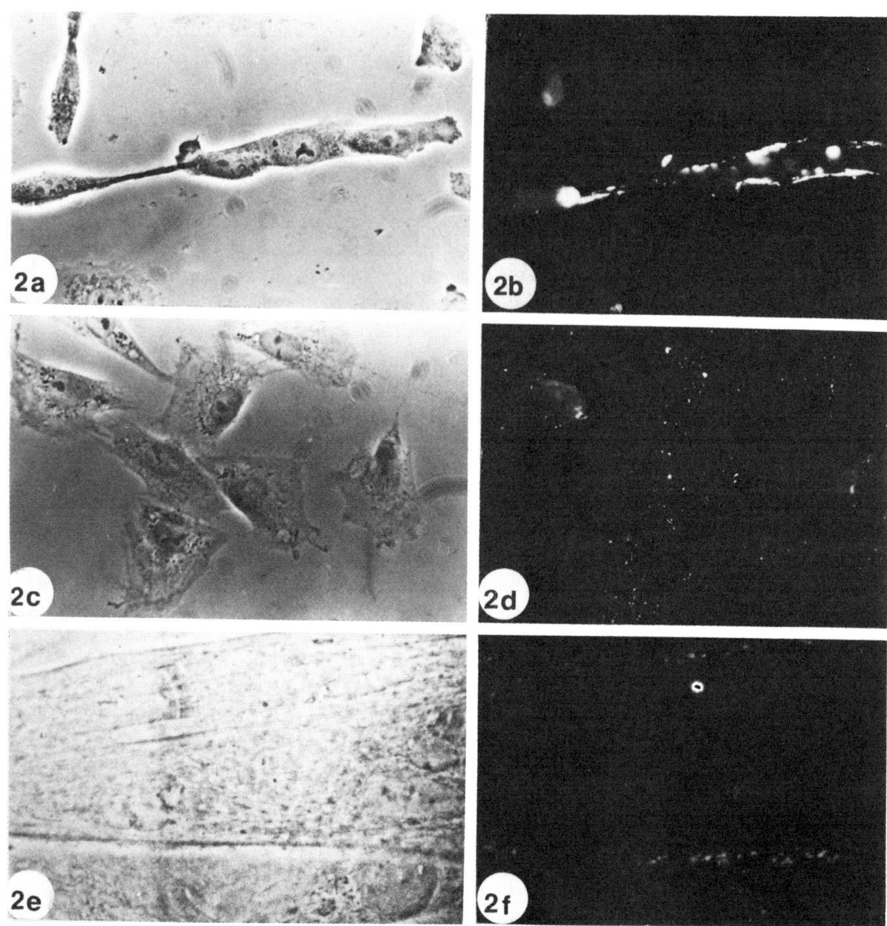

Figure 2--Phase and the corresponding immunofluorescence
(anti-CAF) micrographs of L_6 myoblasts treated as described in
Figure 1: without EGTA (a and b); with EGTA (c and d); and with
A23187 (e and f). X1200. Normal presumptive myoblasts (a and
b) display a peripheral distribution for CAF; however, cells
blocked with EGTA (c and d) show a scattered localization of
CAF. CAF assumes a peripheral location in myotubes formed after
A23187 treatment (e and f).

preventing fusion. A23187 appears not only to promote myoblast
fusion by facilitating a rapid uptake of Ca^{2+}, but also to
increase the rate of mitosis in prefusion myoblasts. Both EGTA

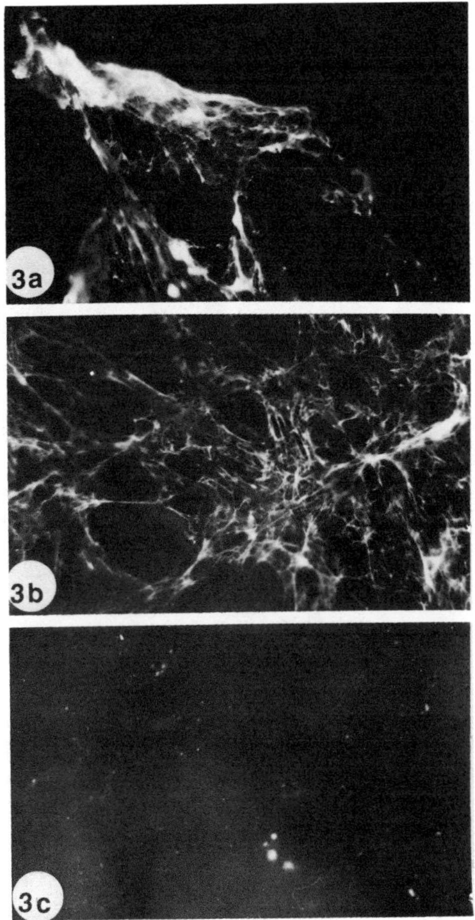

Figure 3--Immunofluorescence (anti-FN) of L_6 myoblasts treated as in Figure 1. X1200. Cells grown without EGTA exhibit reduced amounts of FN as they progress towards fusion (a). EGTA-arrested cells display a well-developed FN-containing extracellular matrix (b), whereas fusing myoblasts resulting from A23187 treatment possess reduced amounts of surface-associated fibronectin (c).

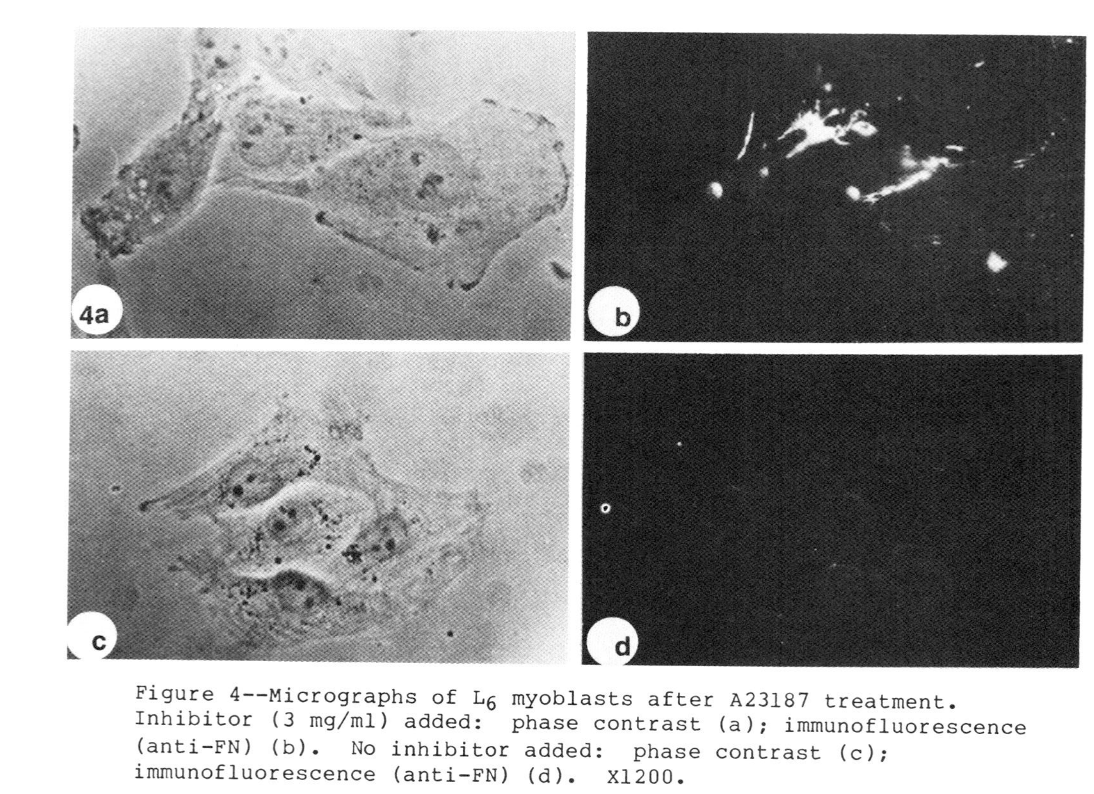

Figure 4--Micrographs of L_6 myoblasts after A23187 treatment. Inhibitor (3 mg/ml) added: phase contrast (a); immunofluorescence (anti-FN) (b). No inhibitor added: phase contrast (c); immunofluorescence (anti-FN) (d). X1200.

Table 1. Comparison of $^{45}Ca^{2+}$ uptake and the percentage of fusion at various culture times

Hours after plating	% Fusion[a]	Net $^{45}Ca^{2+}$ Uptake[b] (cpm/10^4 nuclei/hour)
24	5	100
48	10	200
60	30	300
72	60	450
120	80	700

[a] % Fusion = (fusion of experimental/fusion of control) x 100
[b] In the presence of 1 μM A23187.
Average of three experiments.

and A23187 were used to manipulate the Ca^{2+} concentration both outside (EGTA) and inside (A23187) the myoblast in order to assess the response of the myoblast to these Ca^{2+} fluxes. Specifically, we examined how those Ca^{2+} fluxes might influence the distribution of a calcium-activated neutral protease, CAF, and ultimately, the distribution of fibronectin. Results from the CAF immunofluorescence studies reveal that CAF apparently relocates or is redistributed in response to Ca^{2+} flux. We have seen that, in EGTA-arrested myoblasts, CAF is present in the cytoplasm in a dispersed, random distribution. However, in cells that have been released from EGTA-arrest or cells that have been incubated in A23187, the predominant CAF distribution is peripheral. Fibronectin, like CAF, displays a transient distribution with EGTA-arrested cells. These cells possess an impressive array of extracellular fibronectin whereas A23187-treated or EGTA-released myoblasts retain a diminished quantity of surface-associated fibronectin. The coordinate nature of these two phenomena, i.e., the arrival of CAF at a membrane location at the time fibronectin is lost from the membrane, circumstantially suggests a cause and effect relationship between them. An important question evolves from these observations. How does the relocation of CAF promote the release of fibronectin from the myoblast surface? Studies from other cell types (24) indicate that fibronectin may be released in a transient manner from the surface of a number of cells in an unhydrolyzed form. This also appears to be the case with the fibronectin released from differentiating myoblasts (6). These findings suggest that although fibronectin need not be hydrolyzed in order to be released, it is likely that structural proteins comprising the transmembrane-linkage system that connects the extracellular fibronectin to intracellular

constituents may be altered. Although the precise identity and
organizational relationships of the constituents of the
transmembrane-linkage system are not yet established, it is
generally agreed that actin and its associated proteins (24),
constituents of microfilament-membrane attachments, are involved.

Because fibronectin does not appear to be hydrolyzed upon
release from the myoblast, it seems reasonable to suggest that
CAF, once activated, may act to release fibronectin by
hydrolyzing specific constituents of the linkage complex.
Unpublished observations from this laboratory have documented
the ability of purified activated CAF to hydrolyze two proteins,
a 55-kDa and a 230-kDa protein, both of which are components of
microfilament-membrane attachments. These data support the idea
that CAF may act from within the cell by dismantling the
scaffolding holding the fibronectin on the cell surface.

The possibility that CAF may be elaborated extracellularly
at some time in the cell cycle, or in response to a Ca^{2+}
signal, cannot be ruled out. In fact, preliminary results shown
here (Figures 4a-d), demonstrate that exogenously added CAF
inhibitor can prevent the loss of fibronectin in A23187-treated
myoblasts. These findings suggest that a portion of the CAF
molecule responsive to the inhibitor is accessible to bind to
the inhibitor.

At present, we have little evidence delineating how CAF
might be controlled in the cell. Preliminary results suggest
that two forms of CAF may exist, with one form requiring only
micromolar quantities of Ca^{2+} for maximal activation (25).
Whether the CAF molecule interacts directly with Ca^{2+} or
whether this interaction, like many others in biology, depends
upon calmodulin as an intermediary is also not yet known.
Moreover, recent reports of an endogenous inhibitor (23), that
has been shown to be highly specific for inhibiting CAF, raise
additional questions concerning the _in vivo_ control of CAF
activity.

REFERENCES

1. O'Neill, M.C., and Strohman, R.C. (1969) J. Cell Physiol.
    73:61-68.
2. Koningberg, I.R. (1971) Develop. Biol. 26:133-152.
3. Yaffe, D. (1969) Curr. Topics Develop. Biol. 4:37-77.
4. Stockdale, F.E., and O'Neill, M.C. (1972) In Vitro
    8:212-225.
5. Knudsen, K.A., and Horowitz, A.F. (1977). Develop. Biol.
    58:328-338.
6. Hynes, R.O. (1976). Develop. Biol. 48:35-46.
7. Moss, M., Norris, J.S., Peck, E.J., Jr., and Schwartz, R.J.
    (1978) Exp. Cell Res. 113:445-450.
8. Pauw, P.G., and David, J.D. (1979) Develop. Biol. 70:27-38.

9. Friedlander, M., and Fischman, D.A. (1975) J. Cell Biol. 67(2, Pt. 2):124a.

10. Prives, J., and Shinitzky, M. (1977) Nature (London) 268:761-763.

11. Herman, B.A., and Fernandez, S.M. (1978) Develop. Biol. 48:35-46.

12. Van Der Bosch, J., Schudt, C., and Pette, D. (1973) Exp. Cell Res. 82:433-438.

13. Knudsen, K.A., and Horowitz, A.F. (1978) Develop. Biol. 66:294-307.

14. Paterson, B., and Strohman, R.C. (1972) Develop. Biol. 29:113-138.

15. Nameroff, M., and Munar, E. (1976) J. Cell Biol. 58:107-118.

16. Papahadjopoulos, D. (1978) Cell Surface Rev. 5:766-790.

17. Papahadjopoulos, D., Poste, G., and Vail, W.J. (1979) Meth. Memb. Biol. 10:1-121.

18. David, J.D., See, W. M., and Higginbotham, C.A. (1981) Develop. Biol. 82:297-307.

19. Yamada, K.M., Yamada, S., and Pastan, I. (1977) J. Cell Biol. 74:649-654.

20. Yamada, K.M., Schesinger, D. H., Kennedy, D.W., and Pastan, I. (1977) Biochemistry 16:5552-5559.

21. Wheelock, M.J., and Schollmeyer, J.E. (1982) J. Biol. Chem. 257:12471-12474.

22. Dayton, W.R., and Schollmeyer, J.V. (1981) Exp. Cell Res. 136:423-433.

23. Cottin, P., Vidlenc, P.L., and Ducastaing, A. (1981) FEBS Letters 136:221-224.

24. Hynes, R.O. (1982) In: Cell Biology of the Extracellular Matrix. E.D. Hay (ed.) Plenum, New York, pp. 295-333.

25. Dayton, W.R., Schollmeyer, J.V., Lepley, R.A., and Cortes, L.R. (1982) Biochim. Biophys. Acta 659:48-61.

CHAPTER 10

ELECTROFUSION OF CELLS

U. ZIMMERMANN AND J. VIENKEN
Arbeitsgruppe Membranforschung am Institut für Medizin
Kernforschungsanlage Jülich GmbH, Postfach 1913
D-5170 Jülich, and Lehrstuhl für Biotechnologie, University of
Würzburg, Röutgeuriug 11, 8700 Würzburg,
Federal Republic of Germany

In the last few years, electrofusion of cells has become a
promising new method for genetic engineering and somatic
hybridization. The electrofusion method was developed in this
laboratory at the end of the 1970s (1-4) as a result of
extensive studies of reversible permeability increases in the
membrane brought about by electrical fields of high intensity
and short duration and the discovery of reversible electric
breakdown of cell membranes (5-8).

Electrofusion of cells is a striking example of the
continuing need to carry out broadly based interdisciplinary
research in order to achieve new technological applications.
Biotechnology has traditionally been the domain of
microbiologists, chemists, and more recently immunologists, if
the far-reaching aspects of hybridoma research and the
associated production of monoclonal antibodies are included
(9). The development of the electrofusion method demonstrates
that physicists can also make--as we believe--a valuable
contribution, the result being a new discipline in biotechnology
that we will term <u>biophysical engineering</u>.

The electrofusion technique and the diversity of
applications have been extensively reviewed from various
viewpoints (10-14). We will therefore restrict ourselves to the
fundamental principles of the electrofusion technique and
consider in more detail the latest developments in this field,
particularly in respect to plant protoplast fusion and hybridoma
production.

REVERSIBLE AND IRREVERSIBLE BREAKDOWN

Electrofusion is based on reversible electrical breakdown of the cell membrane in response to electric field pulses of high intensity and short duration (2, 5-8, 10). Breakdown leads to a dramatic, but reversible, increase in the permeability of the biological membrane. The life-span of this field-induced increase in permeability is dependent on the temperature. At low temperatures (about 4°C) it may be 30 minutes or more. At higher temperatures (37°C), the field-induced perturbations in the membrane reseal in a matter of seconds to minutes. After this time, the original membrane impermeability and the normally high electrical resistance of the membrane are virtually completely restored.

The reversibility of these perturbations is critically dependent on the exposure time of the cells in the electrical field (10). The application of field pulses of high intensity lasting nanoseconds to a few microseconds leads to reversible changes in the membrane, whereas excessively long pulse durations (exposure times) cause irreversible changes in the membrane (2, 15-17). It is well known in physics that the action of physical forces on matter is a function of the time during which the forces are effective.

This principle can be demonstrated by a simple example. Microwaves and ultrashort waves can be described by the same wave equation, but because of the different frequencies, the interactions with matter are variable. Microwaves are in the gigahertz range. At this frequency, interactions with water molecules result in the heating of the solution. The frequencies of ultrashort waves are in the kilohertz range, i.e., in a range in which radio and television operate. Therefore, it is useful to divide the frequency range of alternating fields into certain ranges. Conversely, when investigating the interactions of high electrical fields with biological matter, it is also useful to identify various ranges on the basis of the reactions elicited by the fields. In the 100 microsecond to second range, the pulse duration of the electrical fields is usually in the range of the time constant for ion movement in the membrane and polarization effects of the counter-ion cloud around the cell. The potential to which the cell membrane can be polarized does not usually exceed 200 to 300 mV.

Above this threshold value, punch-through effects and irreversible destruction (dielectric breakdown) of the membranes are observed (10). In the range of nanoseconds to a few microseconds, on the other hand, the duration of the electrical fields is so short that only pure electric field effects on the membrane structure can occur. The membrane potentials that can be established across the membrane by application of external fields in this range are also correspondingly higher but do not cause irreversible destruction of the membrane. The following example will serve to eludicate the preceding considerations. The maximum voltage to which planar artificial lipid membranes faced by electrolyte solutions can be charged is about 200 to 350 mV (depending on the composition of the lipid bilayer), if the charging times are longer than about 10 μs. At this critical voltage level, the bilayer membrane is destroyed (so-called irreversible mechanical breakdown). With shorter charging times, on the other hand, it is possible to achieve voltage levels of up to 1 to 2 V (depending on pulse duration, temperature, and composition).

At these voltage levels reversible electrical breakdown is observed, i.e., the permeability of the membrane is dramatically increased, leading to a very rapid discharge of the membrane below the critical voltage value at which irreversible breakdown is normally observed. As the name implies, in this type of electrical breakdown the membrane is not destroyed. In fact, the membrane can repeatedly be polarized to high voltage levels and subsequently discharged by electrical breakdown without any detectable changes in the membrane structure, provided that there is sufficient time between the application of each pulse to allow resealing of the structural changes brought about by the reversible electrical breakdown (15). In a lipid bilayer the resealing times are about 20 μs (18), i.e., they are considerably shorter than those required by lipid-protein membranes (2, 10).

It is hoped that this short discussion of reversible and irreversible breakdown will help to clarify some of the confusing reports of reversible and irreversible electric field effects on biological material (for a detailed discussion, see Zimmermann et al., 19).

The exposure time (or pulse duration of the electrical field) also plays a crucial role in the electrofusion of cells where electrical breakdown is induced in the membrane contact zone between two adjacent cells. Field applications lasting 5 ms, such as those used by Senda et al. (20) for the fusion of plant protoplasts, lead to intermingling of the membranes and, under certain conditions, to the rounding of the fused cells, but they also result in irreversible changes in the membrane structure and in the cell itself. Consequently, these fused cells will neither grow nor divide. Production of viable fusion products requires exposure times that do not exceed 40 to 50 μs.

The reversibility of field-induced permeability changes of the membrane depends not only on the duration of the field but also its intensity. This experimental observation can be easily explained by the integrated Laplace equation, which describes the membrane potential profile of spherical cells exposed to an external electric field (21):

$$V = 1.5 \cdot a \cdot E \cdot \cos \theta \qquad \text{(Eq. 1)}$$

where V = membrane potential, a = radius, E = field strength, and θ is the angle between a given membrane site and field direction. The membrane potential generated across the membrane is proportional to the field strength and the radius of the cell.

For a given field strength and cell radius, the membrane potential is greatest in the field direction (at the so-called poles of the cell), whereas the membrane potential is always equal to zero at sites oriented perpendicular to the field direction (equator). The membrane potential thus declines progressively from the poles to the equator. With increasing field strength (kilovolts per centimeter range, depending on cell radius), the breakdown voltage is first reached at the poles (Figure 1a). Electrical breakdown and the formation of pores occur as a result of local electrochemical compression of the membrane. When the field strength is raised still further (to the supracritical field range), the breakdown voltage is also reached at those membrane sites oriented at an angle of 0° to 89° to the field direction (Figure 1b).

Even more and larger pores are induced in the membrane, leading to an ever increasing permeability change in the cell. So long as the surface area of the pores remains small in relation to the surface area of the cell, the pores can be closed again by diffusion of lipids (and possibly proteins) at higher temperatures. This process is reversible. Secondary reactions (e.g., water influx, local heating in the membrane, and electrophoresis of membrane compounds) are not dominant. However, at very high field strengths, the number of pores and the area of the individual pores may exceed a certain critical value, so that the secondary reactions come into effect. Reversible electrical breakdown then turns into irreversible breakdown.

The angular dependence of the generated membrane potential, V, as described by the integrated Laplace equation (Eq. 1) was proven by Korenstein and colleagues (22) in an elegant experiment. These authors used hypotonically swollen particles (so-called blebs) originating from chloroplasts of spinach and investigated the stimulation of delayed luminescence by external field pulses of 1 to 2 kV/cm intensity and 50 to 500 μs duration.

The angular dependence of the local electrical field shows higher light emission in regions near the pole, where the field strength is maximum, than in equatorial regions, where the local field interaction vanishes. Light emitted by "chlorophyll a" molecules has its transition moment in the long wave range aligned parallel to the membrane plane. Therefore, when a low external electrical field is applied (with all transition moments perpendicular to the field), only emittors in the pole area will be activated, yielding a highly polarized emission. However, with increasing external electrical field, emittors located between the poles and the equator will also be activated. This will result in a reduced degree of

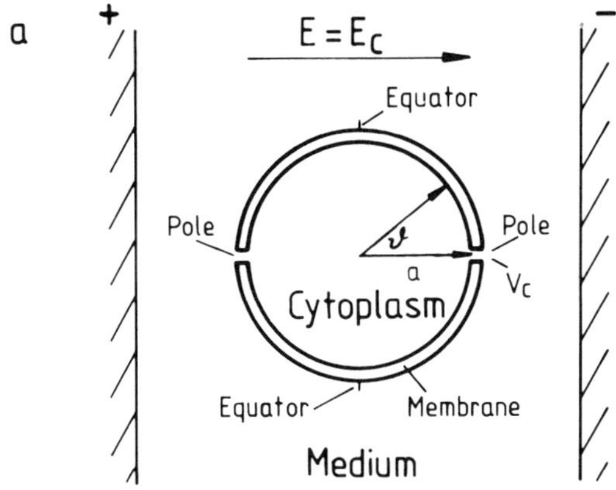

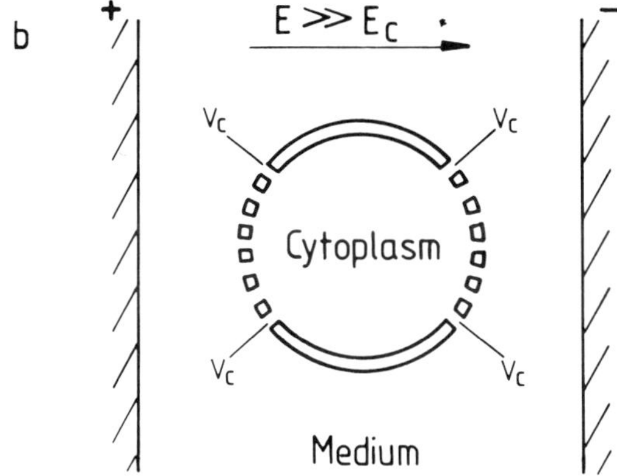

$$V_C = 1.5 \cdot a \cdot E \cdot \cos \vartheta$$

Figure 1--Schematic diagram of a cell exposed to high electrical field strength. The membrane potential, V, which is built up across the membrane in response to the external field, has its highest value at membrane sites oriented in field direction (poles) and progressively decreases towards the equator (perpendicular to the field direction). For membrane sites in perpendicular orientation to the field direction, the potential, V, is zero. The breakdown voltage, V_C, is therefore first reached in field direction, $E = E_C$ (a). In membrane sites oriented at a certain angle to the field direction, it is only reached if supracritical field strengths ($E \gg E_C$) are applied (b). Breakdown of the membrane is indicated by the formation of transmembrane pores.

polarization, because the emittors now are only partially
perpendicular to the electrical field. Both the polarized
emission and its increase with field strength demonstrate the
angular dependence of the local electrical field. In analogy to
photoselection (selection of randomly oriented molecules with
transition moment parallel to the polarized light), an
electroselection procedure could therefore be established taking
advantage of the fact that the chemical reaction is induced only
in the field direction.

The assumption that field-induced pores are formed once
breakdown has occurred is only a model concept that allows a
theoretical calculation of the molecular processes.
Experimental findings, e.g., that organelles (23) and even whole
cells (14) can cross the membrane with the aid of field pulses
of sufficient intensity, indicate that this model concept cannot
be correct because the large pores necessary for this process
would cause the membrane to tear. It is more likely that
changes occurring in the lipid structure lead to an increase in
the local permeability. In this context, it is interesting that
Dressler et al. (24) have demonstrated a field-induced flip-flop
of the lipid molecules in the erythrocyte membrane (see
Zimmermann et al., 14). Recent investigations on the
electrofusion of cells in our laboratory, in collaboration with
Zagury and coworkers, tend to favor a mechanism of this kind.
The term _electroporation_, introduced by Neumann et al. (25) for
reversible electrical breakdown, therefore seems inappropriate.

Field-induced exchange of materials between cells in
suspension and the surrounding solution can be induced
experimentally in a number of different ways. Cells can either
be briefly passed through a stationary electrical field (e.g.,
in the orifice of a Coulter Counter), or a field pulse can be
injected into a cell suspension between two electrodes. The
materials--particles, genes, or even organelles--that are to be
introduced into the cell by means of reversible electrical
breakdown have to be added to the solution before field
application. The application of this technique to controlled
drug release for genetic engineering and to other technological
and pharmaceutical areas is described elsewhere (10, 26). The
reader is referred to this extensive literature.

As mentioned above, electrical breakdown leads also to cell
fusion, provided that a close contact between the cells exists.
In the following section we will therefore consider how intimate
membrane contact can be established between freely suspended
cells.

AC-FIELD-MEDIATED CELL CONTACT

Intimate membrane contact between suspended cells is
established by the application of a non-uniform alternating
electrical field (a.c. field). The electrical field creates a

dipole in the cell by charge separation in the membrane.
Because the field strengths acting on these induced positive or
negative charges are different due to the field inhomogeneity, a
net force arises that pulls the cells in the direction of the
highest field intensity (i.e., generally towards the
electrodes). The movement of a cell in a non-uniform field is
independent of the polarity of the field and is therefore also
observed in an a.c. field (frequency range 1 kHz to 1 MHz).
This phenomenon, which has been known since the 1920s (27-30)
when it was investigated in detail, has been termed
<u>dielectrophoresis</u> by Pohl (31), in contrast to electrophoresis
in a d.c. field. Cells approaching each other during their
movement towards the region of highest field intensity attract
each other, provided that the field strength of the non-uniform
field is sufficiently high to overcome Brownian motion (1 to 400
V/cm, depending on the cell diameter). The attractive forces
arising from the generated dipoles in neighboring cells exceed
the electrostatic repulsion between the apposed negatively
charged cell membranes and the repulsive hydration forces, which
become effective when the distance between the membranes of two
cells is small (say 3 nm). The attraction between at least two
cells at very close quarters is further augmented by the fact
that the second cell encounters an enhanced local field
inhomogeneity in the vicinity of the first cell and tends to
move towards it preferentially. The weak field inhomogeneity
arising from the electrical field between two cylindrical wires
arranged in parallel thus results in the formation of cell
chains (so-called pearl chains) that are aligned along the field
lines emanating from the electrode surfaces. The number of
cells in a pearl chain depends on the suspension density. At
very low suspension densities the formation of two- and
three-cell chains is favored.

Schwan and colleagues (30, 32, 33) have provided a
semi-quantitative theoretical analysis for special cases of
pearl-chain formation. In this theory it is assumed that the
medium and the particles (cells) show no dielectric losses. The
analysis is based on relations for the body forces that are
derived from energy principles in electrostatics. However, as
pointed out by Sauer (34), these relations cannot be applied
when dielectric losses occur. Taking dielectric loss into
consideration, Sauer calculated the trajectories of the
particles in the process of pearl-chain formation on the basis
of the electromagnetic momentum balance. This leads to
meaningful expressions of the body force.

As shown both theoretically and experimentally, the time
constant for pearl-chain formation is in the order of seconds to
minutes, depending on the particle size, viscosity of the
medium, and field strength of the a.c. field.

For the establishment of intimate membrane contact, it is
not necessary for the sinusoidal a.c. field to be applied for
the entire time period. An intermittent sinusoidal field or

even a pulsed field is usually sufficient to prevent dissolution
of the chains by Brownian motion or electrostatic repulsive
forces between the cells in a chain. It is, therefore, worth
noting that the commercial electrofusion devices can adjust the
duty cycle of the alternating field (GCA, Boston/USA; Krüss,
Hamburg/FRG).

The use of an intermittent sinusoidal field and/or of the
duty cycle is advantageous for the following reason. It was
found empirically in the electrofusion of cells of different
species that the viability of the cell hybrids is critically
dependent on the exposure time of the cells in the a.c. field.
It is evident that the a.c. field induces changes in the
membrane structure and transport processes that become
irreversible if the exposure times are too long, say 10
minutes. It is known from experiments with steady fields that a
redistribution of intramembranous particles (proteins, perhaps
also lipid molecules) is induced by the tangential components of
the field (35, 36).

As demonstrated elsewhere (11, 12), in theory a.c. fields
can also elicit movement of intramembranous particles in the
membrane surface because the field vector not only changes with
time but also locally on the membrane surface of a spherical
cell. The possibility is being discussed that such effects may
be responsible for the emergence of particle-free lipid
domains. The emergence of particle-free lipid domains is also
regarded as a prerequisite for chemically and virally induced
fusion (37). It is thus possible that a brief application of
the a.c. field is also a prerequisite for electrofusion, in
order to create particle-free lipid domains, but that
excessively long exposure times, on the other hand, lead to a
marked segregation of the membrane lipids and proteins, thus
jeopardizing the vital transport processes across the membrane.
Access to these processes may soon be provided by the dielectric
spectroscopy method developed by Kell for cell membranes (38).

In any case, the a.c. field should be applied for as short
as time as possible prior to the application of the breakdown
pulse. For example, for the fusion of a single myeloma cell
with a single lymphocyte, exposure times of 30 s seem to be
optimal, because the highest yield of dividing hybridoma cells
can be achieved under these conditions (39, and Zagury,
Fouchard, Vienken, Zimmermann, in preparation). After the
application of the breakdown pulse and the first step of
intermingling of the membranes, the a.c. field should be reduced
exponentially to zero after about 30 s. This function can be
preset in the commercially available equipment (GCA and Krüss),
so that the greatest possible reproducibility of the fusion
process can be achieved. Adverse side effects of the a.c. field
on the cellular functions and membrane integrity can also be
caused by inappropriate frequencies and field strengths.

The membrane acts as a capacitor, and like any capacitor,
it is short-circuited at higher frequencies, i.e., the field

lines go through the cell interior. As a rule, this effect
becomes apparent at frequencies above 1 MHz, so that cell
alignment should be carried out in the range of 10 kHz to 1 MHz.

After application of the breakdown pulse, the membrane
becomes permeable and highly conductive (12). Under these
conditions, current can flow through the cell interior even in
the kilohertz range. During the fusion process the use of the
a.c. field should therefore be restricted as much as possible.

The processes elicited in cells and membranes in response
to a.c. fields of different frequencies will undoubtedly have to
be studied in greater detail to ascertain the optimal fusion
conditions for cells of different species.

We know that use of a.c. fields of appropriate frequency
and intensity does not necessarily lead to adverse side-effects
and that, on the contrary, they may have a stimulating effect on
ion uptake, DNA synthesis, and cellular transcriptions (40-42).
These observations thus open up the possibility of increasing
growth and cell division in fused hybrids by selecting the
appropriate frequency and intensity of the a.c. field during the
entire electrofusion process as well as during the application
of a.c. fields following the rounding stage. This is
undoubtedly an interesting aspect of the action of electrical
fields on biological systems that could gain importance far
beyond the scope of electrofusion.

A further aspect that should be borne in mind for the
alignment of cells by a.c. fields is the correct choice of field
intensity. The distance between two apposed membranes in a
pearl chain, and hence the extent of membrane contact, depends
on the field strength of the a.c. field. An increase in the
field strength leads to a flattening of the spherical cells in
the contact zone, so that greater areas of the two parallel
apposed membranes come into contact. This configuration leads
to optimum fusion. However, when selecting the appropriate
field strength for the establishment of a large area of intimate
membrane contact, care must be taken not to exceed the field
strength, which would cause the breakdown voltage to be exceeded
in the membrane sites oriented in the field direction. Because
breakdown occurs during every half-wave, reversible breakdown
goes over into irreversible destruction of the membrane. In
this range of field strengths, the cells may also undergo
deformation, which may result in vesicle formation at
supracritical field strengths (12). For reasons as yet unknown,
although the cells revert to their original shape when the a.c.
field is switched off, they can no longer be fused.

FUSION

As soon as the breakdown pulse has been applied, apposed
pores are generated in the adjacent membranes. The pores give
rise to a cytoplasmic continuity between the two adjacent cells

and lead to the formation of bridges between the two membranes
(12, 43). The dumbbell-shaped configuration formed in this
manner then assumes a spherical shape for thermodynamic reasons,
ultimately resulting in a fused cell. The speed with which the
dumbbell-shaped configuration turns into the spherical cell,
following the intermingling of the membranes, depends on various
factors (see also next section). The higher the field strength
of the applied breakdown pulse, the faster the formation of a
spherical cell (44). Erythroleukemic Friend cells, for example,
fuse and attain a spherical shape within a few seconds if the
field intensity of the pulse is 5 kV/cm. This field strength is
twice as high as the field strength required to reach the
breakdown voltage in the membrane contact zone. At a field
strength of about 1 kV/cm, on the other hand, the time taken to
reach the rounding-off stage is about 2 to 3 minutes. At higher
field strengths the membrane perturbations are considerably
greater and distributed over a larger membrane area (see above),
so that cell fusion can proceed more rapidly. Fusion times of
the order of some minutes have been found to give higher yields
of growing and dividing hybrids, so that excessively high field
strengths should be avoided when applying the breakdown pulse.

Figure 2 shows for plant protoplasts from Kalanchoë
daigremontiana different stages of electrofusion up to the
rounding-off stage. Electrofusion of protoplasts of K.
daigremontiana and Petunia has shown that excessive membrane
material is removed by vesicle formation in the membrane contact
zone (43, 45) where the pores and the cytoplasmic continuity are
formed (Figures 3 and 4).

The availability of various techniques for preselecting the
number of cells to be fused is of particular interest from the
point of view of producing hybrids of technological interest
(11, 12, 46). With strains of yeasts with different genetic
markers, it was possible to obtain high yields of yeast hybrids
by this method (47-49). The production of human and mouse
hybridoma cells by electrofusion of myeloma cells with
lymphocyte cells has been described in the literature (50-51).
These hybridoma cells are of particular interest to industry and
pharmaceutical houses because of the production of monoclonal
antibodies. Sea urchin eggs also have been fused electrically
(52). The fused eggs could be fertilized and made to undergo
division. Fused plant protoplasts exhibited regeneration of the
cell wall (12, 53).

Liposomes (54) as well as two planar lipid bilayer
membranes that were brought into close contact by hydrostatic
pressure (55) could also be fused by an electric field pulse.

Fusion of animal cells of permanent cell lines is also
possible with electrofusion (11-14). Pronase pretreatment of
the cells or the presence of pronase during field application is
very often required to obtain optimum fusion (however, see
below) (44, 56).

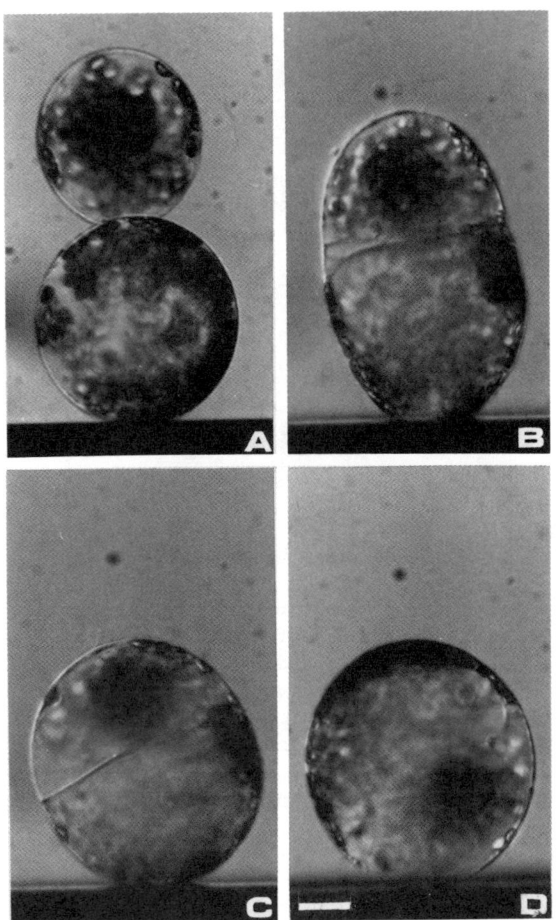

Figure 2--Electrofusion of mesophyll protoplasts of <u>Kalanchoë</u> <u>daigremontiana</u>. Dielectrophoretically collected cells suspended in mannitol solution were fused by the application of an electric pulse of 20-μs duration and 500-V/cm field strength. Pictures were taken before the pulse and 30, 60, and 120 s after the pulse (the electrode appears as a dark line). Bar = 20 μm.

All these examples demonstrate that electrofusion is a universal method for the fusion of cells and liposomes. This is to be expected, given that dielectrophoresis occurs within all polarizable material and that reversible electrical breakdown is a common feature of all biological membranes, including both lipid-protein and pure lipid membranes.

The universality of the electrofusion method becomes particularly obvious if interkingdom fusion is considered.

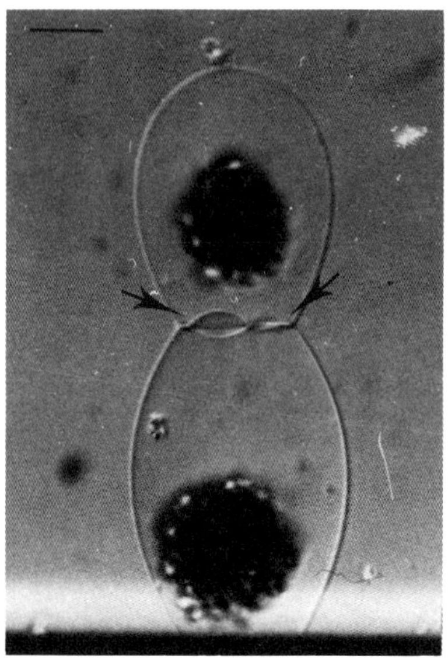

Figure 3--After electrofusion of plant protoplasts
(K. daigremontiana), the volume of the fused cell is the sum of
the volume of the parent cells (43). The excess in membrane
material arising from the reduction in membrane area is removed
by the formation of vesicles, which are visible in the light
microscope. These vesicles are observed in the contact zone of
the fusing cells (arrows). Photograph was taken 1 minute after
the application of the breakdown pulse (field strength, 1 kV/cm;
15µs duration). Bar = 20 µm.

Figure 5 shows fusion between a Friend cell (erythroleukemic
cell transformed by the Friend virus) and a plant protoplast of
Petunia. The fused cells exhibit the properties of both animal
and plant cells. Cell wall regeneration is observed, and
dimethyl sulfoxide-induced hemoglobin synthesis can be
demonstrated in these hybrids (Salhani, Zimmermann, Ward,
Cocking, in preparation).
    Electrofusion also permits the simultaneous, controlled
fusion of thousands of cells to form giant cells (11, 12, 14).
Giant cells can be obtained by injecting suspensions of high
cell density into the electrode gap. Under these conditions,
many cell chains are formed, and the cells in a given chain make
intimate membrane contact not only with other cells in the same
chain but also with cells from neighboring chains. The
application of a breakdown pulse high enough in intensity to
induce breakdown over a large membrane area associated with the
formation of a large number of pores then leads to fusion, not
only between cells along the field lines but also between cells

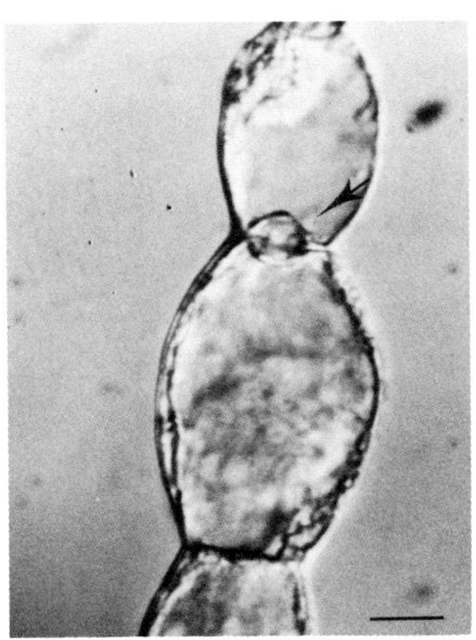

Figure 4--Emergence of a vesicle in the contact zone of fusing
Petunia protoplasts (arrow). Such vesicles can be easily
observed if the fusing cells have their internal vacuoles close
to the contact zone of the cells; otherwise any vesicle that may
be created is hidden from the observer by cell organelles.
Interference phase micrograph. Bar = 10 µm. Taken from
Zimmermann et al. (45).

lateral to the field lines (so-called three-dimensional
fusion). Within several minutes, giant cells are formed that
may reach a diameter of 0.5 to 1.0 mm. Giant cells have been
produced from red blood cells, Friend cells, 3T3 cells, and
Avena protoplasts (Figure 6). These giant cells are large
enough that microelectrodes can be inserted into them, making
possible direct measurement of the electrical properties of the
membrane. In the future, it may even be possible to make
technological use of giant cells for the growth of cells on
their surface, in analogy to Cytodex particles (57). Giant
cells would be suitable for this process because they are stable
and can be stored for long periods of time.

LARGE-SCALE PRODUCTION OF FUSED CELLS

A special chamber, the so-called helical chamber, has been
developed for the large-scale production of hybrids by
electrofusion (19, 46). Figure 7 shows a schematic diagram of
the helical chamber. In principle, the helical chamber consists

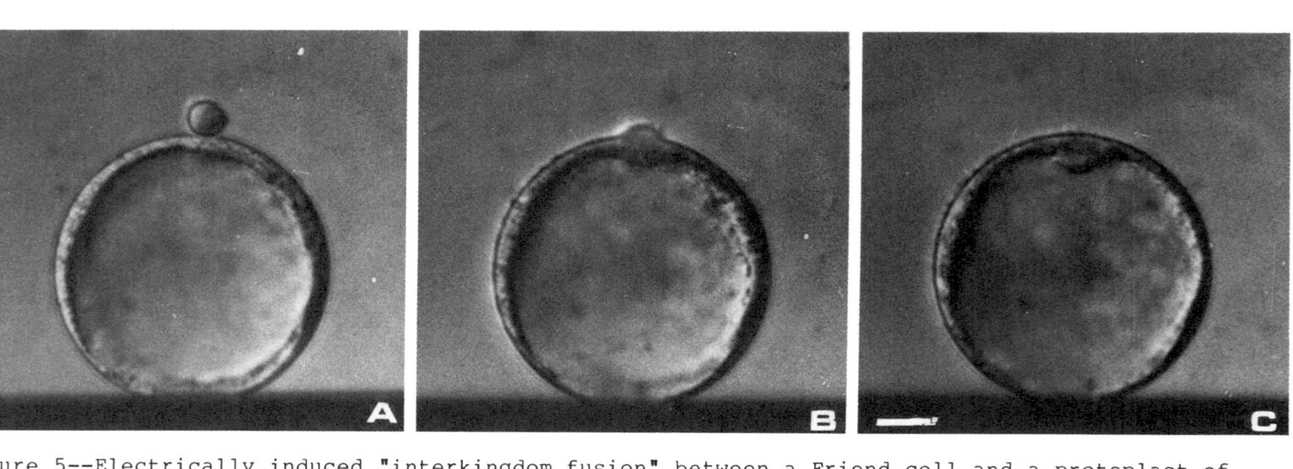

Figure 5--Electrically induced "interkingdom fusion" between a Friend cell and a protoplast of _Petunia_ _hybrida_ _albino_ _commenci_. Micrographs show the time course of fusion before and 30 s (B) and 60 s (C) after the application of two breakdown pulses (2 kV/cm, 20 µs duration). Prior to fusion, both Friend cells and _Petunia_ protoplasts were incubated in a 0.4 M mannitol solution supplemented with 1 mg/ml pronase (Serva GmbH, Heidelberg). One of the electrodes appears as a black line. Interference phase contrast micrographs by N. Salhani. Bar = 20 µm.

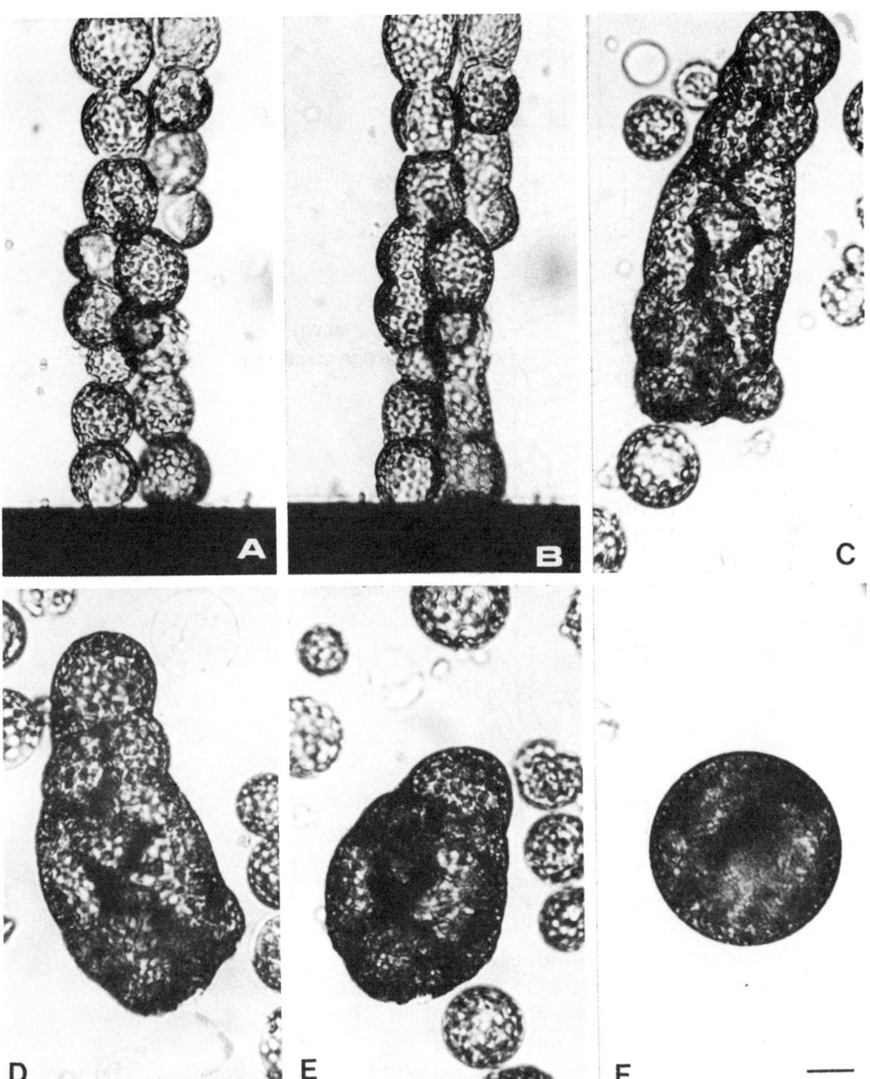

Figure 6--By use of large electrode gaps and high suspension densities, more than one pearl chain of plant protoplasts (<u>Avena sativa</u>) can be arranged in parallel by dielectrophoresis. Vertical as well as lateral fusion between adjacent cells was achieved by the injection of a supracritical electrical field pulse of 600 V/cm and 20μs duration. During the induced fusion process, micrographs were taken in time intervals of 30 s (B), 2 min (C), 5 min (D), 10 min (E), and 30 min (F) after field application. Bar = 20 μm.

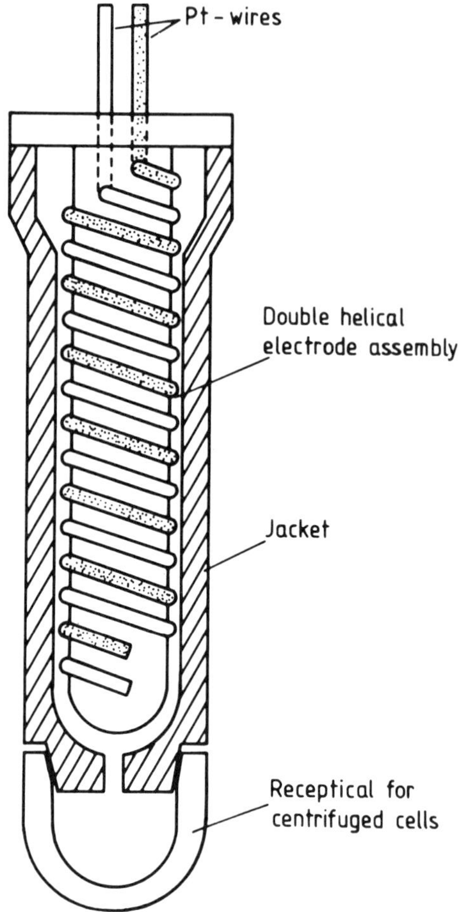

Figure 7--Large-scale production of fused cells by a helical
fusion chamber. The helical chamber consists of a Perspex
mandrel around which two cylindrical platinum electrodes are
wound at a distance of about 200 µm (total length of 1 m).
The mandrel can be filled with cooling medium in order to avoid
temperature increases during the field application. It is
introduced into a cylindrical jacket that already contains the
cell suspension. The cell suspension rises up through the
intervening space. Cell chains are formed between the parallel
electrode wires in response to an alternating field. After
application of a breakdown pulse and subsequent fusion, the
cells can be collected by centrifugation into a lower tube
through a valve (not shown). The receptacle is disposable and
can very easily be removed for cloning the fused hybrids. The
whole procedure can be performed under sterile conditions
(Matschke, Buechner, and Zimmermann, unpublished results).

of a hollow Perspex tube onto which are wound two parallel
electrode wires (at a distance of 200 to 500 µm). The Perspex
tube can be filled with coolant. As in the conventional fusion
chamber, which consists of two parallel electrode wires glued on
to a microslide, fusion proceeds between the electrode wires in
each turn. Because the electrode wires are about 1 m long,
enormous numbers of fused cells are produced. For the fusion
process the Perspex tube is carefully introduced into a
cylindrical jacket containing the cell suspension. The cell
suspension rises up into the space between the tube and the
jacket. Application of the a.c. field causes the cells to align
themselves between the electrodes. After fusion the cells are
centrifuged through a valve into a receptacle that is fixed to
the base and can easily be removed. The receptacle is
disposable and contains nutrition or selection medium. With the
helical chamber, fusion can be carried out under sterile
conditions and also be observed under the microscope. This
chamber is particularly suitable for cloning experiments.

SINGLE-CELL STUDIES

     The main advantage of electrofusion, i.e., the ability to
monitor the entire fusion process (including the identification
of fused cells) under the microscope, offers the additional
possibility of carrying out biochemical and biophysical
investigations of single fused cells. The insertion of
microelectrodes into giant cells produced by three-dimensional
(lateral) fusion of thousands of cells (11) will soon enable the
membrane potentials and resistances to be measured so that we
can expect to acquire a wealth of new information on the
membrane properties of microscopic cells in the not too distant
future. Many biochemical assays are now so sensitive that they
can be applied to individual microscopic cells. An interesting
example was recently reported by Verhoeck-Köhler et al. (58),
who determined the ATP and ADP content in individual cell
hybrids that had been obtained by the electrofusion of two,
three, or four mesophyll protoplasts of <u>Avena</u> <u>sativa</u>. These
authors were able to show that the ATP and ADP content increased
in proportion with the number of fused cells and that the
ATP/ADP ratio thus remained constant (Table 1). These results
also show that there is no reduction in the ATP and ADP content
during fusion.
     Verhoeck-Köhler et al. (58) were further able to
demonstrate that the ATP content of the cell determines the
speed of the fusion process after the first (primary)
intermingling step of the cell membranes. For this purpose,
different energy states were established within the mesophyll
protoplasts before the initiation of the fusion process. This
was achieved by a timed preincubation of the protoplasts with
specific effectors of phosphorylation. The rounding-off process

Table 1--ATP and ADP levels in intact single _Avena_ protoplasts and fusion products and in single osmotically lysed protoplasts

Protoplasts	ATP (fmol)	ADP (fmol)	ATP/ADP
Single	7.0 $\pm$ 2.3	5.0 $\pm$ 0.7	1.4
Hybrid out of two	13.7 $\pm$ 4.5	10.2 $\pm$ 4.6	1.34
Hybrid out of three	21.0 $\pm$ 3.4	15.2 $\pm$ 2.8	1.37
Hybrid out of four	29.1 $\pm$ 3.1	16.6 $\pm$ 4.2	1.87
Single osmotically lysed	4.9 $\pm$ 0.6	7.0 $\pm$ 0.6	0.70

Values are means $\pm$ SD, N = 6. From Verhoeck-Köhler et al. (58).

slows down as the ATP/ADP ratio decreases (Table 2), i.e., the
time required for the rounding-off of the fused cells is closely
linked to the cellular energy state. On the other hand, the first
step in the intermingling process between adjacent cells, initiated
by the application of the fusion pulse, is unaffected by the level
of adenylates within the cell. The energy of this step depends
only on the electric field intensity.
   Low Ca^{2+} concentrations also slow down the rounding-off
process of the fused product (12). At these concentrations (14)
the fusion process is virtually arrested at the first intermingling
stage when the fusion product takes on a dumbbell-shaped configuration. ATP and Ca^{2+} apparently control the membrane fluidity
and the restoration of the cytoskeleton. These aspects undoubtedly
deserve more attention in the future. However, the present results
demonstrate the great value of electrofusion for membrane
research. This type of investigation was previously impossible
with the conventional chemical and viral fusion techniques.

Table 2--Rounding-up time and ATP, ADP-levels of fused cells, treated with various effectors

Effector	Time for rounding up (s)	ATP (fmol)	ADP (fmol)	ATP/ADP
- -	54 $\pm$ 30	16.6 $\pm$ 3.8	7.2 $\pm$ 5.2	2.31
Antimycin	64 $\pm$ 31	12.4 $\pm$ 7.6	11.4 $\pm$ 4.0	1.09
DBMIB	76 $\pm$ 21	11.8 $\pm$ 4.2	12.0 $\pm$ 3.8	0.98
Antimycin/DCMU	153 $\pm$ 59	4.8 $\pm$ 3.2	19.2 $\pm$ 7.6	0.25
FCCP	148 $\pm$ 49	3.8 $\pm$ 2.4	20.0 $\pm$ 9.2	0.19
CCCP	128 $\pm$ 60	5.8 $\pm$ 3.8	18.0 $\pm$ 10.2	0.32

DBMIB = dibromothymoquinone, DCMU = dichloro-phenyl-dimethyl-urea,
FCCP = carbonyl cyanide p-trifluoromethoxy-phenylhydrazone,
CCCP = carbonyl cyanide m-chlorophenylhydrazone.
Values are means $\pm$ SD; rounding-up time: N=30; ATP, ADP-levels:
N=6. From Verhoeck-Köhler (58).

ELECTROFUSION OF CELL PAIRS

Because electrofusion can be viewed under the microscope, it is also possible to fuse a single pair of cells under the microscope and to immediately isolate the hybrid for cloning after the rounding-off stage. Vienken et al. (39) have recently developed a technique for the electrofusion of single cell-pairs that is not only promising for the production of single hybridoma cells but may also help to elucidate the effect of pronase treatment on cells in electrofusion. A microslide, onto which were glued two parallel electrode wires, was covered with a layer of poly-L-lysine. After being washed with distilled water, the slide was rinsed with a solution containing pronase. Pronase partly compensates for the net positive charge of poly-L-lysine. The microslide was then dried, placed in a Petri dish, and completely covered with sterile mineral oil (Vaseline) (Figure 8).

A droplet of 0.3 M mannitol with a few myeloma cells (or a lymphocyte and a myeloma cell) was injected into the electrode gap under the mineral oil cover so that a cavity filled with solution was formed. Sterility was provided by closing the Petri dish with a covering plate. Application of an alternating sinusoidal field of 200 V/cm and a frequency of 1 MHz caused the cell pair to align itself along the field lines. Fusion was achieved by injecting a field pulse with a field strength of 2.5 kV/cm and a duration of 20 μs. Just after the breakdown pulse had been applied, the alternating field was switched off. Because the adhesion between the pretreated surface of the microslide and the cells, the movement of the cells was restricted. Anyhow, the two cells were often observed to move about 1 to 2 μm apart after the application of the breakdown pulse. They then approached each other again under the same orientation and fused. This effect was probably due to the local non-uniform field arising from the anisotropic membrane potential after breakdown. Under the conditions described above, pronase is only present between the glass surface and the cells, not between the cells themselves. One can therefore conclude that, when pronase is present in the solution and hence in the membrane contact zone between adjacent cells, it prevents the cells from moving too far apart (for further details see Vienken et al., 39).

This modified technique allows electrofusion of cells to be carried out without the danger of pronase molecules being taken up during the fusion process, which should of course increase the yield of viable hybrids. The single cells can easily be removed with the aid of micropipettes, particularly if electrode gaps of 500 μm are used in combination with the high output voltage of 270 V (delivered by the device of GCA and Krüss). Single cell pairs of myeloma cells, but also pairs of single myeloma cells and lymphocytes were fused under these conditions and subsequently cloned. This modified technique, combined with

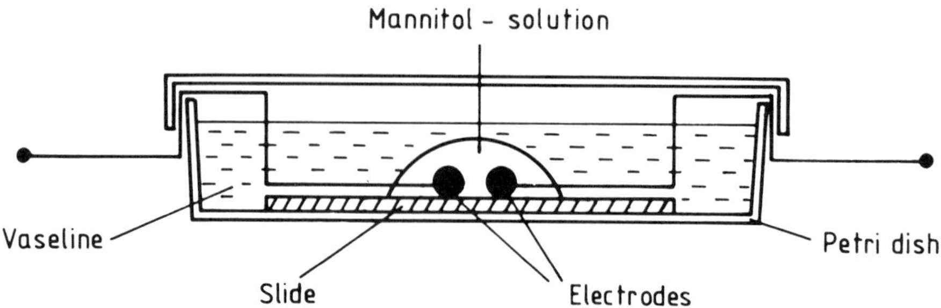

Figure 8--Schematic cross section of the experimental
arrangement used for electrofusion under sterile conditions.
Two platinum electrodes (diameter, 100 μm) are glued in
parallel on a slide, which is then treated with
poly-L-lysine/pronase. The microslide is placed in a Petri dish
that is then filled with sterile mineral oil (Vaseline). By
means of a micropipette a droplet of a solution containing 0.3 M
mannitol and both 0.5 mM $CaCl_2$ and $MgCl_2$ is injected into
the Vaseline in order to form a small cavity over the
electrodes. By this means the Vaseline prevents the mannitol
solution from evaporation. The cells to be fused are introduced
into the mannitol droplet with a micropipette. The electrodes
are then connected to the electronic equipment and the Petri
dish is closed with a covering plate. For reasons of clarity,
cells are not shown. Taken from Vienken et al. (39).

the selection set-up for antibody secreting lymphocytes, could
thus represent a very efficient procedure for the production of
hybridoma cells.

APPLICATION OF CELL ROTATION TO FUSION

     The successful production of hybridoma cells depends on the
preselection of actively secreting lymphocytes. This was
formerly a difficult and tedious procedure because only a very
small proportion of the lymphocyte population is activated.
     It has recently been shown (Arnold, Wendt, Zimmermann, and
Zagury, in preparation) that actively secreting cells can
readily be distinguished from nonsecreting cells by means of the
technique of cell rotation (13, 59-63). When a rotating
electric field of variable frequency is applied to a cell
suspension, maximum rotation rate occurs at a specific frequency
that depends (among other things) on the cell membrane
characteristics (for details, see 60, 61).
     The differences in the specific frequencies detected
between secreting and nonsecreting lymphocytes may therefore
reflect changes in membrane capacitance or resistance, or in
internal conductivity of the cell, to all of which this

technique is sensitive (60, 61). It may be expected that
preselection of other physiological stages of cells may be
possible with this technique, allowing more successful and
precise cell fusion.

MAGNETOELECTROFUSION

     Until recently, electrofusion of cells generally required
the use of non-electrolyte solutions. At 10^{-4} S/cm, the
conductivity of these solutions is low enough to prevent heat
development by the a.c. field. Heat development would
undoubtedly lead to local turbulences which, in turn, would
disrupt the intimate membrane contact between adjacent cells in
a pearl chain. On the other hand, the use on nonelectrolyte
solutions has the disadvantage that cells can only be exposed to
solutions of low conductivity for a very limited time because of
the danger of adverse side effects that could strongly influence
the viability of the fused cells. Plant protoplasts, which are
normally obtained from tissue subjected to enzymatic degradation
in a nonconducting environment, are relatively resistant to the
withdrawal of electrolytes and are able to survive for hours to
days suspended in such solutions without any detectable
deterioration of the cellular and membrane functions. In
contrast, animal cells can only be stored for half an hour to an
hour in solutions of low ionic strength. It is therefore
advisable to perform the entire electric somatic hybridization
process in as short a time as possible (say 10 to 15 minutes).
Electrofusion in conducting solutions has been made possible, in
principle, by the development of the helical and other new
fusion chambers that allow extensive cooling. It will soon be
possible to carry out comparative investigations of the
viability of hybrids obtained by electrofusion in nonconducting
and conducting solutions. An alternative method of performing
fusion in conducting solutions, recently developed in this
laboratory, will now be described briefly, for it may well open
up new areas of application for the electrofusion technique.
     In this new, modified technique, a magnetic field is used
instead of an a.c. field for the formation of pearl chains.
With few exceptions cells are nonmagnetic, but they can be made
artificially magnetic by the absorption of small magnetic
particles (diameter about 10 nm) onto the external membrane
surface. It is also possible to entrap magnetic particles
within cells with the aid of the electrical breakdown technique
(26). The magnetic particles are or can be made biocompatible
by an appropriate coating. Human red blood cells that were
magnetized by the absorption of magnetic particles (ferrofluid)
could be shown to form pearl chains. Membrane contact was
sufficiently close for these cells to be fused into giant cells
by an electric field pulse (Kramer, Vienken, Vienken,
Zimmermann, in preparation). The conductivity of the solutions

used for magnetoelectrofusion was 10^{-3} S/cm. Higher
conductivities are possible, but it is important to bear in mind
that agglomeration of the magnetic cells and particles occurs at
certain electrolyte concentrations. The use of magnetic
particles may also have the advantage that some of the particles
absorbed onto the external surface are incorporated into the
newly fused cells during the fusion process so that the
subsequent separation of these fused cells would be very easy.

ELECTROFUSION AS A TOOL OF EVOLUTION

It has recently been demonstrated (64, 65) that electrical
field induced fusion of cells is possible under conditions that
mimic those that may have occurred during evolution.
Instead of the metal electrodes normally used, pieces of
conducting ore (FeS) 200 μm apart were used. With one
"electrode" connected to ground, the other to a 50-cm length of
wire, it was found that pearl chaining could be induced by means
of induction from a 2-m-long conductor (driven by a frequency
generator) some 5 m distant. Fusion resulted when a spark
discharge between two 5-cm electrodes connected to a standard
interval (combustion engine ignition coil) was operated some 5
cm away.
Within the rocks of the Earth's surface there exist veins
of conductive ore, often considerably longer than 50 cm, that
could have acted as the necessary antennas. The voltage
necessary for pearl-chaining could arise from the influx of
radio energy from the sun, although the extent of ionospheric
absorption during earlier times is unknown. It is even possible
that random collection of cells within the interstices of
surface rocks occurs sufficiently often on an evolutionary
time-scale that the dielectrophoretic step is unnecessary.
The voltage necessary for membrane breakdown would readily
result in a pick-up of the electric field from a lightning
discharge, or the direct effect of the currents radiating from a
nearby stroke to ground. The 1- to 100-μs duration typical of
lightning (66) is suitable to give occasional viable hybrids.
Large voltage potentials can be also built up during the
freezing of aqueous solutions, the so-called Workman-Reynolds
effect (67, 68). For example, a voltage potential of 20 V
(unfrozen solution with respect to ice) can be generated during
the freezing of 1 mM solution of NaCl. Experimental evidence
supports the view that the generation of such high voltage
potentials arises from charge separation that seems to be the
result of selective rejection from the ice rather than selective
inclusion of ions in the ice crystals. If the generation of the
voltage potentials results from charge separation, the external
electric fields may extend considerable distances from the
ice-solution interface. Thus, it is conceivable that
electrofusion of cells could operate during evolution in

response to a freeze-thaw cycle. In this context it is
interesting to note that experimental proof is available that
freezing and thawing lead to fusion of liposomes (69). Freezing
and thawing may be therefore an alternative source for the
generation of short-lived electrical fields of high intensity
which are required for fusion of cells.

CONCLUSIONS

     The electrofusion method undoubtedly has a number of
advantages over the traditional chemical and viral fusion
techniques. The most significant of these advantages are the
complete visual control of the fusion process, the possibility
of preselecting the fusion conditions as well as the number of
cells to be fused, and the high yield of fused cells, which is
also associated with a markedly increased yield of karyogamy, at
least as far as we are able to ascertain at the moment (48, 49).
     The mechanism of electrofusion is reasonably well
understood. In principle, it is possible to elucidate
individual molecular stages in the process in combination with
optical and electron microscopical techniques (70).
Electrofusion of cells, therefore, undoubtedly represents an
important tool both in membrane research and in the research of
differentiation processes. With the aid of this technique it
should be possible to elucidate membrane structure and transport
as well as the regulatory processes in the membrane and in the
cell itself. The importance of electrofusion for the study of
processes that occurred during evolution cannot be fully
evaluated at the present time. However, the finding that cell
fusion can be induced by means of electromagnetic waves and
spark discharge indicates that this technique and its
application to the solution of a variety of biological problems
could lead us to new insights with regard to the origin of life.
     The great potential of the electrofusion method undoubtedly
lies in the production of hybrids of technological interest.
The production of new yeast strains for the brewing industry or
of hybridoma cells represents only two possible applications and
serves to emphasize the variety of laboratories interested in
this new method. The extent to which this electrical method
will replace the conventional techniques of chemical and viral
fusion in the future still remains to be seen. As reported
elsewhere (12), the chemical and viral techniques are ultimately
attributable to changes in the intrinsic electric field in the
membrane which, in turn, elicits the fusion process. If this
assumption is correct, the controllable electrical method should
also help to elucidate the mechanism of viral and chemical
fusion, particularly when combined fusion methods (e.g.,
dielectrophoresis and polyethylene glycol) are examined in
detail. It is thus conceivable that investigations utilizing
the electrical fusion method could increase our understanding of

the chemically or virally induced fusion procedures to the point where these methods could be improved and made more controllable.

The rapid growth of literature in the past two years on the subject of interactions of electric fields with biological matter raises the hope that a considerably more detailed picture of the electrically induced processes in the cell membrane will be available in the very near future.

ACKNOWLEDGMENT

The author is supported by a grant from the Deutsche Forschungsgemeinschaft, Sonderforschungsbereich 160.

REFERENCES

1. Zimmermann, U. and Pilwat, G. (1978) Sixth Int. Biophys. Cong. Kyoto, Abstr. IV-19(H), p. 140.

2. Zimmermann, U., Vienken, J., and Pilwat, G. (1980) Bioelectrochem. Bioenerg. 7:553.

3. Zimmermann, U., Vienken, J., and Scheurich, P. (1980) In: Biophysics of Structure and Mechanism, Vol. 6, Gersonde, K. (ed.) Springer International, p. 86.

4. Zimmermann, U., and Scheurich, P. (1981) Planta 151:26.

5. Zimmermann, U., Schulz, J., and Pilwat, G. (1973) Biophys. J. 13:1005.

6. Zimmermann, U., Pilwat, G., and Riemann, F. (1974) In: Membrane Transport in Plants. Zimmermann, U. and Dainty, J. (eds.) Springer-Verlag, p. 146.

7. Zimmermann, U., Pilwat, G., and Riemann, F. (1974) Z. Naturforsch. 29c:304.

8. Zimmermann, U., Pilwat, G., and Riemann, F. (1974) filed on February 2, 1974; German Patent No. 24 05 119 of May 5, 1977; British Patent No. 1.481 480 of November 23, 1977; US CIP Patent No. 4 081 340 of March 28, 1978; French Patent No. 75 02743 of October 16, 1978; U.S. Patent No. 4.154.668 of May 15, 1979.

9. Galfrè, G. and Milstein, C. (1981) Meth. Enzymol. 7:3.

10. Zimmermann, U., Scheurich, P., Pilwat, G., and Benz, R. (1981) Angew. Chem. 93:332; Int. Ed. 20:325.

11. Zimmermann, U., and Vienken, J. (1982) J. Membrane Biol. 67:165.

12. Zimmermann, U. (1982) Biochim. Biophys. Acta 694:227.

13. Arnold, W.M., and Zimmermann, U. (1983) In: Biological Membranes, Vol. 5. Chapman, D. (ed.) Academic Press, London (in press).

14. Zimmermann, U., Vienken, J., and Pilwat, G. (1984) In:
     Investigative Microtechniques in Medicine and Biology,
     Vol. 1. Chayen, J. and Bitensky, L. (eds.) Marcel
     Dekker, New York, Chapt. 3.
15. Benz, R., Beckers, F., and Zimmermann, U. (1979) J.
     Membrane Biol. 48:181.
16. Benz, R., and Zimmermann, U. (1980) Bioelectrochem.
     Bioenerg. 7:723.
17. Benz, R., and Zimmermann, U. (1981) Planta 152:314.
18. Benz, R., and Zimmermann, U. (1981) Biochim. Biophys.
     Acta 640:169.
19. Zimmermann, U., Büchner, K.-H., and Arnold, W.M. (1984)
     Proc. ISBB, Blackie and Son, Ltd., Nottingham (in press).
20. Senda, M., Takeda, J., Abe, S., and Nakamura, T. (1979)
     Plant Cell Physiol. 20:1441.
21. Jeltsch, E., and Zimmermann, U. (1979) Bioelectrochem.
     Bioenerg. 6:349.
22. Farkas, D.L., Korenstein, R., and Malkin, S. (1980) FEBS
     Letters 120:236.
23. Zimmermann, U., Küppers, G., and Salhani, N. (1982)
     Naturwissenschaften 69:451.
24. Dressler, V., Schwister, K., Haest, C.W.M. and Deuticke,
     B. (1983) Biochim. Biophys. Acta 732:304.
25. Neumann, E., Schaefer-Ridder, M., Wang, Y., and
     Hofschneider, P.H. (1982) EMBO J. 1:6.
26. Zimmermann, U. (1983) In: Targeted Drugs. Goldberg,
     E., Donaruma, L., and Vogl, O. (eds.) John Wiley and
     Sons, New York, p. 153.
27. Muth, E. (1927) Kolloid Z. 41:97.
28. Krasny-Ergen, W. (1936) Hochfrequenztechnik
     Elektroakustik 40:126.
29. Liebesny, P. (1939) Arch. Phys. Ther. 19:736.
30. Saito, M., Schwan, H.P., and Schwarz, G. (1966) Biophys.
     J. 6:313.
31. Pohl, H.A. (1951) J. Appl. Phys. 22:869.
32. Schwarz, G., Saito, M., and Schwan, H.P. (1965) J. Chem.
     Phys. 43:3562.
33. Schwan, H.P., and Sher, L.D. (1969) J. Electrochem. Soc.
     116:170.
34. Sauer, F.A. (1983) In: Coherent Excitations in
     Biological Systems, Froehlich, H. and Kremer, F. (eds.)
     Springer-Verlag, p. 134.
35. Poo, M. (1981) Annu. Rev. Biophys. Bioeng. 10:245.
36. Sowers, A.E., and Hackenbrock, D.R. (1981) Proc. Natl.
     Acad. Sci. USA 78:6246.
37. Poste, G., and Nicolson, G.L. (1978) Cell Surface
     Reviews, Vol. 5, Membrane Fusion. Elsevier/North
     Holland, Amsterdam.
38. Kell, D.B. (1984) Bioelectrochem. Bioenerg. (in press).
39. Vienken, J., Zimmermann, U., Fouchard, M., and Zagury, D.
     (1983) FEBS Letters 163:54.

40. Basset, C.A.L., and Herrmann, J. (1968) J. Cell Biol.
     3a:9a.
41. Korenstein, R., Somjen, D., Laub, F., Danon, A.,
     Fischler, H., and Binderman, I. (1983) In: Biological
     Structures and Coupled Flows. Oplatka, A., and Balaban,
     M. (eds.) Academic Press, New York, p. 401.
42. Goodman, R., Basset, C.A.L., and Henderson, A.S. (1983)
     Science 220:1283.
43. Vienken, J., Zimmermann, U., Ganser, R., and Hampp, R.
     (1983) Planta 157:331.
44. Zimmermann, U., Pilwat, G., and Richter, H.-P. (1981)
     Naturwissenschaften 68:577.
45. Zimmermann, U., Vienken, J., Pilwat, G., and Arnold, W.M.
     (1984) In: Cell Fusion. Pitman Books, London, p. 60.
46. Zimmermann, U. (1983) Trends Biotechnol., 1:149.
47. Halfmann, H.J., Roecken, W., Emeis, C.C., and
     Zimmermann, U. (1982) Curr. Genetics 6:25.
48. Halfmann, H.J., Emeis, C.C., and Zimmermann, U. (1983)
     Arch. Microbiol. 134:1.
49. Halfmann, H.J., Emeis, C.C., and Zimmermann, U. (1983)
     FEMS Microbiol. Letters 20:13.
50. Vienken, J., and Zimmermann, U. (1982) FEBS Letters
     137:11.
51. Bischoff, R., Eisert, R.M., Schedel, I., Vienken, J., and
     Zimmermann, U. (1982) FEBS Letters 147:64.
52. Richter, H.-P., Scheurich, P., and Zimmermann, U. (1981)
     Develop. Growth Diff. 23:479.
53. Bates, G.W., Gaynor, J.J., and Shekhawat, N.S. (1983)
     Plant Physiol. 72:1110.
54. Büschl, R., Ringsdorf, H., and Zimmermann, U. (1982) FEBS
     Letters 150:38.
55. Melikyan, G.B., Abidor, I.G., Chernomordik, L.V., and
     Chailakhyan, L.M. (1983) Biochim, Biophys. Acta 730:395.
56. Pilwat, G., Richter, H.-P., and Zimmermann, U. (1981)
     FEBS Letters 133:169.
57. Maroudas, N.G. (1975) J. Theor. Biol. 49:417.
58. Verhoeck-Köhler, B., Hampp, R., Ziegler, H., and
     Zimmermann, U. (1983) Planta 158:199.
59. Zimmermann, U. Vienken, J., and Pilwat, G. (1981) Z.
     Naturforsch. 36c:173.
60. Arnold, W.M., and Zimmermann, U. (1982) Z. Naturforsch.
     37c:908.
61. Zimmermann, U. and Arnold, W.M. (1983) In: Coherent
     Excitations in Biological Systems. Fröhlich, H., and
     Kremer, F. (eds.) Springer-Verlag, p. 211.
62. Küppers, G., Wendt, B., and Zimmermann, U. (1983) Z.
     Naturforsch. 38c:505.
63. Pilwat, G., and Zimmermann, U. (1983) Bioelectrochem.
     Bioenerg. 10:155.
64. Zimmermann, U. and Küppers, G. (1983)
     Naturwissenschaften, 70:568.

65. Küppers, G., and Zimmermann, U. (1983) FEBS Letters, 163:323.
66. Feynman, R.P. (1971) The Feynman Lectures on Physics, Vol. 2. Addison-Wesley, London.
67. Workman, E.J., and Reynolds, S.E. (1950) Phys. Rev. 78:254.
68. Lodge, J.P., Baker, M.L., and Pierrard, J.M. (1956) J. Chem. Physics 24:716.
69. Hui, S.W., Stewart, T.P., Boni, L.T., and Yeagle, P.L. (1981) Science 212:921.
70. Sowers, A.E. (1982) J. Cell Biol. 95:240a.

CHAPTER 11

NEW DEVELOPMENTS IN HYBRIDOMA TECHNOLOGY

INTERSPECIFIC HYBRIDS FOR THE PRODUCTION OF
IMMUNOGLOBULINS OF VETERINARY INTEREST

S. SRIKUMARAN, A.J. GUIDRY*, and R.A. GOLDSBY
Department of Biology, Amherst College, Amherst, MA 01002
and *Laboratory of Milk Secretion and Mastitis Research
Agricultural Research Service, U.S. Department of Agriculture
Beltsville, MD 20705

Bovine immunoglobulins (Ig) have been studied for decades but their immunochemical characteristics have not been detailed as completely as their counterparts in humans and mice (1). This has been attributed mainly to the rare occurrence of myeloma proteins in cattle. The one report (2) of Bence-Jones proteins in cattle has not been pursued further. Apart from that, there have been no other reports of myelomas in cattle to date. In cattle and in other domestic animals such as sheep, goats, and pigs, the lack of availability of myeloma proteins has hindered detailed immunochemical characterization of Ig. This problem can be circumvented by the construction of interspecific hybrids between a nonsecreting murine hybridoma and normal B lymphocytes from a desired animal species. Even though chromosome loss is an established phenomenon in interspecific hybrids, we and others (3-5) find that if hybridomas can be generated at a sufficient frequency, then rigorous selection techniques can compensate for the high rate of chromosome loss and permit isolation of the minority population of stable hybrids secreting a desired Ig. We report here the interspecific fusion of murine hybridoma with normal bovine spleen cells, resulting in stable hybrid cell-lines that secrete monoclonal bovine Ig.

Bovine x murine hybridomas were prepared by polyethylene glycol (PEG)-assisted fusion of normal bovine lymphocytes to the mouse cell line SP-2/0 (6), a murine x murine hybridoma that does not secrete or internally produce mouse Ig. The fusion was performed as outlined by Goldsby et al. (7). Bovine lymphocytes were obtained from the spleen of an adult female Holstein cow by first mincing six randomly selected cubes of spleen (~ 2 cm) with scissors and grinding the mince between the frosted faces of microscope slides in serum-free growth medium [50% RPMI 1640 and 50% Dulbecco's modified Eagle's medium (DME)]. The lymphocytes were washed

twice in serum-free growth medium, mixed in a 1:1 ratio with
1×10^8 SP-2/0 cells, and centrifuged at 400 x g for 15 min
at room temperature. The pellet was then slurried in 2 ml of
52% polyethylene glycol 1540 in serum-free growth medium and
incubated for 2 min at 37°C. The mixture was then diluted
over the course of 10 min to 50 ml by the addition of
serum-free growth medium and centrifuged for 15 min at 400 x g
at room temperature. The fused cells were resuspended in HAT
(8, 9) medium (45% RPMI 1640; 45% DME; 10% γ-globulin-free
horse serum; 2 mM glutamine; penicillin, 10 U/ml;
streptomycin, 100 µg/ml; 1×10^{-4} M hypoxanthine; 4 x
10^{-7} M aminopterin; and 3×10^{-5} M thymidine) and
distributed in 0.2-ml portions to the wells of microtiter
dishes (10). We obtained 63 single hybrid clones between 14
and 21 days after fusion.

Supernatants of the hybrid clones were assayed for
production of bovine Ig by a competition radioimmunoassay
(cRIA) (11), with affinity-purified polyclonal rabbit antibody
to bovine Ig light chain (Figure 1). Twenty-one of 63 initial
hybrid clones produced bovine Ig. After three cycles of
subcloning, three cell-lines (LHRB-1, LHRB-2, and LHRB-3)
continued to secrete bovine Ig.

A panel of monoclonal antibodies with known specificities
for bovine Ig isotypes was used to determine that LHRB-1,
LHRB-2, and LHRB-3 were secreting bovine Ig heavy chains and
to establish the isotype of the heavy chains. The monoclonal
antibody DAS-1 (12), which binds to bovine IgG1, IgG2, and IgM
but fails to react with IgA, was used to establish that the
products of these cell lines were bovine Ig. As reported in
Table 1, [^{35}S]methionine-labeled products of these
cell-lines bind to DAS-1-conjugated Sepharose 4B. This
binding was inhibited by whole bovine serum but not by whole
mouse serum. In a similar experiment, the labeled products
did not bind above background levels to affinity-purified
sheep antibody to mouse Ig (all isotypes) coupled to Sepharose
4B. These results indicated that the hybridoma products were
bovine Ig.

Two other monoclonal antibodies--DAS-2 (13), which is
specific for bovine IgG2, and DAS-6 (14), which is specific
for bovine IgM--were used in concert with DAS-1 in a cRIA that
established the isotypes of Ig secreted by LHRB-1, LHRB-2, and
LHRB-3 as bovine IgG1, IgG2, and IgM, respectively (Table 2).
The monoclonal antibody DAS-1 bound more strongly to bovine
IgG1 and IgG2 than to IgM in indirect solid-phase RIA. Also,
DAS-1 coupled Sepharose 4B adsorbed bovine IgM. However, in
cRIA, bovine IgM could not competitively inhibit the binding
of ^{125}I-IgG1 or IgG2 to DAS-1 immobilized onto PVC
microtiter plates. This was attributed to the relative lower
affinity of DAS-1 for bovine IgM than for IgG1 or IgG2.

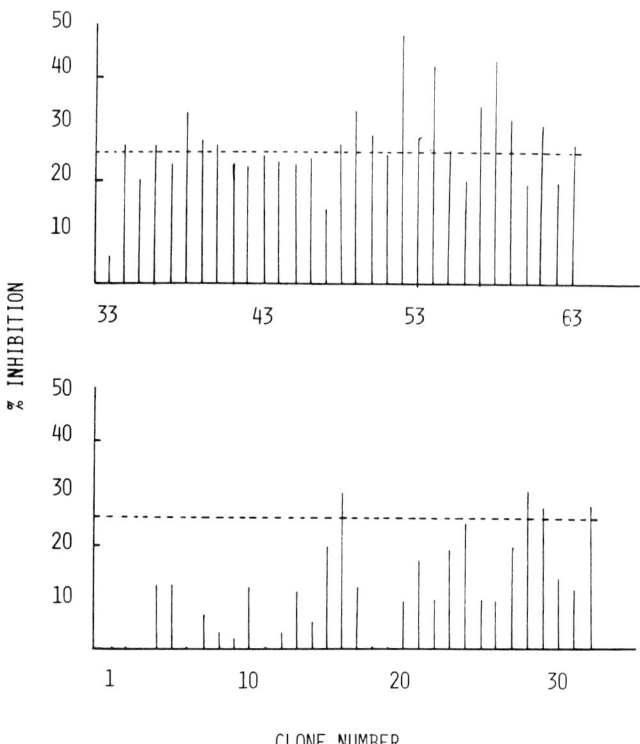

Figure 1--Competition radioimmunoassay of culture supernatants from bovine x murine initial clones. Polyvinyl chloride microtiter plates were coated with rabbit antibody to bovine Ig light chain. A solution of 20 µl of antibody (50 µg/ml) in phosphate-buffered saline (PBS) was added to each of the wells, and the dishes were placed in an incubator at room temperature for 1 hour and then washed three times with PBS containing 1% bovine serum albumin (PBS-BSA). Supernatants were tested for the presence of bovine Ig by adding 20 µl of the supernatant and 20 µl of the ^{125}I-labeled bovine IgG (specific activity, 10 µCi/µg) to the antibody-coated wells. Wells that received 20 µl of culture supernatant from the parent cell-line SP-2/0 and 20 µl of ^{125}I-labeled bovine IgG were used as controls and % inhibition calculated. The horizontal slashed line represents 25% inhibition.

Table 1--Specific immunoadsorption of ^{35}S-methionine-labeled
monoclonal bovine Ig by DAS-1-conjugated Sepharose 4B.

Sample treatment	CPM $\pm$ SD		
	LHRB-1	LHRB-2	LHRB-3
No preincubation	41,610 $\pm$ 60*	9302 $\pm$ 22	6245 $\pm$ 305
Preincubation with bovine serum	3,473 $\pm$ 41	1214 $\pm$ 107	2891 $\pm$ 180
Preincubation with mouse serum	40,226 $\pm$ 370	8948 $\pm$ 127	6216 $\pm$ 299

*The hybrid cells (1 x 10^7) were cultured in 2 ml of
methionine-free DME medium containing 1% fetal bovine serum and
0.5 mCi of [^{35}S]methionine for 6 hours. The culture
supernatant was used in a binding assay. Six milligrams of the
affinity-purified mouse monoclonal antibody DAS-1, specific for
bovine IgG and IgM, was coupled to 1 g of CNBr-activated
Sepharose 4B and suspended in 15 ml of PBS containing 0.05%
polyoxyethylene sorbitan monolaurate (Tween 20). To 1 ml of
this suspension was added 100 µl of the labeled culture
supernatant, and the mixture was agitated slowly for 30
minutes. The ability of whole bovine serum or whole mouse serum
to block the binding of the labeled product in the culture
supernatant to DAS-1 was tested by adding 50 µl of each of
these sera, respectively, to the DAS-1-conjugated Sepharose
prior to the addition of the culture supernatant. The treated
Sepharose was then washed three times by the addition of 10 ml
of PBS containing 0.05% Tween 20 and centrifuging at 400 X g for
5 minutes. The radioactivity of the Sepharose pellet was then
counted in a scintillation counter.

The monoclonal Igs secreted by LHRB-1, LHRB-2, and LHRB-3
were isolated from serum-free growth medium in which these
hybridomas were grown by affinity chromatography with polyclonal
guinea pig antibody to bovine Ig light chains coupled to
Sepharose 4B. The affinity-purified monoclonal Igs were
subjected to sodium dodecyl sulfate polyacrylamide gel
electrophoresis (SDS-PAGE) (15), on 10% polyacrylamide gels,
before and after reduction with 2-mercaptoethanol. Before
reduction (Figure 2), LHRB-1 Ig gave a single band, indicating
that it was an unreduced IgG1 molecule (200,000 MW). After
reduction it gave two bands with molecular weights
characteristic of IgG1 heavy (63,000 MW) and light (25,800 MW)
chains. Also, reduced and unreduced LHRB-1 Ig migrated

Table 2--Competition radioimmunoassay to determine the isotype of monoclonal bovine Ig

Immobilized monoclonal antibody	Specificity	Tracer antigen	Percentage inhibition		
			LHRB-1	LHRB-2	LHRB-3
DAS-1	α IgG1, α IgG2, α IgM	^{125}I-IgG1	60*	67	0
DAS-2	α IgG2	^{125}I-IgG2	0	42	0
DAS-6	α IgM	^{125}I-IgM	0	0	81

*Duplicate samples of culture supernatants (20 µl) from either the hybridomas or the parent cell-line SP-2/0 were added to microtiter wells coated with affinity-purified anti-bovine Ig antibody (DAS-1, DAS-2, or DAS-6). Simultaneously, ^{125}I-labeled bovine IgG1, IgG2, and IgM (20 µl, specific activity 10 µCi/µg) were added to the microtiter wells coated with DAS-1, DAS-2, or DAS-6, respectively. After 1 h of incubation and three washing cycles with PBS-BSA, the wells were cut and radioactivity counted. The percent inhibition of binding of the ^{125}I-labeled antigens by the hybridoma Ig (LHRB-1, LHRB-2, and LHRB-3) in the culture supernatant was calculated.

similarly to the polyclonal IgG1 indicating that it was an intact bovine IgG1 molecule.

LHRB-2 Ig (Figure 3) responded similarly, showing apparent molecular weights of 56,200 and 26,000 Da for the heavy and light chains, respectively. The reduced and unreduced LHRB-2 Ig (200,000 MW) migrated similarly to the polyclonal IgG2 indicating that LHRB-2 was an intact bovine IgG2 molecule. Unreduced LHRB-1 Ig and LHRB-2 Ig and polyclonal IgG1 and IgG2 migrated similarly in 4% polyacrylamide gels (16) crosslinked with DATD (Figure 4). After reduction, LHRB-3 Ig gave two bands that migrated similarly to the heavy and light chains of polyclonal IgM (Figure 5). The LHRB-3 Ig heavy and light chains showed an apparent molecular weight of 76,000 and 29,400 Da, respectively. Neither the unreduced LHRB-3 Ig, nor the unreduced polyclonal IgM migrated into the 10% polyacrylamide gel but migrated similarly in 4% polyacrylamide gels crosslinked with DATD (Figure 4), indicating that LHRB-3 Ig was a pentameric IgM.

Affinity-purified LHRB-1 Ig, LHRB-2 Ig and LHRB-3 Ig were subjected to Öuchterlony immunodiffusion analysis (17) with class- and subclass-specific guinea pig anti-bovine IgG1, IgG2, IgM, and IgA and affinity-purified sheep anti-mouse Ig (Figure 6). LHRB-1 Ig, LHRB-2 Ig, and LHRB-3 Ig gave precipitin lines with guinea pig anti-bovine IgG1, IgG2, and IgM, respectively, but not with the guinea pig anti-bovine IgA or sheep anti-mouse Ig.

The LHRB-1, LHRB-2, and LHRB-3 have continued to secrete bovine Ig for over 16 months. Frozen cells of these cell-lines brought back to culture have continued to secrete Ig.

From a recent fusion between bovine spleen cells and the mouse cell-line SP-2/0, we obtained 353 hybridomas, 49 of which secreted bovine Ig. After two cycles of subcloning, 24 cell-lines have continued to secrete bovine Ig (Figure 7). This demonstrates the reproducibility of our earlier results. Table 3 summarizes the bovine x murine hybridomas derived in our laboratory and the isotypes of the Ig secreted by them.

Continuous cultures of bovine x murine hybridomas will supply material that will open areas of investigation of the bovine immune system heretofore inaccessible. These hybridomas will provide monoclonal bovine Ig for a) serological standards, b) production of polyclonal and monoclonal antisera to bovine Ig isotypes, c) sequencing studies, d) serological and structural studies of bovine Ig isotypes and allotypes, and e) defining bovine Ig classes and subclasses (IgA, IgM, IgG1, IgG2a, IgG2b, and possibly IgE and IgD if they exist in the bovine species). Also, these hybridomas will provide mRNA for the production of cDNA probes for the cloning of bovine Ig genes and for the determination of the organization of Ig genes in the bovine genome.

Table 3--Library of monoclonal bovine immunoglobulins

LHRB designation	Isotype	LHRB designation	Isotype
1	IgG1	8	IgM*
2	IgG2	9	IgM*
3	IgM	10	IgM*
4	IgG1*	11	Light chain
5	IgG1*	12	Light chain
6	IgM*	13-29	Unclassified
7	IgM*		

*Tentative isotype.

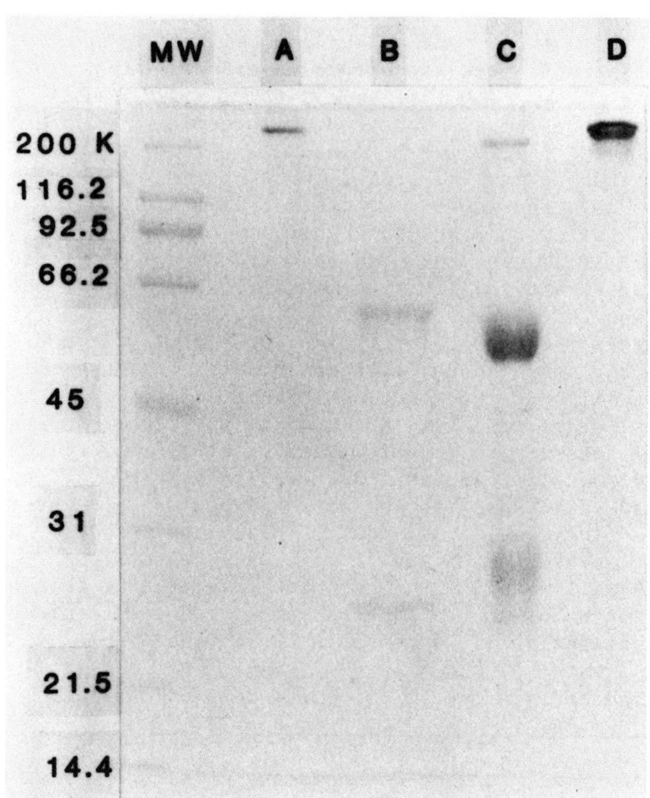

Figure 2--SDS-PAGE of affinity-purified LHRB-1 Ig.
Affinity-purified LHRB-1 Ig was applied on a 10% polyacrylamide
gel before (A) and after (B) reduction. Polyclonal IgG1 was
co-electrophoresed before (D) and after (C) reduction.
Molecular weight standards (MW) were also included.

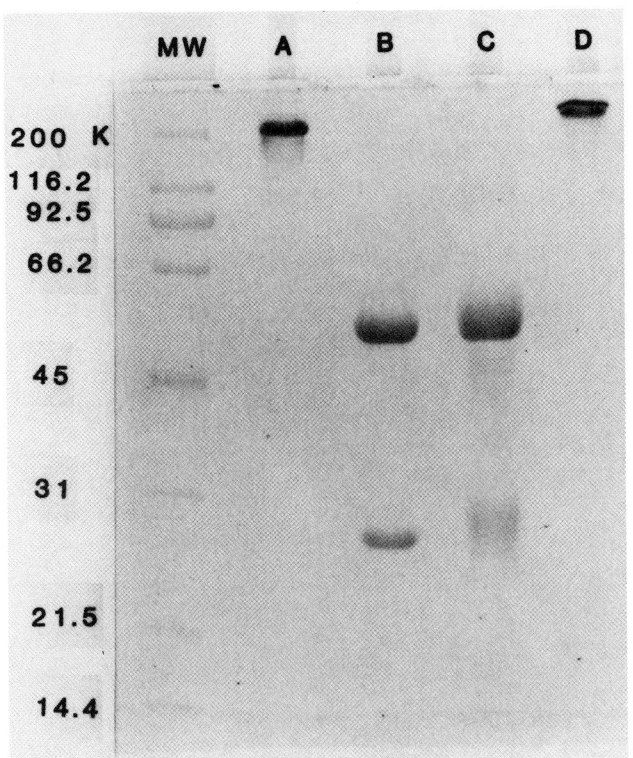

Figure 3--SDS-PAGE of affinity-purified LHRB-2 Ig.
Affinity-purified LHRB-2 Ig was applied on a 10% polyacrylamide
gel before (A) and after (B) reduction. Polyclonal IgG2 was
co-electrophoresed before (D) and after (C) reduction.
Molecular weight standards (MW) were also included.

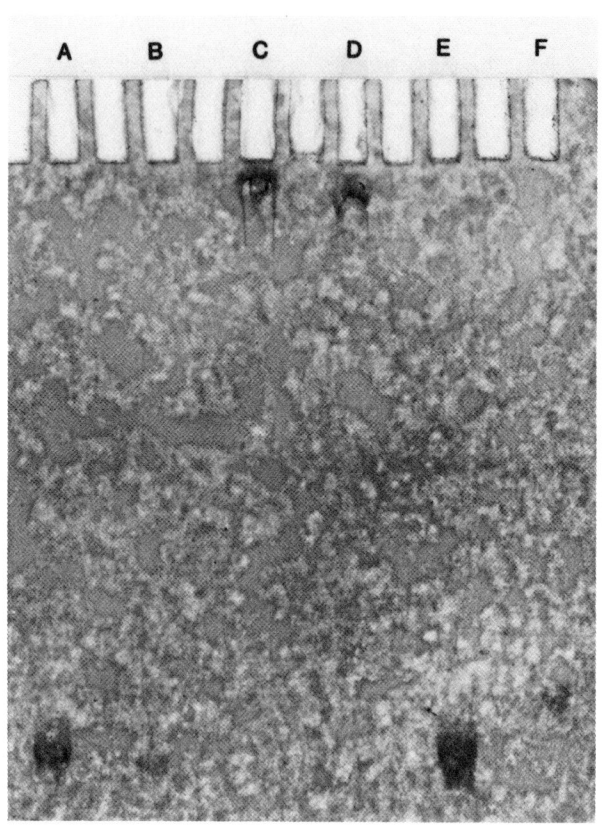

Figure 4--SDS-PAGE of affinity-purified LHRB-1 Ig, LHRB-2 Ig, and LHRB-3 Ig. Affinity-purified, unreduced LHRB-1 Ig (B), LHRB-2 Ig (F), and LHRB-3 Ig (D) were co-electrophoresed with polyclonal IgG1 (A), IgG2 (E), and IgM (C) on a 4% polyacrylamide gel.

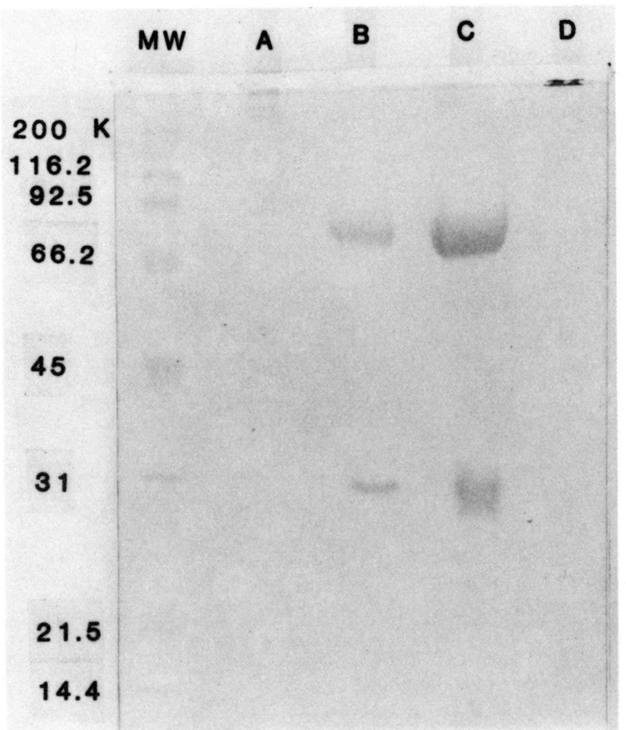

Figure 5--SDS-PAGE of affinity-purified LHRB-3 Ig.
Affinity-purified LHRB-3 Ig was applied on a 10% polyacrylamide
gel before (A) and after (B) reduction. Polyclonal IgM was
co-electrophoresed before (D) and after (C) reduction.
Molecular weight standards (MW) were also included.

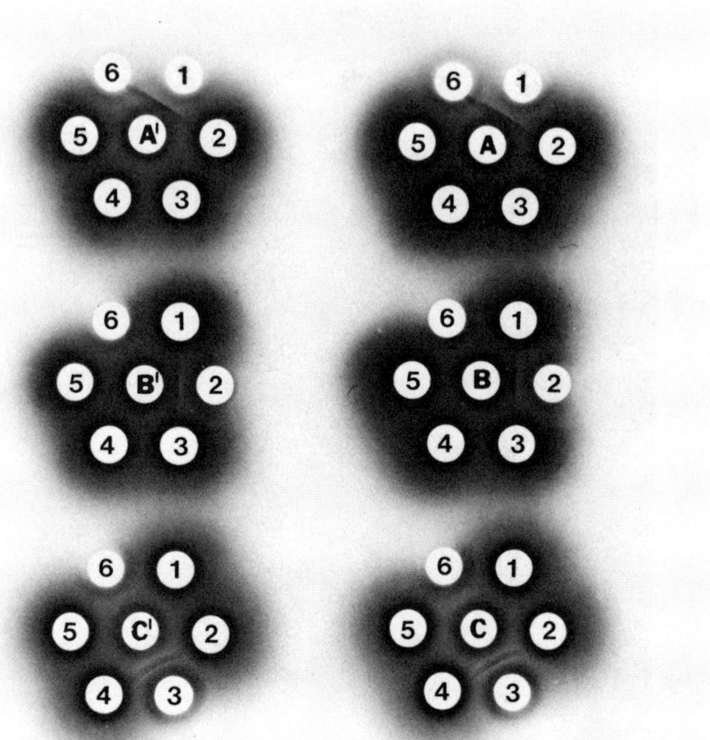

Figure 6--Öuchterlony immunodiffusion analysis of monoclonal
bovine Ig. Affinity-purified monoclonal bovine Ig were tested
against guinea pig antisera to bovine IgG1 (1), IgG2 (2), IgM
(3), IgA (4), and against sheep antibody to mouse Ig (5) and PBS
(6). The center well in A, B, and C contained affinity-purified
LHRB-1, LHRB-2, and LHRB-3, respectively, while the center well
in A', B', and C' contained polyclonal bovine IgG1, IgG2, and
IgM, respectively.

Figure 7--Bovine x mouse hybridomas secreting bovine Ig.

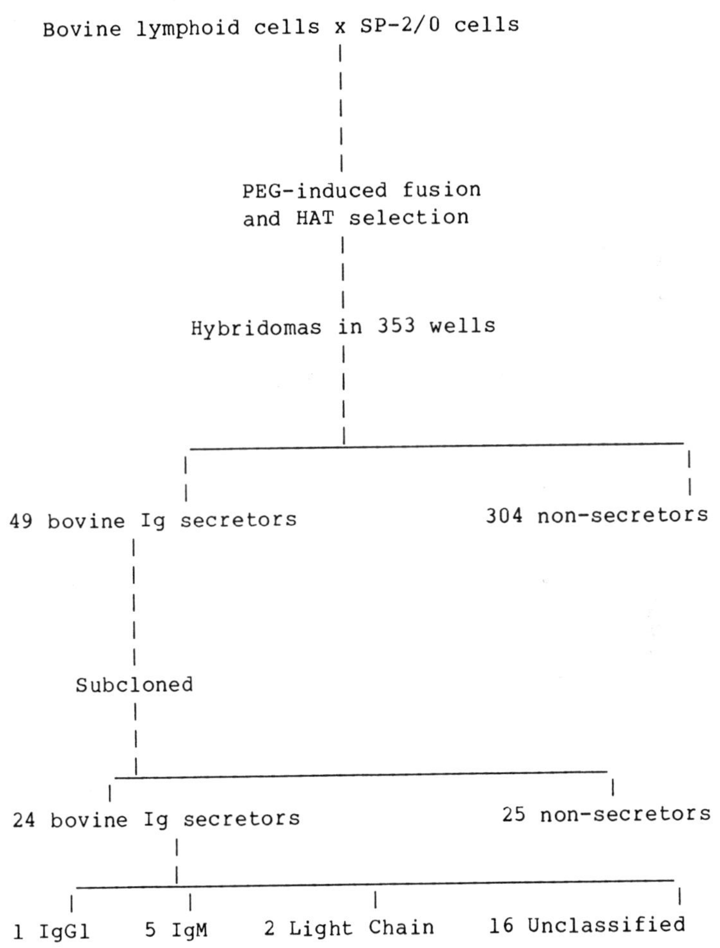

The construction of bovine x murine hybridomas secreting bovine Ig demonstrates the feasibility of obtaining Ig isotypes in general and antibodies of predefined antigenic specificity by the use of spleen cells, from appropriately immunized donors or spleen cells immunized _in vitro_, from any desired animal species. This technology also provides for the production of antibodies "tailored" for specific purposes. For example, situations may arise when there is a requirement for monoclonal reagents that carry out isotype-associated specific effector functions, i.e., that do or do not fix complement, that do or do not bind Fc receptors on certain kinds of cells, or that have long or short serum half-lives. Construction of interspecific hybridomas will be the method of choice for the derivation of such tailored antibodies.

REFERENCES

1. Butler, J.E. (1983) Vet. Immunol. Immunopathol. 4:43-152.
2. Rodkey, S., and Kimmel, A.T. (1972) Immunochemistry 9:23-28.
3. Nowinski, R., Berglund, C., Lane, J., Lostrum, M., Bernstein, I., Young, W., Hakomori, S. I., Hill, L., and Cooney, M. (1980) Science 210:537-539.
4. Yarmush, M.L., Gates, F.T., Weisfogel, D.R., and Kindt, T.J. (1980) Proc. Natl. Acad. Sci. USA 77:2899-2903.
5. Srikumaran, S., Guidry, A.J., and Goldsby, R.A. (1983) Science 220:522-524.
6. Shulman, M., Wilde, C.D., and Kohler, G. (1978) Nature 276:269-270.
7. Goldsby, R.A., Osborne, B.A., Simpson, E., and Herzenberg, L.A. (1977) Nature 267:707-708.
8. Szybalska, E.H., and Szybalska, W. (1962) Proc. Natl. Acad. Sci. USA 48:2026-2034.
9. Littlefield, J.W. (1964) Science 145:709-710.
10. Goldsby, R.A., and Zipser, E. (1969) Exp. Cell Res. 54:271-275.
11. Oi, V.T., and Herzenberg, L.A. (1979) Mol. Immunol. 16:1005-1017.
12. Srikumaran, S. (1981) Production and characterization of monoclonal antibodies to bovine immunoglobulin G_2. M.S. Thesis, University of Maryland.
13. Srikumaran, S., Guidry, A.J., and Goldsby, R.A. (1982) Am. J. Vet. Res. 43:17-21.
14. Srikumaran, S. (1982) Monoclonal bovine immunoglobulins and their antisera: Production, isolation and characterization. Ph.D. Thesis, University of Maryland.
15. Jones, P.P. (1980) In: Selected Methods in Cellular Immunology. Mishell, B.B., and Shigii, S.M., (eds.) W.H. Freeman and Company, San Francisco, pp. 398-440.

16. Ziegler, A., and Hengartner, H. (1977) Eur. J. Immunol.
     7:690.
17. Öuchterlony, O., and Nilsson, L.A. (1978) In: Handbook of
     Experimental Immunology, 3rd edition. Weir, D.M. (ed.)
     Blackwell Scientific Publication, London, pp. 19.16-19.21.

<u>IN</u> <u>VITRO</u> IMMUNIZATION FOR HYBRIDOMA PRODUCTION

YVONNE E. MCHUGH
Hana Biologics, Inc.
2323 5th Street
Berkeley, California   94710

Significant progress is being made in the development and use of <u>in</u> <u>vitro</u> immunization as a means of obtaining immunized lymphocytes for producing hybridomas (see Reading (1) for review). A breakthrough in the technique of immunizing B lymphocytes <u>in</u> <u>vitro</u> was achieved by Mishell and Dutton in 1967. The culture system they developed permitted dissociated normal murine spleen cells to be immunized <u>in</u> <u>vitro</u> with heterologous mammalian erythrocytes (2). Using a similar <u>in</u> <u>vitro</u> immunization system, Hengartner et al. successfully produced hybridomas against sheep red blood cells (3). The procedure most widely used at present for <u>in</u> <u>vitro</u> immunization against soluble protein antigens was first reported by Lubin and Mohler (4), who made hybridomas against human osteoclast-activating factor. Their procedure utilizes thymocyte-conditioned medium as a source of lymphokines (and perhaps thymic epithelial cell factors) that promote antibody production in the presence of antigen. Their experiments suggested that with this particular antigen, <u>in</u> <u>vitro</u> immunization was more effective than <u>in</u> <u>vivo</u> immunization and at the same time required less antigen. This technique has been modified by Pardue et al. to produce monoclonal antibodies to the highly conserved cellular protein calmodulin (5) and by Miner et al. for hybridomas against cell surface components on metastatic variants of murine tumor cells (6).

In <u>vitro</u> immunization of human peripheral blood lymphocytes and subsequent human monoclonal antibody production has been reported by Cavagnaro and Osband (7). By eliminating suppressor T cells and using autologous serum and mixed lymphocyte culture supernatants, they were able to produce human-human hybridomas secreting IgG specific for $Rh_o(D)$ antigen on human erythrocytes. In <u>vitro</u> immunization of human peripheral blood lymphocytes with antigen in the presence of mitogen may also prove useful for expanding specific B-cell clones for fusion from the circulation of immune individuals (8, 9).

In this paper, different approaches that may be taken to successfully perform both primary and secondary <u>in</u> <u>vitro</u> immunization for specific hybridoma production are presented.

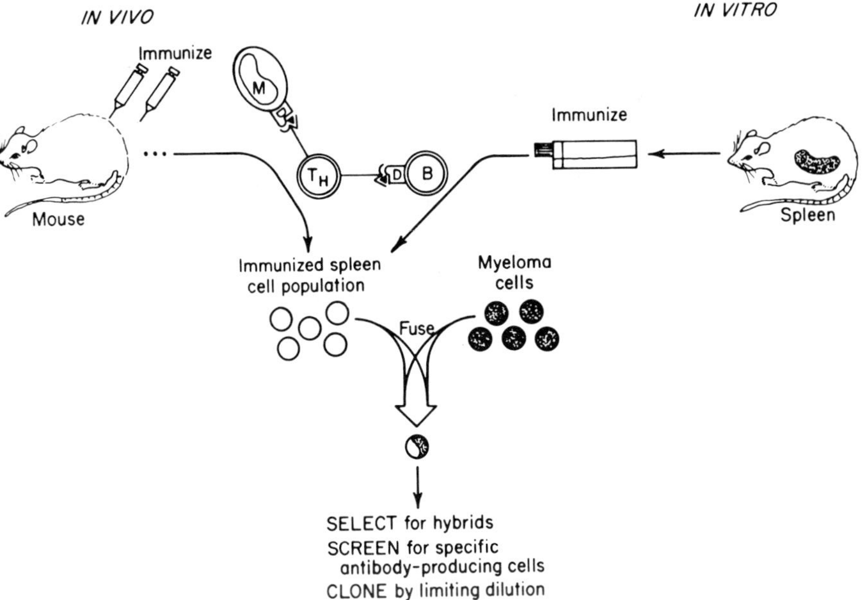

Figure 1--Schematic protocol for <u>in vivo</u> and <u>in vitro</u>
immunizations.

METHODS

     Schemes for <u>in vivo</u> and <u>in vitro</u> immunization are compared
in Figure 1. <u>In vitro</u> immunization of B cells may be
particularly useful in the following situations: 1) when only
microgram (or less) quantities of antigen are available, 2) when
the antigen is autologous or evolutionarily highly conserved,
3) when antigen is toxic at the organismal level, 4) when Ig
products from human hybridomas are sought and 5) when no
positive or otherwise satisfactory hybrids can be obtained from
conventional <u>in vivo</u> immunization regimens.

<u>Primary Immunization</u> In Vitro

     The methods of Lubin and Mohler (4) and Pardue et al. (5)
have been successfully used in our laboratory to produce
monoclonal antibodies against soluble protein and hapten-protein
conjugates. Figure 2 depicts approaches that may be taken when
the antigen of interest is part of a complex surface structure
the antigenic properties of which may be altered upon isolation.

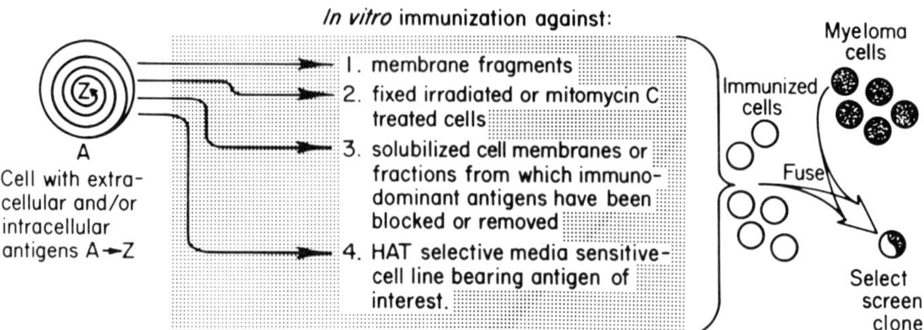

Figure 2--Approaches for <u>in vitro</u> immunization using membrane antigens.

Analysis of the Thymus-Conditioned Medium Component

For testing the effectiveness of a given thymocyte-conditioned medium preparation, a battery of assays was sequentially performed on cells from 1 to 14 day <u>in vitro</u> immunization cultures. These assays included specific (direct and indirect) and polyclonal plaque assays, ^{3}H-thymidine uptake, blast enumeration, ^{35}S-methionine labeling of cultures followed by isolation of antigen-specific antibody produced, measurement of specific and polyclonal levels of IgM and IgG by the enzyme-linked immunosorbent assay, and finally, the determination of the antigen-specific clones after fusions. Because the antigen-specific fractions of all the responses measured were obscured by the polyclonal activation afforded by fetal calf serum and thymus-conditioned medium, specific hybridoma production remains the most reliable source of information under these <u>in vitro</u> conditions (McHugh and Walthall, unpublished).

Antigen Concentration

Too high or low an antigen concentration in an <u>in vitro</u> immunization will result in no positive hybrids. A titration of antigen through the nanogram and microgram ranges is a wise course. The effect of <u>in vitro</u> antigen concentration on the affinity of monoclonals ultimately being produced has not been determined.

Fetal Calf Serum

Some fetal calf serum is thought to contain suppressor factors and cannot be used for in vitro immunization (10). Although serum-free in vitro immunizations generally result in a lower percentage of clones growing in HAT selection (thus lowering the yield overall), this approach may sometimes be the only means of obtaining positive clones (McHugh and Walthall, unpublished).

Secondary Immunization In Vitro

B lymphocytes must be at the correct stage of immunization/differentiation for hybridization to result in specific antibody-secreting hybridomas. After conventional in vivo and in vitro immunizations failed to produce satisfactory anti-ferritin secreting hybridomas, the experiment in Figure 3 was performed. With this approach, initial in vivo immunizations might be performed with antigen in a semipurified form followed by in vitro immunization with more limited amounts of purified material. In addition, this regimen of immunization may increase the ratio of IgG- to IgM-antigen-specific hybridomas produced.

CAVEATS

Although in vitro immunization of B lymphocytes is often helpful in making hybridomas against otherwise nonimmunogenic targets, there are limitations: 1) The majority of hybridomas made after in vitro immunization secrete IgM whereas IgG is often more desirable for many applications. Preliminary results indicate that extension of the in vitro immunization period with reduced amounts of antigen may lead to a higher proportion of IgG secretors (R. Pardue, personal communication). 2) It remains to be determined whether in vitro immunization generally leads to production of monoclonal antibodies of sufficiently high affinities. 3) No general method has as yet emerged that is satisfactory for primary in vitro immunization of human peripheral blood lymphocytes; B-cell subsets in this lymphoid compartment are more difficult to immunize than splenocytes. 4) The components of in vitro immunization systems need to be fully defined by elimination of serum and determination of active species in the thymus-conditioned medium. At present, unknown variation in either can prevent successful immunization.

Finding solutions to these problems will enhance the usefulness of in vitro immunization as a tool in cancer diagnosis and treatment, immunological manipulation, and immunopharmacology.

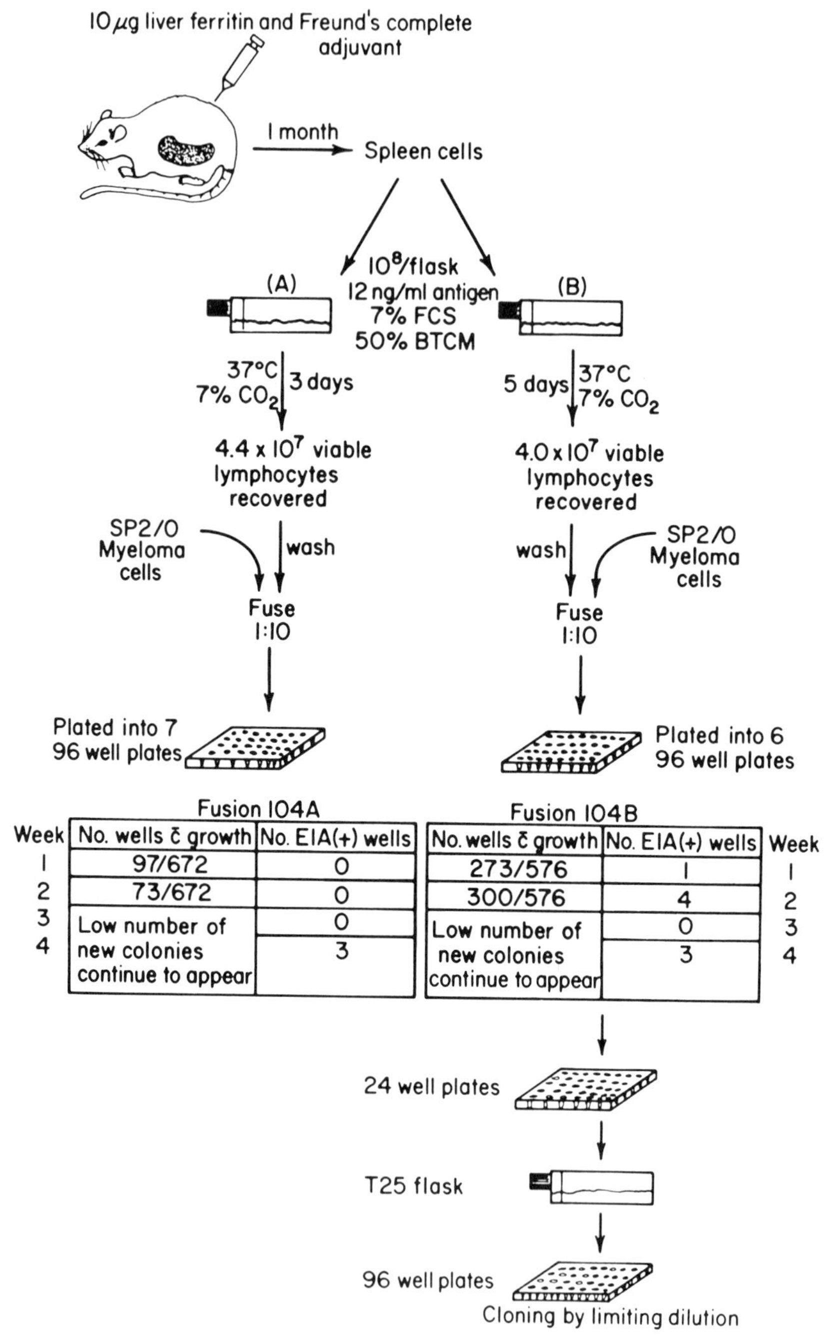

Fusion 104A				Fusion 104B		
Week	No. wells c̄ growth	No. EIA(+) wells		No. wells c̄ growth	No. EIA(+) wells	Week
1	97/672	0		273/576	1	1
2	73/672	0		300/576	4	2
3	Low number of new colonies continue to appear	0		Low number of new colonies continue to appear	0	3
4		3			3	4

Figure 3--Protocol for production of anti-ferritin monoclonal antibodies.

ACKNOWLEDGMENTS

Ben J. Walthall designed the thymocyte-conditioned medium system used in our laboratory. Thanks are due to Sue Hirabayashi and Linda Kiehnau for excellent technical assistance.

REFERENCES

1. Reading, C.L. (1982) J. Immunol. Meth. 53:261.
2. Mishell, R.I., and Dutton, R.W. (1967) J. Exp. Med. 126:423.
3. Hengartner, H., Luzzati, A.L., and Schreier, M. (1978) Curr. Top. Microbiol. Immunol. 81:92.
4. Lubin, R.A., and Mohler, M.A. (1980) Mol. Immunol. 17:635.
5. Pardue, R.L., Brady, R.C., Perry, G.W., and Dedman, J.R. (1983) J. Cell Biol. 96:1149.
6. Miner, K.M., Reading, C.S., and Nicolson, G.L. (1981) J. Invasive Metastasis 1:158.
7. Cavagnaro, J., and Osband, M. (1983) Biotechniques 1:31.
8. Khansari, N., Fudenberg, H.H., and Merler, E. (1983) Immunobiology 104:42.
9. Astaldi, G.C.B., Wright, E.P., Willems, C.H., Zeijlemaker, W.P., and Janssen, M.C. (1982) J. Immunol. 128:2539.
10. Edwards, P.A.W., Smith, C.M., Neville, A.M., and O'Hare, M. (1982) Eur. J. Immunol. 12:641.

IN VIVO TREATMENT WITH ANTI-IDIOTYPE: A POSSIBLE ALTERNATIVE TO
CONVENTIONAL IMMUNIZATION

SUZANNE L. EPSTEIN
Transplantation Biology Section, Immunology Branch
National Cancer Institute, National Institutes of Health
Building 10, Room 4B17, Bethesda, MD 20205

     Deliberate immunization usually involves direct exposure of
an animal to an antigen. However, immunity to an antigen can
also be selectively induced in some cases by anti-idiotypic
reagents that recognize the receptors on cells of appropriate
specificity. This approach offers an alternative when
conventional immunization has practical disadvantages. In
addition, such idiotype-specific manipulation provides an
approach to analysis of the immune network and regulation of
immune responses.
     The development of hybridoma technology has greatly
facilitated the analysis of antibody idiotypes (Ids),
anti-idiotypes (anti-Ids), and their biological effects. A case
in point is the analysis of antibodies directed to major
histocompatibility complex (MHC) antigens. The complexity of
both the antigens and antibody responses to them had led to
great difficulties when polyclonal alloantisera were used as
sources of Id. Anti-Id reagents could not be prepared
reproducibly, and results concerning anti-MHC receptors were
controversial (1, 2). However, monoclonal antibodies provide a
source of pure Id in large quantity, and their use has resulted
in much new information on the anti-MHC receptor repertoire and
its expression in response to various stimuli.
     In this report, I will first briefly review induction of
immunity by anti-Id (MHC class II) in the anti-Ia systems I have
studied. Models to explain the results will be discussed, and
results in a variety of other idiotypic systems compared. The
possible applicability of this approach to the general problem
of immunization will be discussed.

IN VIVO TREATMENT WITH ANTI-Id IN AN ANTI-MHC SYSTEM

     The Ids of a number of monoclonal antibodies in both
anti-H-2 and anti-Ia systems have been studied in this

Table 1--The 14-4-4 monoclonal antibody*

Immunization	C3H.SW anti-C3H/HeJ
Fusion	With SP2/0
Isotype	IgG2a, kappa
Reactivity pattern	Specific for I-E^k, cross-reactions with the d, p, and r haplotypes
Specificity	Corresponds to classic antigenic specificity Ia.7, a public specificity common to all strains of mice expressing I-E antigens.

*Summarized from Ozato et al. (11).

laboratory (3-7) and others (8-10). Some monoclonal antibodies represent public Ids, expressed widely in alloantisera and among monoclonal antibodies of the same fine specificity. Other monoclonal antibodies have private Ids that are very infrequent or undetectable on antibodies other than the particular monoclonal antibody. An example of a strikingly public Id is 14-4-4, a monoclonal antibody derived by Ozato et al. (11) (Table 1), that recognizes the Ia.7 determinant on the I-E antigen. This idiotypic system will be described as a case of successful use of anti-Id as a means of immunization, because immunity is induced efficiently.

The anti-Id reagents used for characterization of the 14-4-4 Id were raised in rabbits and miniature swine and purified by affinity chromatography. When the anti-Id was used to test alloantisera raised by skin grafting and boosting with spleen cells, virtually all mice of the C3H.SW strain (from which 14-4-4 was derived) expressed this Id in their responses to Ia.7 (5). Affinity-purified anti-Id was then used for _in vivo_ treatment of naive mice to assess its effect on Id expression and anti-Ia immunity (for details, see Epstein et al., 12). Such treatment was found to induce molecules reactive with anti-Id in enzyme-linked immunosorbent assays (ELISA). In addition, although the mice had never been exposed to alloantigen, they expressed specific anti-I-E activity in their sera (Figure 1).

Essentially all C3H.SW individuals responded to the treatment with readily detectable levels of anti-I-E activity, and this activity persisted for long periods of time without boosting. IgG1 predominated over IgG2 in the antigen-specific response, a point that might be relevant in situations where protective immunity is conferred by certain isotypes but not others.

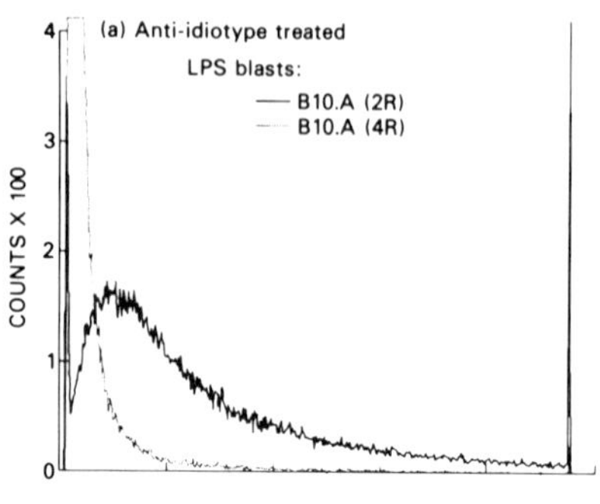

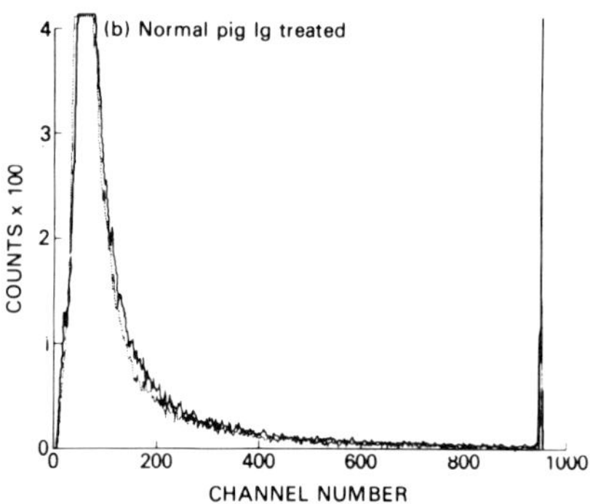

Figure 1--Detection by flow microfluorometry (FMF) of specific
serum anti-I-E activity after _in vivo_ treatment with anti-Id.
Histograms of number of cells detected vs. fluorescence
intensity are shown. Target cells are LPS blasts from B10.A(2R)
(bold line) and B10.A(4R) (thin line). Sera tested are from
C3H.SW mice treated with pig anti-14-4-4 Id(a), or normal pig
Ig(b). Reproduced from Epstein et al. (12).

     Treatment of various other strains of mice showed
differences in the responses of anti-Id. All strains produced
material reactive with anti-Id in ELISA tests, but large
differences were seen in levels of specific anti-I-E immunity
(Figure 2). The genetic control of these differences is
currently under study. At least a portion of the response
appears to be controlled by heavy-chain allotype-linked genes,
as would be expected for a structural V region gene.
     Although the Id in these experiments is monoclonal, the
anti-Id has generally been polyclonal xenogeneic anti-Id.
Monoclonal anti-Ids would have the advantage of specificity for

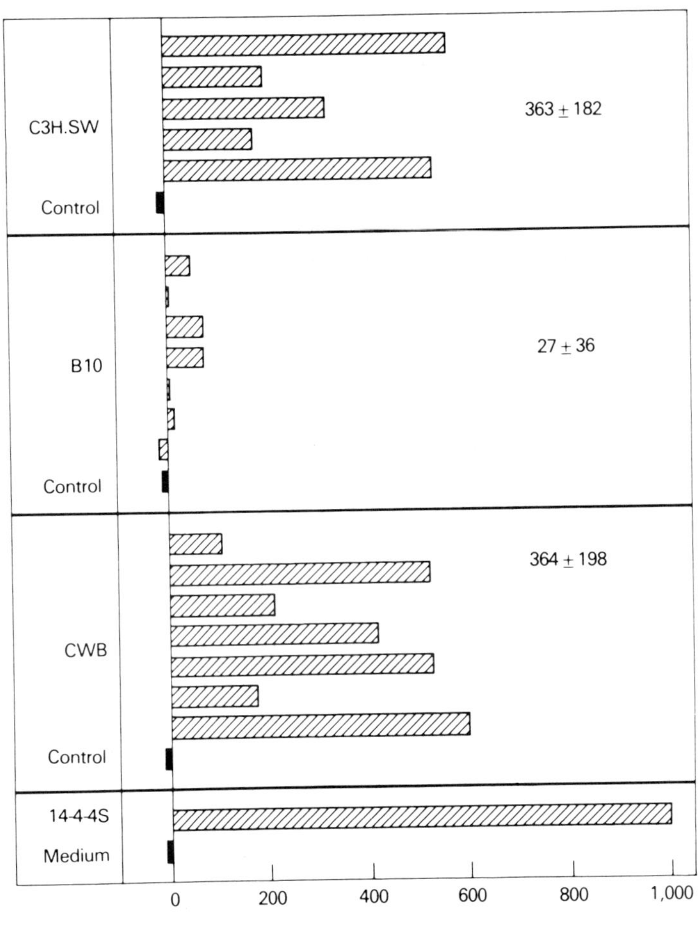

Figure 2--Induction of anti-I-E by pig anti-14-4-4 Id in different strains of mice. Mice were bled 4 weeks after treatment. Specific fluorescence units calculated as units on B10.A(2R) - B10.A(4R) cells. Test sera: individual sera from mice treated with pig anti-Id (shaded bars), mean values for three sera from mice treated with normal pig Ig (solid bars). Controls: 14-4-4 culture supernatant (shaded bar), medium (solid bar). Mean ± SD for each group of anti-Id treated mice is shown to the right of bars. Reproduced from Epstein et al. (12).

single idiotopes rather than a mixture of idiotopes, allowing the resolution of possibly differing biological effects. In addition, preparation of large amounts of anti-Id would be greatly simplified. Monoclonal anti-Ids of syngeneic origin would have the additional advantage of representing components that might occur naturally in the responding mice and would thus represent candidates for components of a physiological network.

Unfortunately, syngeneic anti-Ids in the anti-Ia.7 system fail to detect the public Id, and detect only private idiotypic determinants (13; Epstein and Ozato, unpublished results). Rat monoclonal anti-Ids detect public idiotopes (Devaux and Pierres, manuscript in preparation), but to date we have been unable to detect any induction of immunity by such reagents. Different conditions may be required than are required with the polyclonal anti-Ids, or Id suppression may possibly result from treatment.

MODELS FOR EFFECTS OF ANTI-Id _IN VIVO_

Anti-Id, called Ab2 in the terminology of the immune cascade (14), would be expected to react with several types of molecules, the total set of which can be referred to as Ab3 (15, 16). First of all, Ab3 would include Id-positive, antigen-binding molecules that would likely have structural similarity to the original Id (Ab1). Secondly, Ab3 could include molecules not binding the antigen in question but sharing idiotopes and presumably some structural homology with Ab1. Finally, Ab3 could include anti-anti-Id, that is, antibodies directed against the combining site of the injected Ab2 and constituting a response to it as a foreign antigen. This last set of molecules would not necessarily be expected to have structural homology with Ab1. Ability to make this type of Ab3 would not be expected to be limited to strains of particular allotypes, since a variety of V regions could be employed to make such a response.

In the anti-Ia.7, antibodies of both antigen-binding and "non-antigen-binding" types (i.e., not specific for I-E, although they may bind some other unknown antigen) are produced. In fact, by far the majority of the Ab3 molecules do not bind to the I-E antigen. Sequence data would be needed to determine the possible representation of the two non-I-E binding sets, and such studies are in progress employing monoclonal Ab3 antibodies. Monoclonal Ab3 antibodies have been obtained and partially sequenced in an anti-H-2K^k system, with one of four showing V_H homology to Ab1 (16).

Anti-Id could induce antigen-specific immunity by at least two types of mechanisms. First, it could act by recognizing Id, for example by binding to and activating lymphocytes of appropriate specificity that express Id-positive receptors on their surfaces. A more complicated mechanism in this category would involve perturbation of an Id-anti-Id regulatory network

by the binding of anti-Id to some Id-positive component, such as
a helper or suppressor cell or its factor.

Secondly, anti-Id might actually resemble antigenic
determinant(s), and its mode of action might then be by direct
immunization. Anti-Id in this scheme has been termed an
internal image of antigen (17), a related epitope (18), or an
antigen mimic. In such a model, all animals that can respond to
the antigenic determinant would be expected to respond to its
image, and indeed some such systems are known, as reviewed by
Nisonoff and Lamoyi (18). Such a model is unlikely in the
anti-Ia.7 system, because some strains able to respond to the
same epitope on I-E recognized by 14-4-4 do not make an I-E
specific response after anti-Id treatment.

RESPONSE TO ANTI-Id TREATMENT IN OTHER SYSTEMS

The results of anti-Id treatment in systems specific for a
variety of antigens have been reviewed recently (19). The
antigenic systems that have been studied include haptens,
globular proteins, bacterial surface carbohydrates, MHC
antigens, and parasite surface antigens. Antigen-specific
immunity induced ranged from substantial to nonexistent, and Id
suppression instead of induction was seen in some cases.
Results differed markedly, depending on an unknown number of
variables that may include the origin and characteristics of the
anti-Id preparations (including presence of carrier determinants
on xenogeneic immunoglobulins), the dose and route of
immunization, the specificity of the anti-Id for particular
idiotopes (public or private, combining site related or
framework), and the genetic constitution of the responding
animals. The isotype of the anti-Id was previously thought to
be an important determinant of inducing versus suppressing
capability (20), but recent results suggest that this may not be
true (Rajewsky et al., in preparation).

Results in some systems are encouraging for the use of
anti-Id treatment as an alternative to conventional
vaccination. In the case of trypanosomes protective immunity
was achieved by anti-Id treatment (21), and preliminary results
suggest protection in an <u>Escherichia</u> <u>coli</u> system (Stein and
Soderstrom, personal communication). Anti-Id pretreatment has
also been used successfully to potentiate subsequent responses
to hepatitis virus surface antigen (22). However, it is far
from clear that such results are easily generalizable.

For example, although induction of antigen-specific
immunity is easily achieved in the 14-4-4 system, little or no
induction is seen in two other anti-Ia systems (17-3-3 and
10-2.16) under the same conditions (Epstein, Ildstad, and Sachs,
unpublished observations). These systems both represent private
Ids and the difference might be attributed to that factor.
However, in the 14-4-4 system, some anti-Id preparations induce

and some do not, even though they all recognize public, combining-site-related idiotopes. Perhaps anti-Id must recognize particular idiotopes (regulatory idiotopes, 23) in order to be effective. Alternatively, differences in regulatory effects of different idiotopes may simply reflect quantitative differences in frequency of expression, as proposed by Sacks et al. (19). In any case, the outcome of _in vivo_ treatments is greatly influenced by the nature of the anti-Id preparation used, and most likely by the nature of the idiotopes it recognizes. Idiotopes that are relatively frequent even in normal serum may be regulated differently than those occurring very rarely, for example.

CONCLUSIONS

Anti-Id can provide a means of potent immunization without exposure of the responding animals to antigen. In the anti-Ia.7 system described above, this immunity can be reproducibly elicited and detected in most animals and is long-lasting. However, the ability to induce immunity to an antigen by this means depends on a number of factors that are not yet fully defined, and results vary greatly from one idiotypic system to another.

In addition, even if induction can be demonstrated in a system, genetic factors in the responding animals may lead to a reduction or elimination of the antigen-specific response. Thus, application of this technique to outbred animal populations is made difficult, and specialized reagents might be necessary, either mixtures as suggested by Sacks et al. (19) or customized reagents for each of multiple responder types.

In practical terms, anti-Id is thus unlikely to be suitable for all immunizations. Classic vaccines will continue to be the means of choice in many cases. In other cases, polypeptides isolated by cloning of the relevant genes will be useful, and in yet other cases, artificial small peptides corresponding to key parts of an antigen may give rise to appropriate immunity (24). In many cases, though, anti-Id may prove to be a valuable means of immunization, while in addition telling us a great deal about the immune repertoire and the regulation of its expression.

REFERENCES

1. Krammer, P. (1981) Curr. Top. Microbiol. Immunol. 91:179.
2. Bluestone, J.A., Auchincloss, H., Jr., Epstein, S.L., and Sachs, D.H. (1983) In: Köhler, H., Cazenave, P.-A., and Urbain, J. (eds.). Idiotypy. Academic Press, New York, (in press).
3. Sachs, D.H., Bluestone, J.A., Epstein, S.L., and Ozato, K. (1981) Transplant. Proc. 13:953.

4.   Bluestone, J.A., Epstein, S.L., Ozato, K., Sharrow, S.O.,
     and Sachs, D.H. (1981) J. Exp. Med. 154:1305.
5.   Epstein, S.L., Ozato, K., Bluestone, J.A., and Sachs, D.H.
     1981. J. Exp. Med. 154:397.
6.   Ozato, K., Epstein, S.L., Bluestone, J.A., Sharrow, S.O.,
     Hansen, T., and Sachs, D.H. (1982) Eur. J. Immunol., in
     press.
7.   Bluestone, J.A., Auchincloss, H., Jr., Sachs, D.H.,
     Fibi, M., and Hämmerling, G.J. (1983) Eur. J. Immunol.
     13:489.
8.   Devaux, C., Epstein, S.L., Sachs, D.H., and Pierres, M.
     (1982) J. Immunol. 129:2074.
9.   Grützmann, R., and Hämmerling, G.J. (1982) Eur. J. Immunol.
     12:307.
10. Nagy, Z.A., Elliott, B.E., Carlow, D.A., and Rubin, B.
     (1982) Eur. J. Immunol. 12:393.
11. Ozato, K., Mayer, N., and Sachs, D.H. (1980) J. Immunol.
     124:533.
12. Epstein, S.L., Masakowski, V.R., Bluestone, J.A., Sharrow,
     S.O., Ozato, K., and Sachs, D.H. (1982) J. Immunol.
     129:1545.
13. Devaux, C., and Pierres, M. (1983) Scand. J. Immunol.
     17:375.
14. Urbain, J., Wikler, M., Franssen, J.D., and Collignon, C.
     (1977) Proc. Natl. Acad. Sci. USA 74:5126.
15. Moser, M., Leo, O., Hiernaux, J., and Urbain, J. (1983)
     Proc. Natl. Acad. Sci. USA 80:4474.
16. Bluestone, J.A., Krutzsch, H.C., Auchincloss, H., Jr.,
     Cazenave, P.-A., Kindt, T.J., and Sachs, D.H. (1982) Proc.
     Natl. Acad. Sci. USA 79:7847.
17. Jerne, N.K. (1974) Ann. Immunol. (Inst. Pasteur). 125C:373.
18. Nisonoff, A., and Lamoyi, E. (1981) Clin. Immunol.
     Immunopathol. 21:397.
19. Sacks, D.L., Kelsoe, G.H., and Sachs, D.H. (1983) Springer
     Sem. Immunopathol. 6:79.
20. Eichmann, K. (1974) Eur. J. Immunol. 4:296.
21. Sacks, D.L., Esser, K.M., and Sher, A. (1982) J. Exp. Med.
     155:1108.
22. Kennedy, R.C., Adler-Storthz, K., Henkel, R.D., Sanchez,
     Y., Melnick, J.L., and Dreesman, G.R. (1983) Science
     221:853.
23. Bona, C.A., Heber-Katz, E., and Paul, W.E. (1981) J. Exp.
     Med. 153:951.
24. Lerner, R.A. (1982) Nature 299:592.

CHAPTER 12

APPLICATIONS OF MONOCLONAL ANTIBODIES IN PLANT SCIENCES

COLLABORATIVE EFFORTS FOR PRODUCING MONOCLONAL ANTIBODIES
FOR THE NATIONAL BANK OF PLANT VIRUS ANTISERA AT THE
AMERICAN TYPE CULTURE COLLECTION

H. T. HSU
Plant Virus and Antiserum Collection, Department of Virology
American Type Culture Collection
12301 Parklawn Drive, Rockville, MD 20852

Plant virologists have benefited greatly from the
development of serological techniques. These techniques have
proven extremely useful for identification of plant viruses,
for studies of relationships of plant viruses, for routine
testing of plants for freedom from virus infection, and as
quantitative methods for virus assays.

Plant virus antisera are routinely produced in rabbits.
When a "purified" virus preparation is used to immunize
animals, the resulting antiserum is a mixture of many
different immunoglobulins with different antigenic
specificities. The antiserum produced often is limited in
quantities by a single animal. It is virtually impossible to
reproduce the antiserum for a given plant virus antigen.

Hybridoma technology provides methods for producing
antibody preparations of molecular homogeneity. The
usefulness and application of monoclonal antibodies in the
biomedical field are well established. This is reflected by
the exponential increase in the number of hybridoma
publications since the technique was introduced in 1975 (1).
However, monoclonal antibody production has been adapted to
the field of plant pathology only recently (2).

During the last 2 years, monoclonal antibodies to 18
different plant viruses have been produced (Table 1). Of
these plant viruses, half were prepared at the American Type
Culture Collection. The monoclonal antibodies to plant
viruses have been useful for studying the differentiation of
strains and relationships of viruses (3, 12, 15), quantitation
of viral antigens (6, 7), detection and diagnosis of virus
infection (10), neutralization of viral infectivities (5), and
analysis of mapping of epitopes (16, Jordan and Hsu, personal
communication). In the practical application of diagnosis,

Table 1--Plant viruses for which monoclonal antibodies have been produced

Viruses	Investigators	References
Alfalfa mosaic	Halk et al.	3
Apple mosaic	Halk et al.	3
Barley yellow dwarf	Diaco et al., Hsu et al.	4, 5
Carnation etched ring	Hsu et al.	6, 7
Citrus leaf rugose	Halk, Hsu, Garnsey	Hsu, p.c.[1]
Citrus tristeza	Gumpf	Gumpf, p.c.[1]
Lettuce mosaic	Hill et al.	8
Potato leafroll	Gugerli, Martin, and Stace-Smith	9, 10
Potato A	Gugerli	9
Potato Y	Gugerli	9
Prune dwarf	Jordan, Hsu	Jordan, p.c.[1]
Prunus necrotic ringspot	Halk et al.	3
Southern bean mosaic	Tremaine and Ronald	11
Soybean mosaic	Hill et al.	8
Tobacco mosaic	Briand et al., Dietzgen and Sander	12, 13
Tobacco streak	Halk et al.	3
Tomato ringspot	Powell and Marquez	14
Tulip breaking	Hsu et al.	7

[1]p.c.--personal communication

commercially produced monoclonal antibodies are currently being made available for indexing potato viruses (9).

Monoclonal antibodies are also potential tools for other studies in plant virology. They can be utilized in the investigation of antigenic drift of plant viruses and in studies of the biological properties of vector-borne plant viruses (5). By neutralization tests, monoclonal antibodies may be able to specifically select new mutants that are difficult to isolate from a mixture. Monoclonal antibodies also offer potential application to the isolation of plant viruses that occur in low concentration in infected plants and are difficult to purify from plants. Affinity chromatography using monoclonal antibodies may be used to concentrate and purify these viruses for chemical and physical analysis (Slack and Rochow, personal communication).

The ATCC plant virus and antiserum collection is recognized as the national serum bank for plant viruses. Production of antiserum to reference plant viruses has been one of the major tasks of the ATCC-USDA cooperative project since the program was implemented in 1975. Monoclonal antibodies will increase in importance in taxonomic and epidemiological studies of plant viruses. They will also acquire increased applicability and relevance for biological and analytical studies of plant viruses. Development of monoclonal antibody-secreting cell lines will assure the long-term supply of homogeneous and biochemically defined immunological reagents for the assay of plant viruses. Murine monoclonal antibodies together with existing rabbit antisera to plant viruses will provide a powerful combination of tools for investigation of plant viruses.

REFERENCES

1. Kohler, G., and Milstein, C. (1975) Nature 256:495-497.
2. Hsu, H.T., Halk, E.L., and Lawson, R.H. (1983) Proc. 4th Int. Congr. Plant Pathol. p. 25.
3. Halk, E.L., Hsu, H.T., Aebig, J., and Franke, J. (1984) Phytopathology (in press).
4. Diaco, R., Lister, R.M., Durand, D.P., and Hill, J.H. (1983) Phytopathology 73:788.
5. Hsu, H.T., Aebig, J., and Rochow, W.F. (1984) Phytopathology (in press).
6. Hsu, H.T., Aebig, J., Rochow, W.F., and Lawson, R.H. (1983) Phytopathology 73:790.
7. Hsu, H.T., Lawson, R.H., Beijersbergen, J.C.M., and Derks, A.F.L.M. (1983) Proc. 4th Int. Congr. Plant Pathol. p. 117.
8. Hill, E.K., Hill, J.H., and Durand, D.P. (1984) J. Gen. Virology (in press).
9. Gugerli, P. (1983) Proc. 4th Int. Congr. Plant Pathol. p. 25.
10. Martin, R.R., and Stace-Smith, R. (1983) Phytopathology 73:792.

11. Tremaine, J.H., and Ronald, W.P. (1983) Phytopathology 73:794.
12. Briand, J.P., Al Moudallal, Z., and van Regenmortel, M.H.V. (1982) J. Virol. Meth. 5:293-300.
13. Dietzgen, R.G., and Sander, E. (1982) Arch. Virol. 74:197-204.
14. Powell, C.A. and Marquez, E.D. (1983) Proc. 4th Int. Congr. Plant Pathol. p. 121.
15. Halk, E.L., Hsu, H.T., Aebig, J., and Chang, K. (1982) Phytopathology 72:707.
16. Al Moudallal, Z., Briand, J.P., and van Regenmortel, M.H.V. (1982) EMBO J. 1:1005-1010.

PRODUCTION OF MONOCLONAL ANTIBODIES TO THREE ILARVIRUSES:
USE AS SEROTYPING AND DIAGNOSTIC REAGENTS

EDWARD L. HALK
The American Type Culture Collection
12301 Parklawn Drive, Rockville, MD 20852

Immunochemical techniques provide an invaluable tool for plant
virus identification, virus taxonomy, and diagnosis of plant virus
infection. However, a lack of readily available high-titer,
high-quality antisera may limit the widespread usage of sensitive
immunological methods (enzyme-linked immunosorbent assay, or ELISA;
radioimmunoassay, or RIA; etc.) for viral diagnostics or
epidemiological and taxonomic studies. This is particularly true
for many members of the Ilarviruses, which are weak antigens in
rabbits and difficult to purify in quantities sufficient for
quality antiserum production. Production of monoclonal antibodies
to these viruses could potentially provide highly specific
antibodies that could be used as taxonomic and diagnostic
reagents. This study was undertaken to determine the utility of
monoclonal antibodies specific for apple mosaic (ApMV), prunus
necrotic ringspot (NRSV), and tobacco streak (TSV) viruses both as
reference reagents to distinguish strains of these viruses and as
diagnostic reagents.
Hybridoma cell lines secreting antibodies specific for ApMV-F,
NRSV-G, TSV-WC, or alfalfa mosaic virus (AMV) were produced by
somatic cell fusions between mouse myeloma cell lines NS1/1 or
P3X63Ag8.653 and spleen cells of BALB/c mice immunized with a
mixture of the four viruses. Antibody-secreting hybridomas
specific for viral antigens were detected by indirect ELISA on
virus-coated PVC plates. Fourteen stable hybridoma cell lines were
selected by two to three cycles of single cell cloning. High-titer
ascitic fluid was produced from each hybridoma. Antibody titers of
ascitic fluid in indirect ELISA ranged from 6.25×10^4 to 7.5×10^6 for the homologous viruses. Of the fourteen hybridomas, 7,
4, 1, and 2 were of the IgG1, IgG2a, IgG2b, and IgM subclasses,
respectively.
Five TSV-WC specific monoclonal antibodies were produced, only
two of which were able to precipitate homologous virus in double
diffusion tests. Four serotypes of TSV were identified when

strains of TSV from white clover, soybean (from Brazil), tobacco, grape, or rose were assayed in indirect ELISA against a panel of the five TSV specific monoclonal antibodies. Thus, at least four antigenic sites on TSV were identified. None of the TSV specific monoclonal antibodies reacted to all five of the TSV strains.

Two AMV specific hybridomas were produced, both of the IgG2a subclass. The AMV hybridomas have not been further characterized.

Seven hybridomas produced monoclonal antibodies specific for either NRSV-G or ApMV-F. Three of these were ApMV specific, two were NRSV specific, and two were cross-reactive with some isolates of both NRSV and ApMV. The ascites titer of both cross-reactive antibodies was higher for NRSV than ApMV. Only three of the seven monoclonal antibodies precipitated NRSV-G in agar double diffusion assays. When the seven NRSV or ApMV specific monoclonal anti-bodies were used in indirect ELISA to screen nine NRSV isolates from cherry, peach, rose, or hops, three NRSV serotypes were identified. Two hybridomas, N46E10 and NA70C9, produced anti-body that reacted with all NRSV isolates tested. The Danish plum line pattern strain of NRSV was the only NRSV strain that failed to react with N63F10 antibody. None of the NRSV- or ApMV-specific antibodies reacted with TSV-WC, prune dwarf virus, or AMV in indirect ELISA.

When eleven ApMV isolates from apple, plum, hops, rose, or yellow birch were tested against the same panel of monoclonal antibodies, five ApMV serotypes were identified. Antibody A70A5 reacted with all ApMV isolates tested, and monoclonal antibodies A74F11 and A63E10 reacted with ten of the eleven ApMV isolates. Cross-reactive NRSV-ApMV monoclonal antibodies NA70C9 and NA49F8 differentiated four of the five serogroups. The pattern of reactions of NRSV- or ApMV-specific monoclonal antibodies to 21 isolates of ApMV or NRSV indicate that there are at least four antigenic sites on both NRSV and ApMV.

The feasibility of using monoclonal antibodies as diagnostic reagents for NRSV or ApMV in ELISA was examined with antibodies from hybridomas N63F10, N46E10, NA70C9, A70A5, and A63E10. These antibodies represented the broadest spectrum of reactivity to NRSV of ApMV isolates. Alkaline phosphatase antibody-enzyme conjugates were made with hybridoma antibodies purified by affinity chromatography on protein A or goat anti-mouse IgG-IgM columns. Rabbit anti-NRSV (RAN) or rabbit anti-ApMV (RAA) antisera were used for comparisons. Double antibody sandwich (DAS) ELISAs were run in several configurations of coating globulin and antibody-enzyme con-jugate utilizing polyclonal rabbit sera and monoclonal antibodies.

NRSV infection in cucumber and roses was readily detected by monoclonal antibodies or RAN in all combinations of coating globulin and antibody-enzyme conjugate tested. Coating globulin consisted of an equal mixture of affinity purified monoclonal antibody N46E10, NA70C9, and N63F10, whereas the enzyme conjugate was a 1:1 mixture of NA70C9 and N63F10 antibody. The N46E10 antibody (IgM) did not produce a useable antibody-enzyme conjugate and was thus used as coating globulin only.

Results with ApMV-infected plants were variable.
ApMV-infected cucumber, <u>Chenopodium quinoa</u>, and roses were not reliably detected by an ELISA system composed of monoclonal antibody
coating globulins and conjugates (1:1:1, NA70C9/A70A5/A63E10).
However, DAS ELISA consisting of RAA coating globulin and
antibody-enzyme conjugate (1:1:1; NA70C9/A70A5/A63E10) detected
some infected plants. However, intact, purified ApMV at
concentrations as high as 100 ng used as a positive control was not
readily detected by the monoclonal antibody conjugates.

A series of experiments with intact or disrupted ApMV showed
that antibodies NA70C9, A63E10, and A70A5 detected only disrupted
ApMV in DAS ELISA. Disrupted virus was detected from 100 down to
3.7 ng; however, intact purified virus was not readily detected
even at 100 ng. These results indicate that NA70C9, A63E10, and
A70A5 antibodies react with interior antigenic sites inaccessible
on intact virus. Antibody NA70C9 also reacted only with disrupted
NRSV in similar tests.

The ability of antibodies NA70C9, A70A5, and A63E10 to react
with virus coated directly to an ELISA plate but not to intact
virus in DAS ELISA probably results from partial disruption of
these labile viruses as they bind to the plate, thus exposing
interior antigenic sites.

The ELISA data with NRSV indicate that a limited number of
broadly reactive monoclonal antibodies are as good or better than
polyclonal sera in DAS ELISA. However, experiments with the
ApMV-specific monoclonal antibodies show that when designing assay
procedures, the location of epitopes recognized by monoclonal
antibodies needs to be understood.

The author's present address is Advanced Research Division,
Agrigenetics Corporation, 5649 E. Buckeye Road, Madison, WI 53716.

BASAL LOCALIZATION OF PRESUMPTIVE AUXIN TRANSPORT CARRIERS
IN PEA STEM CELLS

MARK JACOBS
Department of Biology, Swarthmore College
Swarthmore, PA   19081

Although the phenomenon of polar transport of the plant
hormone auxin has been well characterized during the last 50
years, the mechanism by which the process occurs has not been
elucidated. The currently attractive "chemiosmotic hypothesis
of polar auxin transport" (1, 2) states that a stem cell expends
energy to maintain a pH gradient across its plasma membrane,
with its cell wall more acidic than its cytoplasm. In the
acidic environment of the cell wall, molecules of auxin will
tend to be in the undissociated form (IAAH). Because cells are
more permeable to IAAH than to auxin anions (IAA$^-$) according
to the hypothesis, the undissociated form can readily enter the
cell by diffusion. Once inside the cytoplasm, auxin molecules
tend to dissociate to IAA$^-$ in the higher prevailing pH, and in
this form cannot freely equilibrate across the plasma membrane.
As the concentration of IAA$^-$ increases, movement out of the
cell is thermodynamically favorable but can only occur via
specific IAA$^-$ carrier molecules that are hypothesized to lie
only, or at a greater concentration, in the basal plasma
membrane of the cell.
The two major points of the chemiosmotic hypothesis are,
therefore: 1) the cell wall must be more acidic than the
cytoplasm in order to drive the uptake of auxin in the IAAH
form; and 2) the auxin anion carrier must be differentially
located at the base of the cells.
There is abundant evidence that the walls of plant stem
cells can be regulated at lower pH values than the cytoplasm can
and that extracellular pH can regulate auxin uptake into a
variety of cells and tissues (3, 4). However, there has been no
direct evidence for the basal location of the auxin anion
carrier. We began our work with monoclonal antibodies in order
to localize this component of the chemiosmotic hypothesis that
donates polarity to auxin movement through plants.
Because we presumed the auxin efflux carrier to be a
protein, we began by considering methods of labeling it _in situ_

with fluorescein-conjugated antibodies. However, in order to
raise normal serum antibodies in an animal in any useful amount
one needs to inject a fairly well purified protein, and we knew
too few of the biochemical characteristics of the auxin anion
carrier to be able to purify it.

    With the monoclonal technique, one does not need a purified
protein. We began by injecting a relatively crude preparation
of pea stem cell membranes into mice. After chopping and
grinding etiolated third internode tissue of peas, we used a
15,000 to 40,000 x g pellet resuspended in a typical plant
hormone binding assay medium (250 mM sucrose, 1 mM DTE, 0.5 mM
$MgCl_2$, and 10 mM MOPS, pH 6.5) to a total protein
concentration of 1 µg/µl (5). We injected 200 µl of this
preparation three times into each test mouse. Spleen cells from
the immunized mice were fused with NS-1 plasmacytoma cells, and
hybrid clones were selected in HAT (hypoxanthine, aminopterin,
thymidine) tissue culture medium (6).

    After the hybrid cells had grown for 20 days, we screened
the culture supernatant from each well, using as our screening
assay the binding of naphthylphthalamic acid (NPA) to its
membrane-associated receptor in pea. We did not use an auxin
binding assay as the screen for antibodies marking the auxin
anion carrier because in any general membrane preparation of the
type we injected into the mice originally there are several
different auxin binding sites. There could, therefore, be
several hybridomas producing antibodies that interfered in a
general auxin binding assay but would not label the auxin anion
carrier _in situ_. On the other hand, NPA has only one known
binding site in plant stem cells and that is believed to be on
the plasma membrane. It also has only one well-documented
effect in the same cells: a specific, strong inhibition of
polar auxin transport that is thought to take place at the
efflux site. It is therefore likely that the NPA receptor is
either identical to, or interacting with, the auxin anion
carrier. Because an NPA binding assay is also relatively easy
to perform rapidly, we used it as our screening test for an
antibody potentially marking the polar efflux carriers.

    After determining which wells contained cells producing
antibodies that significantly inhibited NPA binding to its
receptor in pea membranes, we cloned the positive hybridomas in
soft agarose, separated the easily visible monoclones into wells
on a new plate, and again screened for active antibody. Once
active monoclones were identified, the cell supernatants were
used as a direct source of active monoclonal antibodies.

    Because the NPA receptor in pea stem tissue was apparently
a potent immunogen, several monoclonal hybridomas produced
antibodies recognizing that protein. To use them to label pea
stem tissue, we cut median longitudinal sections of third
internode tissue of dark-grown peas freehand with a razor blade
from segments of fresh material. The sections were about 1 cm
in length, the width of the internode, and about 500 µm thick

on one end tapering to about 20 μm in thickness at the other.
The basal end of each section was marked, and the sections were
rinsed in distilled water. They were incubated for 50 min in
undiluted supernatant from active monoclonal hybridomas, rinsed
three times in a phosphate-buffered saline/bovine serum albumin
solution, incubated for 50 min in fluorescein-conjugated rabbit
antimouse antiserum, mounted in 50% glycerol, and observed under
ultraviolet light.

Control sections were exposed either to antibodies inactive
in inhibiting NPA binding in the screening assay or to a
commercial preparation of monoclonal antibody to an animal
antigenic determinant, GQ ganglioside. Although a relatively
strong blue-grey autofluorescence from xylem cells was observed
with all treatments, we never saw anything but a diffuse
background fluorescein fluorescence with control treatments. On
the other hand, when active antibodies were used, groups of
cells were specifically labeled, showing strong fluorescein
fluorescence only at their basal ends. [Color photomicrographs
of our results have been published elsewhere (5).] To test
whether the appearance of a specific fluorescence at the basal
ends of cells labeled a protein associated with the plasma
membrane or with the cell wall, we plasmolyzed pea stem sections
in 1 M and 2 M sucrose after exposing them to the fluorescent
secondary antibody. The bright fluorescent label remained with
the plasma membrane as it pulled away from the basal cell wall,
indicating that the U-shaped fluorescent zones in the cells
represent a distribution of the NPA binding site that is
primarily on the basal plasma membrane but also extends partway
up the side membranes of the cell.

Only certain cells--elongated parenchyma cells sheathing,
or in close proximity to, vascular bundles--appeared to be
labeled. The mature xylem and phloem of the vascular bundles
were not labeled, but the cells that were labeled could include
differentiating xylem and/or phloem initials. Cells of the
cortex and pith not closely associated with the vascular bundles
were never labeled.

It was bothersome to be so specifically locating the NPA
recorder of pea stem tissue without having direct evidence that
this protein could interact with auxin. Membrane-bound NPA
receptors do not bind auxin, but Sussman and Gardner solubilized
the NPA receptor from corn coleoptile membranes and found that,
in solubilized form, it did bind auxin (7). It was for this
reason that we solubilized the NPA receptor from pea membrane
preparations to test whether it could also bind auxin. As we
reported (5), the solubilized pea NPA receptor does bind auxin
as well as NPA and, furthermore, auxin and NPA binding by the
receptor are both inhibited by our active, but not by our
inactive, monoclonal antibodies. This shows that the protein we
have localized using the active antibodies binds NPS and can
interact with IAA$^-$, two characteristics expected of the auxin
anion carrier. Our results thus suggest that the auxin anion

carriers of the chemiosmotic hypothesis are differentially located at the base of transporting cells.

REFERENCES

1.   Rubery, P.H., and Sheldrake, A.R. (1974) Planta 118:101.
2.   Goldsmith, M.H.M. (1979) Annu. Rev. Plant Physiol. 28:439.
3.   Edwards, K.L., and Goldsmith, M.H.M. (1980) Planta 147:457.
4.   Davies, P.J., and Rubery, P.H. (1978) Planta 142:211.
5.   Jacobs, M., and Gilbert, S.F. (1983) Science 220:1297.
6.   Kohler, G., and Milstein, C. (1975) Nature 256:495.
7.   Sussman, M.R., and Gardner, G. (1980) Plant Physiol. 66:1074.

APPLICATION OF MONOCLONAL ANTIBODIES TO THE CHARACTERIZATION OF
PHYTOCHROME

LEE H. PRATT
Botany Department, University of Georgia, Athens, GA 30602

Plants sense via phytochrome both wavelength distribution
and fluence rate of incident radiant energy. They respond in
many ways, most of which result either in enhanced capability to
convert incident radiant energy to chemical energy via
photosynthesis or in altered partitioning of chemical energy
within the plant. Examples of plant responses to light include
de-etiolation, regulation of seed germination, control of
internode elongation rate, and modification of leaf morphology.
How phytochrome mediates these and other responses is poorly
understood (see Shropshire and Mohr (1) for reviews).

Phytochrome is a large (monomer size near 120 kDa),
water-soluble chromoprotein with an open-chain linear
tetrapyrrole chromophore. It exists in two forms, one
red-absorbing and physiologically inactive (Pr), the other
far-red-absorbing and physiologically active (Pfr). When either
form absorbs light, it is converted to the other. Little is
known about the physicochemical differences between the two
forms that must serve as the basis for the unique biological
activity of Pfr. The low abundance of phytochrome, especially
in plants grown in a natural environment (in which case the
pigment is about 0.01% of extractable protein), has hindered its
investigation at a molecular level. The application of
monoclonal antibody technology to phytochrome (2) provides many
methods for investigating its molecular properties and
function. My participation in this workshop is intended to
highlight some of our recent applications of this technology to
phytochrome.

CHARACTERIZATION OF MONOCLONAL ANTIBODIES

Twenty-five monoclonal antibodies (9 to pea and 16 to oat
phytochrome) have been screened by enzyme-linked immunosorbent
assay (ELISA) against phytochrome from etiolated pea, zucchini,

lettuce, oat, rye, and barley seedlings, as well as against both Pr and Pfr (3). Although all antibodies were prepared against approximately 118-kDa phytochrome, all nevertheless appear to recognize determinants found on the 60-kDa chromophore-containing half of the molecule. None of the antibodies is capable of distinguishing between the inactive Pr and active Pfr forms by this assay. Two bind equally well to phytochrome from all six plants. These two antibodies, which presumably recognize an evolutionarily conserved epitope, may serve as invaluable probes for regions of the chromoprotein that are important to its molecular function.

IMMUNOCYTOCHEMICAL VISUALIZATION OF PHYTOCHROME

Of eight antibodies directed to pea phytochrome, three were found to work well for immunofluorescence staining of formaldehyde-fixed pea tissue (4). In the epicotyl of etiolated pea seedlings, phytochrome was found to be most abundant in cortical cells. With the exception of stomatal guard cells, phytochrome was not detected in abundance in epidermal cells. As Pr, the pigment was diffusely distributed throughout the cytosol. Immediately after photoconversion to Pfr by a 10-s pulse of red light, it was no longer antigenically detectable. If left as Pfr for 10 min at room temperature prior to fixation, however, it again became detectable, but in a new "sequestered" distribution. Photoconversion back to Pr at any time resulted in a rapid redistribution of phytochrome to the original diffuse condition. These observations are neither a consequence of using monoclonal antibodies selective for only a single epitope on phytochrome nor are they a function of selective recognition of only one form of phytochrome. Whether the sequestering of phytochrome as Pfr is related to its molecular function is an intriguing question that merits further attention.

PHYTOCHROME QUANTITATION BY ELISA

We have developed a direct heterologous sandwich assay for phytochrome quantitation in completely crude plant extracts (5). Wells of vinyl assay plates are coated with immunopurified, polyclonal rabbit antibodies to phytochrome. After blocking further nonspecific binding with bovine serum albumin, 50-μl aliquots of crude plant extracts are added. These are then incubated with monoclonal antibody to phytochrome, immunopurified alkaline phosphatase-labeled antibody to mouse IgG, and p-nitrophenylphosphate substrate solution. ELISA activity is linear with respect to phytochrome over a range of less than 100 pg (<1 femtomol) to more than 1 ng. The assay has been used with green, light-grown oat seedlings to confirm or document that (1) phytochrome levels

oscillate threefold on a diurnal cycle, (2) phytochrome
accumulation rate accelerates markedly during prolonged dark
incubation, (3) phytochrome accumulates in mature cells as well
as in young, recently meristematic cells, (4) the herbicide
norflurazon inhibits phytochrome accumulation during prolonged
darkness, and (5) phytochrome from a green oat plant is
antigenically distinct from that found in an etiolated oat plant.

FUTURE

The initial applications of monoclonal antibody technology
to phytochrome described here serve primarily to illustrate the
potential for future applications. Of major interest is the
possibility that monoclonal antibodies may provide a way to
discriminate between Pr and Pfr and perhaps between
"etiolated-plant" and "green-plant" phytochrome, if they are as
different as initial data indicate.

REFERENCES

1.   Shropshire, W., Jr., and Mohr, H. (eds.) (1983)
     Photomorphogenesis. Encyclopedia of Plant Physiology, New
     Series, Vols. 16A & B. Springer-Verlag, Berlin.
2.   Cordonnier, M.-M., Smith, C., Greppin, H., and Pratt, L.H.,
     (1983) Planta 158:369-376.
3.   Cordonnier, M.-M., Greppin, H., Pratt, L.H. (1983) Plant
     Physiol. (in press).
4.   Saunders, M.J., Cordonnier, M.-M., Palevitz, B.A., and
     Pratt, L.H. (1983) Planta (in press).
5.   Shimazaki, Y., Cordonnier, M.-M., and Pratt, L.H. (1983)
     Planta (in press).

MONOCLONAL ANTIBODIES FOR PHYTOHORMONE RESEARCH

DAVID L. BRANDON
Western Regional Research Center, Agricultural Research Service
U.S. Department of Agriculture, Berkeley, CA   94710

Phytohormones are low-molecular-weight substances that have many important regulatory functions in plant growth and development. Their mode and sites of action remain poorly understood. The quantitation and localization of phytohormones are difficult because they occur in low concentrations and because they are highly soluble in solvents used in preparing samples for electron microscopy.

Immunoassays offer a quick and simple approach to the measurement of biologically important compounds, and they have recently been applied to the quantitation of plant growth regulators. They clearly have the potential to replace--or at least supplement--physicochemical methods and bioassays.

In this laboratory, we have prepared and characterized antibodies elicited with a variety of protein-cytokinin conjugates and developed a simple radioimmunoassay that we have used to quantitate naturally occurring cytokinins (1). We have also used immunocytochemical techniques to localize cytokinins in corn root tips (2). We are in the process of extending this work by using monoclonal antibodies (3).

MONOCLONAL ANTIBODY PREPARATION

BALB/c mice were immunized with protein conjugates of cytokinin ribosides. High responders were used for hybridoma preparation, with NS-1 as the parental myeloma cell. Positive cultures were identified by solid-phase radioimmunoassay (RIA) and were expanded and then cloned by limiting dilution. Cells were recloned by growth in soft agar. Quantities of antibodies were obtained by culture of the hybridomas in large flasks or by the growth of the cells as ascites tumors. Ascites fluid and sera from mice bearing the hybridomas were prepared.

Further discussion will pertain to hybridomas obtained from a mouse immunized with isopentenyladenosine (IPA) conjugated to

keyhole limpet hemocyanin (KLH). Several clones were obtained.
The monoclonal antibodies were analyzed by immunodiffusion and
solid-phase enzyme immunoassay (EIA). Each was identified as
IgGl containing a κ light chain.

Antibody CL40 was characterized by both liquid-phase and
solid-phase immunoassays. There was little difference in
binding to IPA, the hydroxylated hormone, _trans_-zeatin riboside
(TZR), and the saturated analog, dihydrozeatin riboside (DHZR).
However, the more elaborate titrations described below reveal
preferential binding to TZR. We are in the process of
characterizing a panel of monoclonal antibodies directed against
cytokinins, as well as production of antibodies directed against
other phytohormones.

SCREENING MONOCLONAL ANTIBODIES FOR PHYTOHORMONES

Techniques for rapid determination of the affinity of
anti-hapten antibodies are important for producing high-affinity
monoclonal antibodies. Ideally, the screening of hybridoma
culture supernatants should be done prior to transfer and
cloning of positive cultures. Early screening could obviate
extensive handling of cells that do not produce useful
antibodies.

Tsu and Herzenberg (4) described the use of hapten-protein
conjugates of various hapten densities in solid-phase RIAs. Low
affinity antibodies bind preferentially to conjugates with high
hapten density.

Van Heyningen et al. (5) have pointed out the utility of
simple antibody titer measurements in ranking monoclonal
antibodies by affinity. However, this method is applicable if
the antibody concentration is known or, at least, if the
antibody concentration is identical among the samples to be
compared. This condition is often approximated in hybridoma
cultures.

We have found that titration of culture supernatants on
assay plates containing various hapten densities (0.1 to 20
moles/100,000 g of protein) reveals differences in specificity
not seen at high concentrations or at high hapten density.

Nevertheless, it would be useful to have a more efficient
screening procedure for determining affinity and specificity.
Most useful would be a selection, rather than a screening,
process. Clearly, the fluorescence-activated cell sorter can be
used. A variety of methods more modest in cost and
instrumentation was discussed by Basch et al. (6). One
promising technique is the use of cell-bound catalase to protect
cells against a toxic environment of hydrogen peroxide.
Presumably, the use of hapten-catalase conjugates at low
concentration and hapten density would selectively protect cells
bearing high-affinity cytokinin-binding immunoglobulin.

Finally, immunization protocols could be optimized to
stimulate appropriate precursor cells selectively.
Mitogen-hapten conjugates may be useful for this approach. It
should be noted that the choice of hapten--i.e., how it is
derivatized--generally has a profound influence on antibody
specificity. We have found that some pyridoxine derivatives can
elicit exquisitely specific antibodies that do not crossreact
with closely related analogues of this vitamin.

THE LOCALIZATION OF CYTOKININS IN PLANT TISSUE

Low-temperature preparative techniques have been used for
preserving the intracellular distribution of labile cellular
components in plant and animal cells. We chose to couple
immunocytochemical and low-temperature preparative techniques to
investigate the distribution of cytokinins in corn root tips
(2). Although this study employed conventional polyclonal
antibodies and was a pilot study, I want to summarize the
results because they point out an important application of
antibodies in the study of plant physiology and could obviously
be extended using monoclonal antibodies.

The localization of cytokinins was investigated in corn
root tips sectioned at -30 to -40°C for immunofluorescence or
freeze-substituted in ethanol or acetone and embedded in plastic
for electron microscopy. We used antibody fragments, rather
than whole molecules, in order to maximize penetration and to
minimize the background.

Preparation of Tissues

For fluorescence studies, root tips were fixed in 2%
formaldehyde; acceptable morphological preservation with low
autofluorescence was obtained. To prepare frozen sections, the
tissue was quench-frozen in Freon, cooled in a liquid nitrogen
bath, and stored in liquid nitrogen until sectioned. Sections
0.25 or 0.5 µm thick were cut at -30 to -40°C. For freeze
substitution, root tips frozen by immersion in Freon slush in
liquid nitrogen were freeze-substituted in ethanol or acetone at
-85°C and finally embedded in plastic. Sections were cut with a
diamond knife.

Antibodies

Antibodies directed against cytokinins were raised in
rabbits and characterized (1, 2). The IgG fraction was
purified, and Fab fragments were prepared. Fab fragments were
conjugated with tetraethylrhodamine isothiocyanate. Colloidal
gold was prepared and Fab fragments directed against
dihydrozeatin riboside were adsorbed onto the gold particles (7
to 10 nm).

<u>Immunocytochemistry</u>

Labeled anticytokinin stained experimental samples with apparent specificity. Rhodamine-Fab stained cells adjacent to but not cells within the quiescent center. The cytoplasm of meristematic and root cap cells was fluorescent; nuclear regions were not. Fluorescence was blocked in cells incubated with DHZR in the medium, indicating that the bound antibody was associated with cytokinin.

It was necessary to pretreat sections with normal rabbit serum to eliminate the otherwise high background staining. Immunogold label was clearly limited to cytoplasmic regions. Label was rarely found in nuclei, vacuoles, or cell walls. Particles were associated with intracellular membranes that might be endoplasmic reticulum. The plasma membrane--when it could be identified--was not densely labeled. Labeling by gold anti-DHZR was blocked by unlabeled anti-DHZR. Sections adsorbed few particles when exposed to control Fab-colloidal gold. Hence, the distribution of particles was dependent on the antibody specificity.

CONCLUSIONS

These results indicate that immunocytochemical techniques when combined with low-temperature histological techniques are useful for studies of phytohormone localization. As a preliminary step toward performing high resolution autoradiography, we have also used ^{125}I-labeled antibody, and replicated the pattern of staining obtained with the other probes.

Because cells of the quiescent center rarely divide and cytokinins are generally considered to be hormones that stimulate division, our observations are consistent with the biology of the quiescent center. The association of the quiescent center with cytokinin biosynthesis may change during root development, so it would be of particular interest to correlate cytokinin localization with physiological and developmental events.

Monoclonal antibodies could be applied to this problem in order to quantitate individual phytohormones and their conjugated forms. Because large amounts of phytohormones exist as conjugates and may be essentially inactive, it is essential to have quantitative information about the distribution of the various forms in tissues and within cells.

ACKNOWLEDGMENTS

   The work discussed here is a collaborative effort with Drs.
Joseph Corse and Maria Elena Zavala. We are grateful to Clark
Craig, Sheri Krams, and Winnie Zing for their assistance.

REFERENCES

1. Brandon, D.L., Hua, S.-S., Corse, J.W., Yang, S.F., and
    Layton, L.L. (1980) Plant Physiol. 65 (suppl.):25.
2. Zavala, M.E., and Brandon, D.L. (1983) J. Cell Biol.
    97:1235-1239.
3. Brandon, D.L., Krams, S.M., Craig, C.W., and Corse, J.W.
    (1983) Twenty-second Midwinter Conference of Immunologists,
    Pacific Grove, CA.
4. Tsu, T.T., and Herzenberg, L.A. (1979). In: Selected
    Methods in Cellular Immunology. Mishell, B., and Shiigi,
    S. (eds.),   W.H. Freeman, San Francisco, pp. 373-397.
5. Van Heyningen, V., Brock, D.J.H., and Van Heyningen, S.
    (1983) J. Immunol. Methods 62:147-153.
6. Basch, R.S., Berman, J.W., and Lakow, E. (1983) J.
    Immunol. Methods 56: 269-280.

PRODUCTION OF MONOCLONAL ANTIBODIES TO CELL WALL ANTIGENS OF
CORYNEBACTERIUM SEPEDONICUM, INCITANT OF THE BACTERIAL RING ROT
DISEASE IN POTATO

S.H. De BOER and A. WIECZOREK
Agriculture Canada Research Station, 6660 N.W. Marine Drive
Vancouver, BC, Canada, V6T 1X2

Two hybridoma cell lines that produce monoclonal antibodies
to Corynebacterium sepedonicum (Spieck. and Kotth.) Skapt. and
Burkh., causal agent of the bacterial ring rot disease of
potato, were developed. The hybridomas were produced by fusing
NS1 myeloma cells with spleen cells from BALB/c mice previously
immunized with whole C. sepedonicum cells. Hybridomas were
screened for specific antibody production with an indirect
enzyme-linked immunosorbent assay (ELISA) procedure performed on
whole cells deposited on nitrocellulose or glass-fiber paper
held in a 96-well manifold. After recloning, two stable clones
were obtained. These clones were designated 3E10 and 2A4, and
the monoclonal antibodies produced by them, McAb 3E10 and
McAb 2A4, respectively. Monoclonal antibodies were purified
from ascitic fluid by protein A affinity chromatography,
adjusted to 0.5 mg/ml protein, and stored in aliquots at -20°C
until required. About 2 mg of immunoglobulin protein was
obtained per milliliter of ascitic fluid.
    Specificity of the monoclonal antibodies for the ring rot
bacterium was tested by indirect immunofluorescence using goat
anti-mouse IgG conjugated with dichlorotriazinylaminofluores-
cein. They were tested against 20 C. sepedonicum strains, 2
strains each of 10 other plant pathogenic corynebacteria
(including the closely related C. michiganense and C. insidiosum
species), and 12 unidentified strains that were isolated from
potato and cross-reacted with rabbit antiserum produced against
C. sepedonicum. McAb 2A4 reacted with all C. sepedonicum
strains but with none of the other plant pathogenic
corynebacteria. However, it did react with three of the
unidentified strains from potato. McAb 3E10 also reacted with
all C. sepedonicum strains and with none of the other plant
pathogenic corynebacteria or the unidentified strains from
potato. In extensive immunofluorescence tests with McAb 3E10 on

diseased but ostensibly ring-rot-free potato stem and tuber
tissue, some fluorescing bacteria-like cells were occasionally
observed. Only one such cross-reacting organism has been
isolated to date. It is a rod-shaped, gram-positive bacterium
and stains poorly with McAb 3E10 in immunofluorescence.

Monoclonal antibodies from both hybridomas reacted with C.
sepedonicum cells in standard, indirect ELISA tests. Both McAb
3E10 and McAb 2A4 also reacted with a cell wall extract obtained
from lithium chloride-treated cells. Separation of the extract
by sodium dodecyl sulfate-polyacrylamide gel electrophoresis and
subsequent staining of gels with Coomassie brilliant blue R-250
and Schiff's reagent revealed the presence of both protein- and
carbohydrate-containing components. Subsequent treatment of the
extract with pronase to digest the proteins rendered the extract
nonreactive with McAb 3E10, but reactivity with McAb 2A4
remained intact. Furthermore, electrophoretic blotting of the
polyacrylamide gel electrophoresis banding pattern onto
nitrocellulose and subsequent staining with an ELISA procedure
indicated that McAb 2A4 reacted with the carbohydrate
component. McAb 2A4, but not McAb 3E10, also reacted in ELISA
and immunodiffusion with an extracellular polysaccharide
purified from C. sepedonicum cells. The evidence suggests that
McAb 3E10 reacts with a cell wall-associated protein and
McAb 2A4 with an extracellular polysaccharide of C. sepedonicum.

ACKNOWLEDGMENT

This research was supported in part by a Farming for the
Future grant from the Agricultural Research Council of Alberta.

CHAPTER 13

DIAGNOSIS OF PLANT AND ANIMAL DISEASE

PLANT VIRUSES AND THEIR GENE PRODUCTS IN THE
DEVELOPMENT OF POLYCLONAL AND MONOCLONAL ANTIBODIES

ROGER H. LAWSON
Florist and Nursery Crops Laboratory
Agricultural Research Service
U.S. Department of Agriculture, Beltsville, MD   20705

Plant viruses have been classified into twenty-three groups
and two families. Viruses with ss-RNA genomes and particles
that have no envelopes account for the vast majority of viruses
infecting plants. They can be broadly divided into viruses with
protein coats that are arranged in helical symmetry, resulting
in rod-shaped particles, and those with icosahedral symmetry,
resulting in isometric particles.

The antigenic properties of virions are the most useful
criteria for reliable virus identification (1). The main
advantages of serological diagnosis of virus infections are
specificity, rapidity, and reliability. Specific antibodies
react in a reproducible fashion with the homologous virus and
are a reference standard for identification. Diagnostic
procedures that utilize serology allow virus identification at
both the group and strain level. Differentiation between virus
strains is usually based on cross absorption with polyclonal
antisera to heterologous strains.

In spite of the successful application of serological test
procedures for routine detection and diagnosis of viral capsid
proteins, there have been many instances where different
laboratories have presented conflicting data on the antigenic
characteristics of a particular virus and the relationship of
the virus to other known viruses. These differences in
reactions may be explained by the greater cross reactivity of
serotypes of certain viruses. Another explanation may be that
the virions of some isolates are less stable and soluble
antigens are formed. Soluble antigens are often detected with
virus antisera, and the subunits produced from the degraded
virus show greater cross reactivity and serological identity
than intact virions. For example, some strains of cucumber
mosaic virus dissociate to produce small proteins antigenically
related to the virus. These so-called soluble antigens may also

be produced in excess in the infected cell and are not
encapsidated. These soluble antigens are readily recognized as
rapidly diffusing antigens in agar gel double diffusion tests.

Two groups of cucumber mosaic viruses are recognized on the
basis of spur formation between preciptin lines caused by intact
capsids in double diffusion tests. Precipitin lines caused by
different strains showed reaction to complete fusion in
reactions with the soluble antigen (2). These results show that
meaningful antigenic comparisons can be made only if the antigen
is stablized before immunization to ensure that antibody is
formed to the intact capsid. Monoclonal antibodies should be
useful in further identifying and distinguishing the epitopes
associated with intact capsid protein and soluble antigen.

In addition to the form of the capsid protein, other
factors influence the "specificity" of an antiserum. The
qualitative and quantitative differences in homologous reactions
are of minor importance, but the relative differences in
reactions with heterologous viruses are of considerable
importance. In general, it is believed that high-titered
polyclonal antisera react less specifically than low-titered
antisera and that the fraction of strain-specific antibodies is
less in high-titered antisera than in low-titered ones. It is
probable that with prolonged immunization the concentration of
strain-specific antibodies becomes a smaller fraction of the
total antibodies. There are many examples in which
high-titered, but not low-titered, antisera react with
heterologous antigens. Monoclonal antibodies should be useful
in clarifying these relationships because the problem of
decreased strain specificity in polyclonal antisera that results
from prolonged immunization should not occur with antibodies to
a single epitope produced in a monoclonal system.

In addition to capsid protein, plant viruses produce other
products that are proteinaceous and capable of producing
antibodies. Virus-induced inclusion bodies formed by some
groups of viruses may be useful for virus diagnosis at the group
level. These inclusions are found in the cytoplasm or nucleus
of virus-infected cells and consist of large aggregates of
virus-coded nonstructural protein. The inclusion proteins of
potyviruses (3-5), tobamoviruses (6), and potato virus X (PVX)
(7) have been shown to be antigenically unrelated to the capsid
protein. In the case of PVX there are at least three
morphologically distinct proteinaceous structures in infected
cells. These included ribosome-like beads, thin sheets in
various configurations, and the virus particles (7). These
proteins show a difference in sensitivity to a proteolytic
enzyme and may be different proteins. The sheets or the beads
may be excess structural protein with a different quaternary
structure than the virus. One or the other may, therefore, have
antigenic properties in common with PVX or with its
antigenically distinct polymerized protein. Neither the sheets
nor the beads are antigenically related to PVX and probably

represent distinctly different gene products. Production of monoclonal antibodies to the proteins associated with these different types of inclusions should be useful in clarifying the relationships among them.

The rod-shaped potyviruses have been studied extensively. Polyclonal antisera prepared to intact virions often show a greater specificity than antisera prepared to their dissociated protein subunits, possibly because of the larger surface area of the viral subunit in the monomer than in the capsid. In another explanation, the partial denaturation of the protein exposes internal sequence homologies that are buried in the native molecule.

In many instances, the dissociated viral proteins react as well with antisera prepared against untreated virus as with antisera prepared against viral proteins dissociated by various treatments. Sometimes, however, antisera to whole virus do not react with the dissociated subunits, and it becomes necessary to immunize the animals with the chemically degraded protein.

The association of large amounts of nonstructural proteins with potyviruses is unique among plant RNA viruses. These proteins may aggregate in host cells to form distinct inclusion structures. Cylindrical inclusions are observed in the cytoplasm of all potyvirus-infected hosts. Inclusions in the nucleus have been observed for only a limited number of potyviruses. Although these nonstructural proteins are constantly associated with potyvirus infection, direct evidence that these proteins are of viral origin had not been found until recently. _In vitro_ translation of tobacco etch virus RNA in the mRNA-dependent rabbit reticulocyte lysate system has demonstrated that a number of translation products, distinct from capsid protein, are useful to study viral relationships among the large potyvirus group (8). Using immunoprecipitation techniques, Dougherty and Hiebert (8) showed that the inclusions associated with potyviruses consist of virus-coded nonstructural proteins.

The six gene products identified for tobacco etch virus have an estimated total product molecular weight of 340,000 (9). Immunoprecipitation analyses of the translations revealed some products that were consistently immunoprecipitated with more than one antiserum. These products were presumed to be gene readthroughs, based on their serological reactions and their size. These readthroughs were useful in linking genes on the potyvirus genome and in constructing a genetic map. One product was an 87,000-Da protein that subsequently was shown to be the helper virus component essential for successful transmission of the virus by aphids. Discovery of these proteins presents new possibilities for immunological detection utilizing polyclonal antisera and specific identification with monoclonal antibodies.

Monoclonal antibodies are particularly useful for determining the complex antigenic structure of proteins.

Proteins, as complex antigens, produce a variety of heterogenous antibody families, each family made up of members able to recognize, with various degrees of fit, one particular antigenic determinant of the protein (10). Polyclonal antisera showing this heterogeneity have been the only antisera available to establish the identity of the epitopes of many proteins.

Attempts to localize the epitopes of proteins with polyclonal antisera have led to disagreement in the interpretation of results that may be ascribed to the uniqueness of each antiserum and differences in methods of immunization in the animal species used. In addition, antibody-antigen interaction was measured by different techniques and this may have led different workers to emphasize different subsets of the total antibody population. For example, techniques that utilize highly diluted antiserum will preferentially measure antibodies of high affinity.

Monoclonal antibodies can be used to overcome the difficulties resulting from the use of heterogenous antibody mixtures. Monoclonal antibody preparations are homogeneous and only one epitope at a time is investigated. Protein determinants can now be investigated at the molecular level.

The ability of monoclonal antibodies specific for tobacco mosaic virus to detect antigenic changes induced by a single amino acid substitution in the protein coat has been demonstrated. Some of the substitutions are located at the outer surface of the viral capsid; others are internal and embedded within the protein coat (10).

Monoclonal antibodies have certain other properties useful for elucidating the antigenic structure of proteins. The ability of monoclonal antibodies to detect residue changes varies. Al Moudallal et al. found that many individual monoclonal antibodies failed to recognize certain exchanges that were detected by polyclonal antisera (10). Thus, monoclonal antibodies have an excellent capacity to discriminate and be useful in determining the presence of viral epitopes in certain regions of the protein and will allow further definition of details of antigenic structure related to conformational differences in the protein.

In this paper, I have presented examples of the current knowledge of the immunological relationships of the capsid proteins of some RNA viruses and nonstructural proteins of these viruses. Similar examples of nonviral inclusion proteins have also been described for dsDNA plant viruses (11).

Monoclonal antibodies will become increasingly important in future studies of the relationships among virus serotypes in the same group and for diagnostic purposes in instances where strain differentiation has not been possible with polyclonal antiserum.

REFERENCES

1. Van Regenmortel, M.H.V. (1982) Serology and Immunochemistry
    of Plant Viruses. Academic Press, New York.
2. Lawson, R.H. (1967) Virology 32:357-362.
3. Hiebert, E., Purcifull, D.E., Christie, R.G., and Christie,
    S.R. (1971) Virology 43:638-646.
4. Hiebert, E., and McDonald, J.G. (1973) Virology 56:349-361.
5. Purcifull, D.E., Hiebert, E., and McDonald, J.G. (1973)
    Virology 55:275-279.
6. Granett, A.L., and Shalla, T.A. (1970) Phytopathology
    60:419-425.
7. Shalla, T.A., and Shepard, J.F. (1972) Virology 49:654-667.
8. Dougherty, W.G., and Hiebert, E. (1980) Virology
    104:174-182.
9. Dougherty, W.G., and Hiebert, E. (1980) Virology
    104:183-194.
10. Al Moudallal, Z., Briand, J.P., and Van Regenmortel, M.H.V.
    (1982) EMBO J. 8:1005-1010.
11. Lawson, R.H., and Hearon, S.S. (1980) Phytopathology
    70:327-332.

EVALUATING THE RELATIVE SPECIFIC ACTIVITIES OF MONOCLONAL
ANTIBODIES TO PLANT VIRUSES

RAMON JORDAN
Florist and Nursery Crops Laboratory
Agricultural Research Service
U.S. Department of Agriculture, Beltsville, MD   20705

Virologists and plant pathologists have found serological
techniques extremely useful for the identification and
quantitative assay of plant viruses, for the routine diagnosis
of virus diseases, and for determining the degree of similarity
between viruses (1). The antigenic properties of virions are
the most useful criteria for reliable virus identification, and
serological diagnosis of virus infections is specific, rapid,
and reliable. Serological diagnosis also is one of the most
specific and rapid means of identifying individual strains of
plant viruses and has been shown to be a valuable tool for
measuring virus relationships.

The use of monoclonal antibodies should have several
advantages over the use of polyclonal antisera in the serology
of plant viruses: for example, the variability encountered with
polyclonal antisera would be eliminated; problems arising from
antisera containing antibodies to plant protein contaminants
should not be encountered; and the need to cross-adsorb antisera
with heterologous antigens to achieve strain-specific
identification would be eliminated.

The ultimate suitability of a monoclonal antibody for a
desired serological function (whether it is detection/diagnosis,
strain differentiation, immunopurification, or histochemistry)
must be assessed individually for each system. The affinity of
a monoclonal antibody for its antigen is one of the more
important properties determining its usefulness. For instance,
high-affinity antibodies are necessary for radio- and
enzyme-linked immunoassays used in virus detection, whereas
low-affinity antibodies are more suitable for immunopurification
procedures. Antibody-specific activity and epitope specificity
are important properties to consider in using monoclonals for
strain differentiation and identification. I intend to
highlight some of the methods and concepts used to evaluate the

relative specific activities of monoclonal antibodies to plant viruses.

CHARACTERIZATION OF MONOCLONAL ANTIBODIES TO SEVERAL ILARVIRUSES

Halk et al. (2) produced fourteen hybridoma cell lines secreting monoclonal antibodies specific for _Prunus_ necrotic ringspot (NRSV), apple mosaic (ApMV), tobacco streak (TSV), or alfalfa mosaic (AfMV) viruses using a mixture of NRSV strain G and ApMV strain F as the immunizing antigen. Subsequent screening of hybridoma colonies consisted of an indirect enzyme-linked immunosorbent assay (ELISA) in which monoclonal antibody culture supernatants were allowed to react in polyvinyl microtiter assay plates coated with purified virus. Seven cell lines secreting monoclonals specific for NRSV and/or ApMV were chosen for further investigation (Jordan, Aebig, and Hsu, manuscript in preparation) and are the focus of this report.

The ilarvirus (isometric labile ringspot virus) group consists of multicomponent viruses, with a divided genome, that are grouped together because of similarities in virion morphology and stability and in symptoms produced in infected plants (3). Three serologically distinct subgroups have been recognized. Subgroup III contains prune dwarf virus (PDV), NRSV, and ApMV. In tests with rabbit polyclonal antisera the latter two have been shown to cross-react. Danish plum line pattern virus (DPLP) has been shown to be closely related to NRSV (3).

Analysis and Comparison of Properties and Reactivities of Monoclonal Antibodies With Immunogen

Once a panel of monoclonal antibodies specific for a given antigen has been chosen, several different properties of these antibodies can be determined to help further classify and/or differentiate them: immunoassay titration, immunodiffusion titration, and isotyping of the monoclonal antibodies are just a few examples. The results of three such tests are summarized in Table 1. An indirect ELISA comparing the endpoint titers of the monoclonal antibodies to each of the immunogens subdivides the seven into three classes: nos. 1 - 3 react only with ApMV-F, nos. 4 and 5 react with ApMV-F and NRSV-G, and nos. 6 and 7 react only with NRSV-G. None of the monoclonal antibodies reacting with ApMV-F (nos. 1-5) precipitate antigen in immunodiffusion tests in agar, whereas nos. 5 to 7 do precipitate NRSV-G; thus, nos. 4 and 5 can be further differentiated. Determination of monoclonal antibody isotype is another property useful in differentiating antibodies of similar immunoreactivities. Monoclonal antibodies 6 and 7, which react specifically with NRSV-G can be differentiated in that no. 6 is an IgM and no. 7 is an IgG1. Nos. 1 - 3 are all IgG1s and have

Table 1--Immunoreactivities of hybridoma antibodies with immunogens

Antibody No.	Hybridoma cell line	Isotype	APMV-F[a] EIA	ID	NRSV-G[a] EIA	ID
1	63E10	IgG1	19	0	0	0
2	74F11	IgG1	18	0	0	0
3	70A5	IgG1	18	0	0	0
4	70C9	IgG2a	20	0	23	0
5	49F8	IgG2a	12	0	16	7
6	46E10	IgM	0	0	21	10
7	63F10	IgG1	0	0	25	11

[a]Endpoint titer ($\log_2$) of ascites antibodies against whole virus in indirect ELISA tests (EIA) and double immunodiffusion tests (ID).

similar endpoint titers in immunoassays with ApMV-F; in these tests they cannot be further differentiated.

Comparative Reactivities of Monoclonal Antibodies with Immunogen and Related Antigens

Analysis of the comparative immunoreactivity of the monoclonal antibodies with other strains or isolates of the viruses can also be useful in determining the relative activities of the monoclonal antibodies. Two different approaches that should give similar results are a) testing the ability of serial dilutions of antibody to differentiate (in an indirect ELISA test, as above) between several different related strains of the immunogen held at a constant concentration (antibody titration) and b) testing serial dilutions of the related viruses against a constant standardized amount of monoclonal antibody (antigen titration).

The seven ilarvirus monoclonal antibodies were tested in an antibody titration assay (Table 2) against five strains representing five different serotypes of ApMV, two strains representing two different serotypes of NRSV, and one isolate of DPLP (see also Halk et al. (4)). Of the three monoclonal antibodies specific for ApMV, only no. 3 reacted with all five serotypes tested. Nos. 1 and 2 reacted with four of the five serotypes, differentiating nos. 1 and 2 from no. 3. Even though

Table 2--Comparative immunoreactivities of monoclonal antibodies (ascites) with immunogens and related viruses: antibody
titration[a]

| Antibody | APMV | | | | | NRSV | | DPLP |
No.	AB	215	YB	F	P	G	R-9	
1	20	21	21	21	0	0	0	0
2	17	18	16	17	0	0	0	0
3	19	19	19	19	18	0	0	0
4	18	14	14	17	16	20	20	20
5	0	0	12	12	12	14	0	12
6	0	0	0	0	0	18	18	16
7	0	0	0	0	0	23	20	0

[a]Endpoint titer (log 2) of ascites antibodies in indirect
ELISA tests against representative strains of each of five
sero-groups of ApMV, two serogroups of NRSV and one serogroup
of DPLP; see Halk et al. (2) and Fulton (3).

nos. 4 to 7 can be differentiated based on antibody isotype,
immunogen specificity, and immunodiffusion reaction
(precipitating vs. non-precipitating) (Table 1), these
monoclonal antibodies react differently (both in specificity and
titer) when tested against a panel of related viruses (Table
2).   The difference in antibody specificity for different
serotypes is possibly an indication of epitope specificity of
the monoclonal antibodies (i.e., no. 6 recognizes a site present
on NRSV R-9 that is not recognized by no. 5 and the site
recognized by no. 5 on NRSV-G is not detected by no. 5 on NRSV
R-9).   It becomes fairly obvious at this point that
characterization of the monoclonal antibodies also involves
characterization of the antigens and, in fact, is usually the
reason for generating the monoclonal antibodies in the first
place.

Antigen titration immunoassays with the same monoclonal
antibodies and antigens gave similar results.   An excellent
example of this type of assay in examining the ability of nine
monoclonal antibodies directed to the common strain of tobacco
mosaic virus to differentiate between ten tobamoviruses is given
by Briand et al. (4, 5).

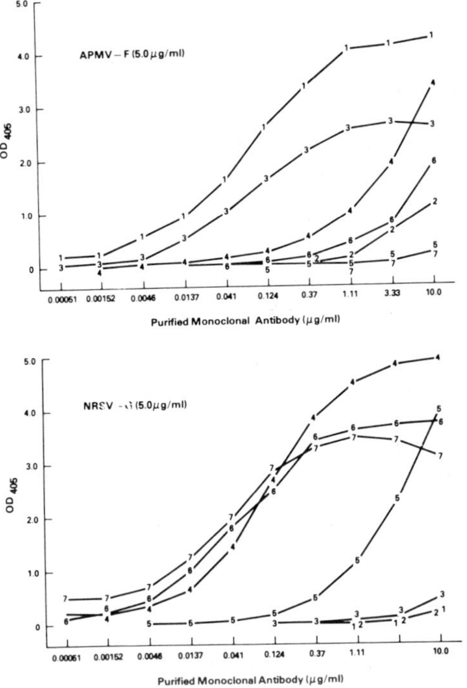

Figure 1--Binding characteristics of seven different monoclonal
antibodies to ApMV-F and NRSV-G by an indirect ELISA test.
Serial threefold dilutions of purified monoclonal antibodies
were incubated in the wells of polyvinyl microtiter plates
containing adsorbed virus (5 µg/ml). Monoclonal antibodies
bound to the virus-adsorbed wells were detected by the addition
of alkaline phosphatase-labeled goat anti-mouse Ig. Properties
of the monoclonal antibodies are summarized in Table 3.

Relative Binding Characteristics

Before attempting to further evaluate the epitope
specificity of monoclonal antibodies, their relative affinities
and specific activities should be determined. For example, in
competitive inhibition assays differences in antibody affinity
can affect the degree of competition among antibodies (6).

The seven monoclonal antibodies described above were first
purified from clarified ascites by ammonium sulfate
precipitation, Sephacryl G-300 chromatography, and DEAE-Affigel
blue chromatography.

The relative binding characteristics of the monoclonal
antibodies were compared by assaying serial dilutions of each on
NRSV-G and ApMV-F virus-coated vinyl assay plates (Figure 1).

The purified monoclonal antibodies generally had the same antigen specificity as the ascites from which they were derived, although no. 6 now appears to also react (albeit weakly) with ApMV-F (Figure 1). The possible reasons for this observation are under investigation.

A convenient way to express the specific activity of an antibody is in units of activity per milligram antibody. The specific activity of polyclonal antisera is usually expressed as the reciprocal of endpoint titer divided by antibody concentration. With purified antibodies in this ELISA test (Figure 1) we can express specific activity in terms of the minimum concentration of monoclonal antibody needed to reach a "positive" reading (two to three times background absorbance). For example, a ranking of the relative specific activities of the monoclonal antibodies reacting to NRSV-G, based on the above criteria, would be 4 = 6 = 7 > 5 >> 3; and to ApMv-F, 1 > 3 > 4 > 6 > 2 >> 3.

In determining the relative binding affinities of the purified monoclonal antibodies, at least two parameters must be taken into consideration: the slope of the linear portion of each graph and the lowest antibody concentration that achieves binding saturation (peak plateau, Figure 1). An increase in ELISA absorbance value is a measurement of increased binding which, in these tests, is proportional to an increase in antibody concentration. The antibody concentration at binding saturation (peak plateau) is a measurement of antibody affinity. The lower the antibody concentration required to achieve plateau binding, the higher the affinity of that antibody for its epitope. The actual peak height (ELISA absorbance value) is a measurement of the relative number of epitopes on the antigen. The apparent order of relative affinities of the ApMV-F-specific monoclonal antibodies are 1 = 3 >> 4 > 6 > 2; and of the NRSV-G-specific monoclonal antibodies the relative ranking is 7 > 6 > 4 >> 5.

The overall relative binding characteristics (Table 3) determined in this manner (solid-phase antibody-binding assay) is not a means of calculating the actual affinity constant, K (1,7). However, it is a more practical method, because the system used to determine the relative affinities is similar in design to the methods used in immunoassays with the monoclonal antibodies.

Evaluating Epitope Specificity

Epitope specificity of the different monoclonal antibodies can be determined by competitive inhibition assays. Competition of antibody binding is based on the premise that if two antigenic sites are identical or overlapping, then an excess of one antibody binding to its epitope would inhibit the binding of a second antibody to its epitope. If the sites are sufficiently distant (and distinct), then simultaneous binding will occur.

Table 3--General provisions of the monoclonal antibodies to ApMV and NRSV

Antibody No.	Isotype	Immunoreactivity to virus serogroups[a]			Binding activity on plastic-adsorbed virus[b]		Site specificity[c]
		ApMV	NRSV	DPLP	ApMV	NRSV	
1	IgG1	4/5	0/2	0/1	High	–	A
2	IgG1	4/5	0/2	0/1	Low	–	A'
3	IgG1	5/5	0/2	0/1	High	–	B
4	IgG2a	5/5	2/2	1/1	Med	High	C
5	IgG2a	3/5	1/2	1/1	Low	Med	D
6	IgM	0/5	2/2	1/1	Low	High	E
7	IgG1	0/5	2/2	0/1	–	High	F

[a]No. of serogroups recognized/no. serogroups tested in an indirect ELISA test (Table 2).

[b]As described in Figure 1 and text.

[c]Relative site specificity determined from analysis of monoclonal antibody isotype, serogroup reactivities, and competitive inhibition assays.

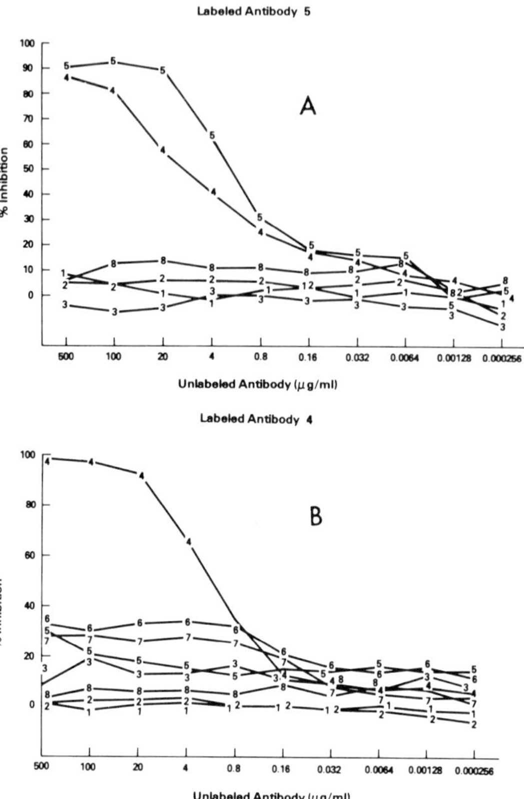

Figure 2--Inhibition of binding of biotinylated monoclonal antibodies 5 (A) and 4 (B) by nonlabeled competitor monoclonal antibodies in a competition inhibition assay. The binding of a constant amount (1 μg/ml) of biotinylated monoclonal antibody to NRSV-G-virus-adsorbed microtiter wells (5 μg/ml coating) in the presence of fivefold serial dilutions of nonlabeled nos. 1 to 7 (see Tables 1 to 3). Inhibition of binding is expressed as percentage of inhibition relative to the binding levels achieved in the presence of reference no. 8 (A and B) as 0% inhibition control, and no. 5 (A) and 4 (B) as 100% inhibition controls.

Other situations may also arise in which steric hindrance (partial inhibition) occurs between monoclonal antibodies that recognize distinct but relatively close epitopes; or the binding of one monoclonal antibody alters the conformation of the epitope (allosteric hindrance) recognized by the second one, resulting in the altered binding of the second monoclonal antibody.

In immunoassays used to determine the degree of competition among the seven above mentioned monoclonal antibodies, polyvinyl assay plates were coated with either ApMV-F or NRSV-G, incubated with unlabeled competitor monoclonal antibody, then (without prior washing) incubated with biotin-labeled second monoclonal antibody. Biotin-labeled monoclonal antibody was detected with avidin-conjugated to enzyme. In attempts to minimize any effect due to differential affinities and to favor the inhibition reaction, at least a 100-fold excess of unlabeled putative inhibitor was incubated overnight before adding the biotinylated monoclonal antibody.

The binding characteristics of two of the labeled antibody preparations for NRSV-G in the presence of serial dilutions of each of the nonlabeled monoclonal antibodies are illustrated in Figure 2. The following controls were included: use of a completely heterologous monoclonal antibody to carnation etched ring virus (no. 8) and 0.1% BSA as inhibitors (negative, 0% inhibition control) and unlabeled homologous monoclonal antibody as inhibitor (positive, 100% inhibition control).

In both examples, 20- to 500-fold excess of unlabeled homologous monoclonal antibody effected 90 to 100% inhibition (Figure 2a and 2b). Less than 10% inhibition was detected with the negative monoclonal antibody control (no. 8) as well as with the ApMV-F-specific monoclonal antibodies (nos. 1, 2, and 3). The binding of the labeled no. 5 was strongly inhibited by unlabeled no. 4 (Figure 2a), whereas the binding of labeled no. 4 was not inhibited to any great extent (<30%) by unlabeled no. 5 (Figure 2b).

Three interpretations of these observations are possible. a) Monoclonal antibodies 4 and 5 recognize the same epitope. This is unlikely; for they have different immunoreactivities with the given serogroups of ApMV and NRSV; i.e., no. 5 does not react with two ApMV serogroups and one NRSV serogroup that no. 4 does react with. It is therefore more likely that the two monoclonal antibodies recognize different epitopes. b) Monoclonal antibodies 4 and 5 recognize non-overlapping but proximal epitopes. The strong (but not complete) inhibition of labeled no. 5 by unlabeled no. 4 suggests this possibility, but the lack of reciprocal inhibition must also be considered. Low affinity antibodies tend to show low competition as competitor antibodies (6) but can be strongly inhibited by competitor molecules. Monoclonal antibody 5 apparently has a lower affinity for its epitope than no. 4 does for its epitope (Table 3; Figure 1). In this situation then, because the dissociation constant of no. 5 for "epitope 5" is greater than its association constant, any "'unbound' epitope 5" would be "blocked" (sterically) by no. 4 binding to the proximal "epitope 4". This is possible, but the competition assays were run in such a manner as to try to minimize any effect due to differential affinities and in fact to favor the inhibition assay (Figure 2). Even at 500-fold excess of unlabeled no. 5,

only 30% inhibition of labeled no. 4 was observed. c)
Monoclonal antibody 4 recognizes "epitopes 4 and 5", whereas no.
5 recognizes only "epitope 5". This explanation of "partial
identity" is a more likely explanation for these observations
than the first two reasons. The lack of reciprocal inhibition
is because of the lack of partial identity of no. 5 for "epitope
4", whereas no. 4 strongly inhibits no. 5 because of the partial
identity of no. 4 for "epitope 5".

The results obtained from the competitive inhibition
assays, coupled with the data generated from analysis of
monoclonal antibody isotype, virus serogroup specificity, and
binding activity in various different ELISA procedures may be
used to attempt to assign site specificity as defined by each of
the monoclonal antibodies. The analysis of these data suggests
at least six different antigenic sites can be delineated by the
seven monoclonal antibodies used in these studies (Table 3), and
that no. 4 apparently recognizes an epitope common to all ApMV,
NRSV, and DPLP isolates tested (Tables 2 and 3). Unfortunately,
no. 4 does not recognize any epitopes on PDV, another member of
ilarvirus subgroup III (3). Monoclonal antibodies generated to
PDV (Jordan, Elliott and Hsu, unpublished) are now being tested
to evaluate their relative immunoreactivities with isolates of
PDV, ApMV, and NRSV.

CONCLUDING REMARKS

These interpretations and explanations are presented mainly
as examples of the concepts and determinations that must be
recognized in attempts to evaluate epitope specificity of
monoclonal antibodies to plant viruses. Another important
factor that should be considered in determining the relative
specific activities of monoclonal antibodies is the assay method
used to measure their activity. For example, in the above
mentioned experiments all the results obtained were by an
indirect ELISA procedure in which whole virus was used to coat
the assay plates. All of the seven monoclonal antibodies
reacted with the appropriate virus antigens as shown. However,
when assay plates were first coated with rabbit anti-ApMV
globulin and then incubated with whole ApMV virus, no. 3 was
found not to bind to virus. These results illustrate the value
of testing several different assay methods in determinations of
monoclonal antibody activity. Other methods might include the
use of homologous and heterologous polyclonal antisera in
indirect and sandwich ELISA, and testing monoclonal antibody
activity against whole and disrupted virus, including peptide
fragments.

The application of monoclonal antibody technology to plant
virology provides an expanded battery of both biological and
analytical methods for the investigation of plant viruses.
Plant virus serology can now become an even more powerful tool

for the detection, assay, differentiation, and topographical and structural analysis of plant viruses and their gene products.

ACKNOWLEDGMENTS

I thank Dr. Hei-ti Hsu and Joan Aebig for their excellent professional and technical assistance.

REFERENCES

1. Van Regenmortel, M.H.V. (1982) Serology and Immunochemistry of Plant Viruses. Academic Press, New York.
2. Halk, E.L., Hsu, H.T., Aebig, J., and Franke, J. (1984) Phytopathology (in press).
3. Fulton, R.W. (1983) Ilarvirus group. CMI/AAB Descriptions of Plant Viruses. Holywell Press.
4. Briand, J.P., Al Moudallal, Z., and Van Regenmortel, M.H.V. (1982) J. Virol. Methods 5:293.
5. Al Moudallal, Z., Briand, J.P., and Van Regenmortel, M.H.V. (1982) EMBO. J. 1:1005.
6. Jackson, D.C., and Webster, R.G. (1982) Virology 123:69.
7. Steward, M.W. and Petty, R.E. (1972) Immunology 22:747.

APPLYING MONOCLONAL ANTIBODIES TO THE DIAGNOSIS OF ANIMAL
DISEASES

RICHARD VAN DEUSEN
National Veterinary Services Laboratory
P.O. Box 844, Ames, IA   50010

If we envision an animal as a DNA expression system for a
given set of genes just as a host cell is the expression system
for a viral genome, we then have a new way of viewing the
disease process. Changes in the environment of any cell then
elicit modifications in the expression of the host system's
genome, resulting in cellular protein change. Monoclonal
antibodies are capable of specifically binding to relatively
small areas, or epitopes, on protein molecules. It is therefore
possible to conceive of producing monoclonal antibodies that
recognize not only epitopes associated with pathogens but also
new or modified epitopes associated with the host system that is
under attack.
Diagnostic applications include the more conventional
identification of pathogens using monoclonal antibody typing
reagents, identification of the isotype (1, 2) of host animal
antibodies against specific antigens as a serologic test to
identify pathogens during acute infection; the immunochemical
identification (3) of vacuole-associated modified host cell
proteins as an approach to developing reagents for the diagnosis
of scrapie; and identification of tumor-cell-associated proteins
(4) in cancer diagnosis. Anti-idiotype antibody (5, 6) is
another potential means of developing immunodiagnostic
reagents. The precision with which a monoclonal antibody
recognizes and bonds to its epitope provides us a new way for
examining disease processes and their causes. Wands et al. (7)
have applied monoclonal antibody in their "signature analysis"
for subtype identification of hepatitis B viruses. Figure 1 is
a hypothetical resultant curve that we would expect when any
antigen is titrated against a constant amount of ^{125}I-labeled
homologous antibody according to antigen-bound-to-solid phase
radioimmunoassay (RIA) methods. If the starting concentration
of antigen is lower or higher, the curve would shift to the left
or right, respectively. If the affinity of the antibody for the

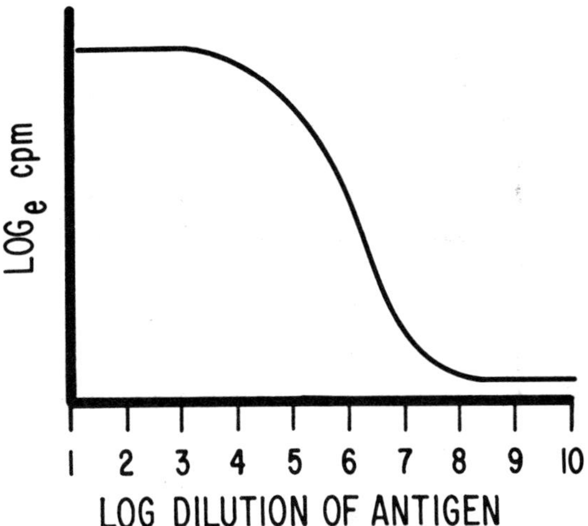

LOG DILUTION OF ANTIGEN

Figure 1--Hypothetical radioimmunoassay with constant ^{125}I labeled antibody concentration. Antigen is bound to a solid phase. The curve is typical for a homologous antigen.

antigen for any reason decreased, the height of the curve would decrease. Signature analysis is based on applying the RIA method with a small panel of labeled monoclonal antibodies each specific for a different epitope. Differences in the relative occurrence of each epitope as well as variations in affinity that result from structural differences within the epitopes when subtypes of the virus are compared are observed. Figure 2 is an example of what one might expect to find when three subtypes of a particular virus are tested with a panel of four monoclonal antibodies. It is essential that the virus dilutions being tested are prepared in sufficient quantity to allow the four curves to be generated with the single set of ten tenfold dilutions. This allows the determination of relative frequency of the four epitopes because the four sub-sets of dilutions are identical. It is also essential that each labeled monoclonal antibody is always used at the same protein concentration and label intensity to allow measurement of affinity differences between virus subtypes. In Figure 2, virus subtype A is depicted as having its four epitopes in descending order of frequency 3, 1, 4, 2; virus B - 3, 4, 1, 2; and virus C - 2, 1, 3, 4. The depicted relative affinities of the antibodies for their epitope on each virus subtype are monoclonal antibody no. 1, A = B > C; no. 2, A = C > B; no. 3, A = B = C; and no. 4, B > A > C. In signature analysis, variation in beginning concentrations of antigen cause the entire set of curves to shift left or right, but the relationship of each curve to the

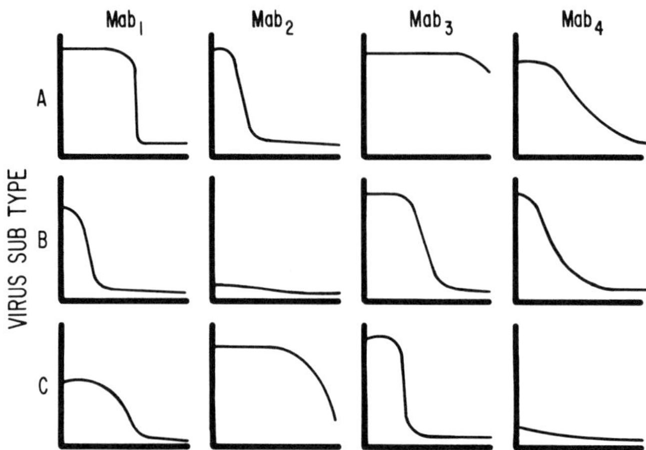

Figure 2--Hypothetical "signatures" of 3 subtypes of a virus.
X axis = log dilution of virus, Y axis = $\log_e$ CPM. Antigen is
bound to a solid phase and ^{125}I labeled monoclonal antibody is
held constant.

other will remain constant, thus providing a reproducible
"signature."

This type of conceptual approach to analyzing differences in
epitopes opens a large field of potential applications for
monoclonal antibodies. It obviates the need for absolute
specificity of each monoclonal antibody in order to identify
differences in antigens. Instead we now have a way of looking
at epitopes as parts of an observable continuum. The extent to
which this idea may be applied appears limitless.

REFERENCES

1. Butler, J.E., Seawright, G.L., McGivern, P.L., and Gilsdorf,
    M. (1981) Exp. Med. Bio. 137:790-791.
2. Paul, P.S., Van Deusen, R.A., and Mengeling, W.L. (1984)
    Vet. Immunol. Immunopath. (in press).
3. Cuello, A.C. (1982) From Gene to Protein: Translation into
    Biotechnology, Miami Winter Symposia, Vol. 19. Academic
    Press, New York.
4. Chang, T.H., Steplewski, Z., Sears, H.F., and Koprowski, H.
    (1981) Hybridoma 1(1):37-45.
5. Gerhard, W., Taylor, A., and Wroblewska, Z. (1981) Proc.
    Natl. Acad. Sci. USA 78(5):3225-3229.

6. Potocnjak, P., Zavala, F., Nussenzweig, R., and Nussenweig,
    V. (1982) Science, 215:1637-1639.
7. Wands, J.R., Wong, M.A., Shorey, J., Brown, R.D., Marciniak,
    R.A., and Isselbacher, K.J. (1984) Proc. Nat. Acad. Sci. USA
    (in press).

APPLICATION OF HYBRIDOMA TECHNOLOGY TO THE DEVELOPMENT OF A
DIAGNOSTIC TEST FOR SWINE TRICHINOSIS

H.R. GAMBLE
Animal Parasitology Institute, Agricultural Research Service,
U.S. Department of Agriculture, Beltsville, MD   20705

    Monoclonal antibodies have found considerable application
in the development of diagnostic tests for parasitic diseases
[see (1) for review]. Hybridoma-derived reagents have been used
a) as antibodies to test for the presence of parasite antigens
in host body fluids and b) to identify and isolate parasite
antigens useful for the detection of parasite-specific host
antibodies. I will describe the use of monoclonal antibodies in
the development of a diagnostic test for swine trichinosis.
    Trichinosis in swine and humans is caused by the parasitic
nematode _Trichinella spiralis_. Infections are acquired by the
ingestion of muscle tissue containing infective larvae. Human
infection results from eating trichinous meat, generally pork or
game, that has not been treated (cooked, frozen, or processed)
to inactivate the parasite. After ingestion, parasite larvae
excyst in the stomach and penetrate the muscosa of the small
intestine, there developing into adults. Adult female worms
produce numerous larvae, which then pass via the lymphatics and
circulatory system to the striated muscle cells, where they
encyst and persist in muscle tissue for the life of the host.
This muscle larval stage (L_1), which is infective for other
hosts, is the stage of diagnostic interest.
    A variety of direct and serologic tests have been used to
diagnose trichinosis in swine and humans [see Ruitenberg et al.
(2) for review]. The enzyme-linked immunosorbent assay (ELISA)
currently is being tested as both an epidemiological and
slaughter-house diagnostic test in the United States and Europe
(3,4). However, the lack of a suitably specific antigen for use
in the ELISA has hindered the adoption of this test. Use of
crude and complex parasite extracts has resulted in significant
rates of false-positive reactions (5-7), due in part to
cross-reactions with other swine nematode infections. The
identification of an antigen unique to _T. spiralis_ would,

therefore, be of considerable value for the serodiagnosis of trichinosis.

PRODUCTION OF ANTI-T. SPIRALIS MONOCLONAL ANTIBODIES

Monoclonal antibodies were produced by standard polyethylene glycol fusion of mouse myeloma cell line P3X63-Ag8.653 with spleen cells from a BALB/c mouse (8). The spleen cell donor received 200 infective T. spiralis muscle larvae (L_1) 35 and 7 days before the cell fusion. Culture supernatants from wells showing growth after 10 to 14 days were screened in a triple-antibody ELISA (9) against a crude extract of T. spiralis L_1. Positive wells were expanded, cloned by limiting dilution, and injected into pristane-primed mice to produced ascites.

Several ELISA screening protocols were used to select monoclonal antibodies of interest for a diagnostic test. First, monoclonal antibodies were screened in an ELISA (as above) against crude extracts and excretory-secretory preparations of various stages in the life cycle of T. spiralis. Excretory-secretory (ES) products, derived from in vitro cultivated L_1, had previously been shown to be a good source of diagnostic antigens (10). Second, monoclonal antibodies were screened against crude extracts of several other common swine nematodes (Table 1). Most of the monoclonal antibodies reacted with several life cycle stages of T. spiralis as well as with extracts of various other nematodes, in particular the closely related Trichuris suis. Only one antibody, $7C_2C_5$, reacted exclusively with T. spiralis L_1. In addition, supernatants from this clone showed specificity for an antigen epitope present in ES products. Because of its specificity, this hybridoma was selected for further studies.

CHARACTERIZATION OF HYBRIDOMA $7C_2C_5$

Hybridoma $7C_2C_5$ secreted an IgM-class antibody molecule as determined by gel diffusion, ELISA, and metabolic labeling. The molecular weights of the antigens recognized by this monoclonal antibody were determined by immuno-blot staining of whole parasite extracts separated in sodium dodecyl sulfate-polyacrylamide gels (11). Three proteins with molecular weights of 53,000, 49,000, and 45,000 (designated Ts.53, Ts.49, and Ts.45) were recognized.

Immunoperoxidase staining of parasite sections demonstrated that the $7C_2C_5$ monoclonal antibody bound to antigens located in the secretory stichocyte cells of the parasite. After incubation of living parasites in $7C_2C_5$ ascites fluid, oral precipitates were formed, confirming the possibility that these proteins were secreted orally from the stichocyte cells.

Table 1--Reactivity of selected monoclonal antibodies with antigens from
T. spiralis life cycle stages and other species of nematodes (_Ascaris suum_,
Strongyloides _ransomi_, _Stephanurus_ _dentatus_, and _Trichuris_ _suis_) in an ELISA.
See Gamble and Graham (12) for details of the assay. + = positive ELISA
result, - = negative ELISA result.

| | _T. spiralis_ | | | _A. suum_ | | | _S._ _ransomi_ | _S._ _dentatus_ | _T. suis_ |
Clone	larvae	adult	ES*	larvae	adult	egg	larvae	adult	adult
1A$_5$C$_3$	+	+	-	-	-	-	-	-	+
1D$_3$C$_3$	+	+	-	+	+	+	-	+	+
7A$_5$C$_1$	+	+	-	-	-	+	-	+	+
7A$_6$C$_1$	+	+	-	+	+	+	+	+	+
7B$_1$C$_3$	+	+	-	+	-	+	+	+	+
7B$_4$C$_5$	+	+	-	-	-	-	-	-	+
7C$_2$C$_5$	+	-	+	-	-	-	-	-	-
7C$_5$C$_2$	+	+	-	+	+	+	+	+	+

Reprinted from Gamble and Graham (12).
*ES = excretory secretory products.

AFFINITY ISOLATION OF Ts.53, Ts.49, and Ts.45

Ascites fluid produced in mice harboring hybridoma
$7C_2C_5$ was partially purified by precipitation in 50%
ammonium sulfate, and the immunoglobulin fraction was coupled to
CNBr-Sepharose CL-4B. The gel was equilibrated in 100 mM Tris
buffer (pH 8.0), and a preparation of crude ES products in the
same buffer was passed over the column. When the A_{280} of the
eluate had returned to baseline, a gradient from 100 mM Tris (pH
8.0) to 0.2 M glycine (pH 2.5) was applied and fractions were
collected (Figure 1). The specifically bound and eluted peak
contained proteins Ts. 53, Ts.49, and Ts.45 exclusively. Ts.45
was recovered in relatively minor amounts.

DIAGNOSIS OF SWINE TRICHINOSIS BY INDIRECT ELISA

Antigens recovered by monoclonal antibody affinity
chromatography were coated onto polystyrene microtiter plates
(Cooke) and used in a triple-antibody ELISA test (12). Briefly,
antigens Ts. 53, Ts.49, and Ts.45 (5 µg/ml) were coated
overnight at 4 C in carbonate-biocarbonate buffer (pH 9.6).
Plates were then washed with phosphate-buffered saline and 0.05%
Tween 20 (PBS-Tween), and 100 µl of swine serum (diluted
1:100) was added to each well. After incubation and washing,
wells were reacted with a rabbit anti-swine IgG reagent, washed
again, and a goat anti-rabbit IgG-horseradish peroxidase
conjugate was added. Bound swine antibodies specific for
T. spiralis were then visualized by the addition of 0.005%
hydrogen peroxide with 5'-aminosalycylic acid (0.8 mg/ml). The
results of the indirect ELISA comparing antigens Ts.53, Ts.49,
and Ts.45 to a crude antigen extract (CWE) are shown in
Figure 2. The monoclonal antibody affinity-isolated antigens
eliminated false-positive reactions and reliably detected all
Trichinella-positive pigs.

COMPETITIVE ELISA WITH MONOCLONAL ANTIBODIES

As an alternative to the indirect ELISA, a competitive
assay was developed which eliminated the need for the affinity
isolation of antigen (13). Briefly, microtiter plates were
coated with crude antigen as before. A mixture of swine serum
and biotin-labeled monoclonal antibody $7C_2C_5$ was then added
to each well. After incubation and washing, an avidin-
horseradish peroxidase conjugate was added, incubated, washed,
and substrate added as before. Reduction of the hydrolysis of
substrate as compared to controls indicated that anti-T. spiralis
swine antibodies were present. Results of this test are
presented in Figure 3. An excellent separation was obtained
between groups of T. spiralis-positive and T. spiralis-negative

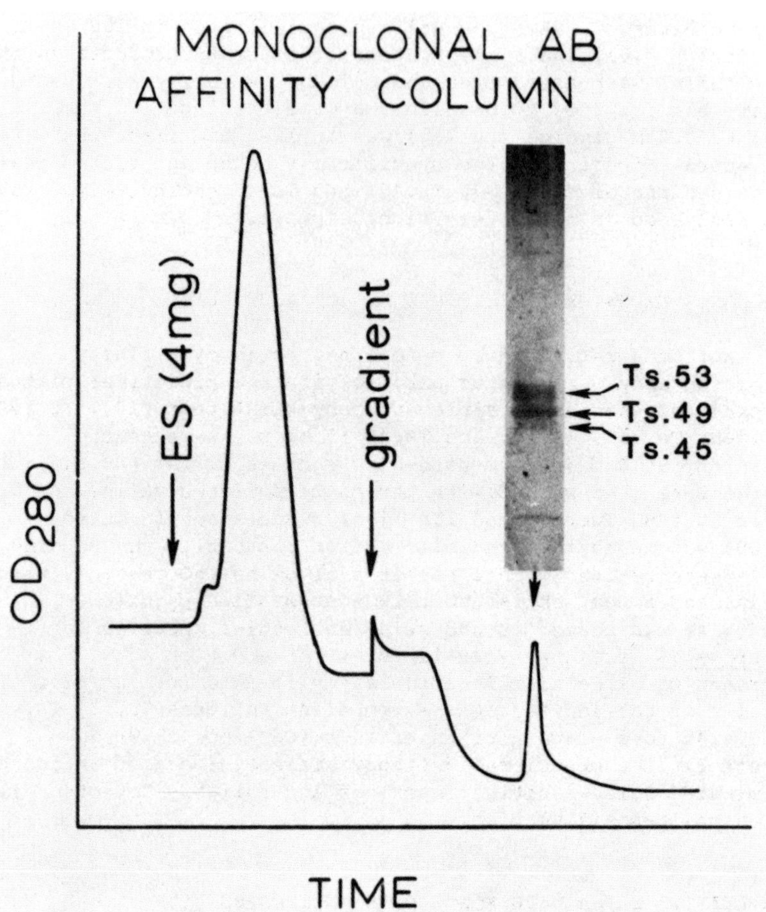

Figure 1--Affinity isolation of <u>Trichinella</u> <u>spiralis</u> proteins
Ts.53, Ts.49, and Ts.45 using monoclonal antibody $7C_2C_5$.
Antibody partially purified by precipitation in 50% ammonium
sulfate was coupled to CNBr-Sepharose CL-4B. A crude
preparation of <u>T. spiralis</u> excretory-secretory products (4 mg)
in 100 mM Tris buffer (pH 8.0) was passed over the column.
When the A_{280} of the eluate had returned to baseline a
gradient of 100 mM Tris buffer (pH 8.0) to 0.2 M glycine
(pH 2.5) was applied to the column and fractions collected.

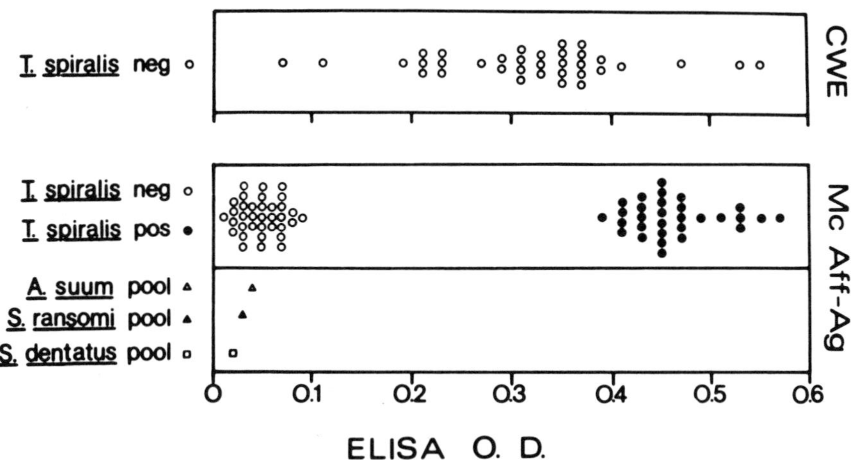

Figure 2--Enzyme-linked immunosorbent assay for the detection of swine antibodies directed against Trichinella spiralis infection. Antigens were crude worm extract (CWE) or excretory-secretory antigens Ts.53, Ts.49, and Ts.45 (McAff-Ag) isolated by affinity chromatography using hybridoma $7C_2C_5$. Swine sera included 35 sera from farm-raised slaughter hogs - o , 30 sera from hogs inoculated with 10,000 T. spiralis muscle larvae and bled after 6 months - • , and pooled samples from animals experimentally infected with Ascaris suum - ▵ , Strongyloides ransomi - ▴ , or Stephanurus dentatus - ▫ . See Gamble and Graham (12) for details. Reprinted from Gamble (1).

serum samples. In addition, no reactivity (inhibition) was obtained with serum from pigs harboring infections of other swine nematodes.

SUMMARY

     With the use of monoclonal antibodies as probes, an antigen epitope unique to the parasite Trichinella spiralis has now been identified. This epitope occurs on three proteins (Ts.53, Ts.49, and Ts.45) found in the parasite's stichocyte cells. Two ELISA tests, an indirect and a competitive assay, based on the recognition of this unique epitope have yielded promising results with regard to specificity and sensitivity for the detection of Trichinella-infected pigs. Studies are currently being conducted to evaluate these diagnostic tests in populations of naturally infected swine.

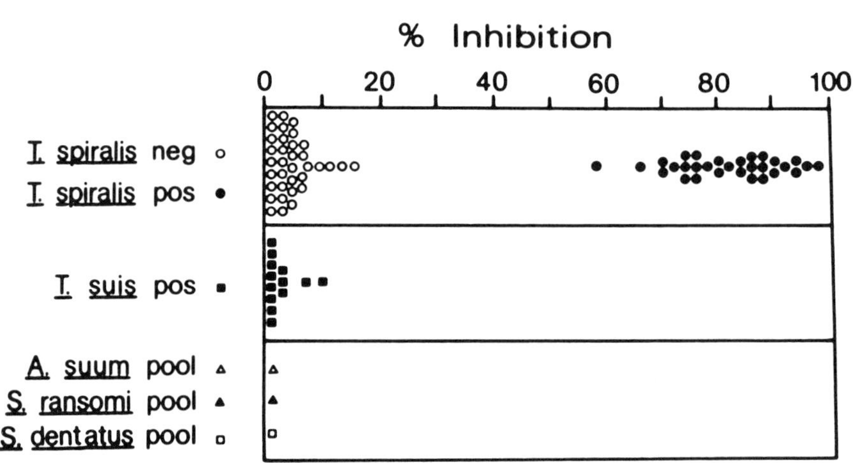

Figure 3--Competitive enzyme-linked immunosorbent assay for the detection of swine antibodies directed against Trichinella spiralis infection. Antigen was a crude excretory-secretory preparation. Swine sera included 35 sera from farm-raised slaughter hogs -O , 30 sera from hogs inoculated with 10,000 T. spiralis muscle larvae and bled after 6 months - ● , 13 sera from hogs with natural infections of Trichuris suis - ■ , and pooled samples from animals experimentally infected with Ascaris suum - △ , Strongyloides ransomi - ▲ , or Stephanurus dentatus - □ . See Gamble and Graham (13) for details. Reprinted from Gamble and Graham (13).

REFERENCES

1. Gamble, H.R. (1984) Vet. Parasitol. (in press).
2. Ruitenberg, E.J., van Knapen, F., and Elgersma, A. (1983)
    In: Trichinella and trichinellosis. Campbell, W.C. (ed.)
    Plenum Press, New York, pp. 529-550.
3. Leighty, J.C. (1983) Food Technol. 37:95-97.
4. Ruitenberg, E.J., van Knapen, F., and Elgersma, A. (1983)
    Food Technol. 37:98-100.
5. Clinard, E.H. (1978) In: Trichinellosis, 4th Int. Conf.
    Kim, C.W., and Pawlowski, Z.S. (eds.) Reedbooks, London,
    pp. 509-517.
6. Ruitenberg, R.J., Lungstrom, I., Steerenberg, P.A., and
    Buys, J. (1976) J. Immunol. Meth. 10:67-83.
7. Van Knapen, F., Franchimont, J.H., Ruitenberg, E.J.,
    Andre, P., Baldelli, B., Gibson, T.E., Gottal, C.,
    Henriksen, S.A., Kohler, G., Roneus, O., Skovgaard, N.,
    Soule, C., Strickland, K.L., and Taylor, S.M. (1981) Vet.
    Parasitol. 9:117-123.
8. Kennett, R.H. (1980) In: Hybridomas: Monoclonal
    Antibodies--A new dimension in biological analysis.
    Kennett, R.H., McKearn, T.J., and Bechtol, K.B. (eds.)
    Plenum Press, New York, pp. 365-367.
9. Gamble, H.R. (1984) In: Hybridoma Technology in
    Agriculture and Veterinary Research. Stern, N.J., and
    Gamble, H.R. (eds.) Rowland and Allenheld, Totowa, New
    Jersey.
10. Gamble, H.R., Anderson, W.R., Graham, C.E., and Murrell,
    K.D. (1983) Vet. Parasitol. 13:349-361.
11. Towbin, H., Staehelin, T., and Gordon, J. (1979) Proc.
    Natl. Acad. Sci. USA 76:4350-4354.
12. Gamble, H.R., and Graham, C.E. (1984) Am. J. Vet. Res.
    45:67-73.
13. Gamble, H.R., and Graham, C.E. (1984) Vet. Immunol.
    Immunopathol. (in press).

STRATEGIES FOR USING MONOCLONAL ANTIBODIES IN
IDENTIFYING IMMUNOGENIC ANTIGENS OF ANIMAL PARASITES

GEORGE CAIN
Department of Zoology
University of Iowa
Iowa City, Iowa

Recent research with several animal and human diseases
caused by animal parasites has suggested that it may be possible
to develop vaccines for many of these diseases. Monoclonal
antibodies have played an important role in this work by helping
to identify antigens that have been and could be utilized in
vaccine trials. Presented here is an overview of how monoclonal
antibodies have been used to identify candidate antigens,
together with a brief consideration of how these antibodies
might contribute further to our understanding of the biology of
the host-parasite relationship.

In general, the strategies used for identifying potential
immunogenic antigens from animal parasites have made use of one
to several of the following approaches:

1) Monoclonal antibodies made against parasites have been
correlated with sera from functionally immune hosts in
immunoprecipitation studies or in a combination of Western blot
and enzyme-linked immunosorbent assay (ELISA); common antigens
are presumed to be immunogenic.

2) Monoclonal antibodies have been screened in
cytotoxicity assays using eosinophils or other elements of the
cellular immune system known to be effective against the
parasite.

3) Monoclonal antibodies have been shown to confer
protection upon passive transfer to an appropriate host.

4) In some systems it is possible to neutralize infection
by prior incubation of the parasites _in vitro_ with monoclonal
antibodies.

5) Monoclonal antibodies have been used in immunoaffinity
chromatography to isolate candidate antigens for immunization
trials.

MALARIA

The earliest example of the use of monoclonal antibodies in the study of parasitic disease is the landmark research of Ruth Nussenzweig and her colleagues at New York University (1). Their initial success was in making a monoclonal antibody to a 44,000 Da surface protein found on the sporozoite, the invasive stage that is injected from the mosquito in the rodent malaria, Plasmodium berghei. This monoclonal antibody was able to neutralize sporozoite infectivity after incubation in vitro. It has become clear from these and other studies from Nussenzweig's laboratory that the circumsporozoite protein, or CSP, is unique to the sporozoite stage and probably represents a general feature of the sporozoites of all species of Plasmodium. Studies on the mechanism of synthesis of CSP, together with studies using monoclonal antibodies to sporozoites of other species, have shown that although these proteins may differ somewhat in detail, they are the same in principle. Thus, P. falciparum contains a 67,000- and a 58,000-Da CSP and P. vivax, a 51,000- and a 45,000-Da protein. By analogy with earlier studies on P. berghei, the larger protein represents a precursor that is processed before appearing on the surface. Serum from human volunteers given sporozoites precipitated these four proteins; furthermore, sporozoites incubated in monoclonal antibodies that immunoprecipitated these proteins, failed to infect chimpanzees (2).

Although Nussenzweig has not yet used monoclonals to isolate CSP for immunoprotection trials, her colleagues at NYU (3) have isolated messenger RNA from P. knowlesi-infected mosquitoes and produced complementary DNA to portions of the CSP gene. Using Nussenzweig's monoclonal antibodies, they were able to show that nucleotide sequences of the gene fragments that expressed this protein specified a rather remarkable array of twelve repeating sequences of twelve amino acids each. These repeating sequences made up about a third of the total CSP. A dodecapeptide containing the repeating sequence was synthesized and shown to bind anti-CSP monoclonal antibodies and to compete with these same monoclonal antibodies in a radioimmunoassay. Interestingly, there is perfect conservation of these 12 amino acids in each repeating unit. Even in other proteins with repeating units (e.g., collagen), there can be conservative substitutions. Godson and his coworkers suggest that, since the dodecapeptide contains hydrophobic amino acids regularly interspersed with hydrophilic amino acids, this structure may provide multiple sites for association with the sporozoite membrane, together with numerous exposed sites that may participate in binding the target cell.

Although possibilities for a vaccine from this protein are obvious, we should realize that if just one sporozoite manages to infect a host liver cell, the infection would not be neutralized because the merozoite, the stage that invades red

blood cells, lacks the CSP. For this reason attention has also
been given to vaccines against the merozoite stage, and here
again monoclonal antibodies have played an important role.
Acting on information provided originally by Cohen and coworkers
in England that merozoite preparations could produce immunity,
Freeman et al. (4) reported that monoclonal antibodies they
developed to bloodstream forms of P. yoelii were specific to
merozoites and protective in passive transfer experiments.
Later, Holder and Freeman (5) showed that these antibodies could
be used to purify a 235,000-Da protein associated with the
merozoite surface and a 230,000-Da protein associated with
schizonts, the forms that rupture to release merozoites. Both
of these purified proteins were successfully used to immunize
mice.

TRYPANOSOMIASIS

     Among the trypanosomes, the African bloodstream forms
provide a formidable challenge to any sort of induced immunity
because of the phenomenon of antigenic variation. Nonetheless
there are points of attack at the beginning and end of the
mammalian infection. Esser at Walter Reed has used monoclonal
antibodies made to metacyclic trypanosomes (the infective stage
that is transmitted by the tsetse fly vector) to show that there
is a limited repertoire of surface antigens, perhaps 15 or 20
antigenic types (personal communication). More monoclonal
antibody studies of this type are needed to determine the extent
of this repertoire; if it is fairly limited, there might be
possibilities for developing a vaccine. Monoclonal antibodies
are doubly valuable in these studies because they are perhaps
the only way to study the antigenic repertoire of the tsetse fly
forms, which are very difficult to obtain in the quantities
necessary for traditional biochemical and immunological studies.
     Recombinant DNA studies on metacyclic forms, now in
progress in Donelson's laboratory at the University of Iowa,
have produced complementary DNA clones from one of the
metacyclic antigen types. By sequencing around the ends of this
and other metacyclic surface antigen genes, it is hoped that
information can be gained about the mechanism of expression of
the surface protein and/or the means by which these genes are
transposed from one portion of the chromosome to another. This
phenomenon of gene transposition may be under the control of
enzymes and hence may be susceptible to a vaccine.
     Pearson (loc. cit.) described experiments with monoclonal
antibodies to the procyclic forms that develop soon after the
bloodstream trypomastigote is ingested by the insect. The
possibility that these antibodies may inhibit development of the
parasite in the insect is an exciting approach to the control of
this disease.

Another area of trypanosomiasis research that has received too little attention is the mechanism by which the surface antigen is attached to the cell membrane. _Trypanosoma_ _brucei_, for example, is instantly lysed by human high-density lipoprotein, as shown by Rifkin (6). The mechanism of this lysis is not understood, but it is intriguing to think of the possibility of a genetically engineered, "universal" lipoprotein that could interfere with the binding of the surface antigen and/or promote lysis in the same way as human high-density lipoprotein.

LEISHMANIASIS

The diseases caused by the several species of _Leishmania_ constitute an important veterinary concern because domestic and wild animals often serve as reservoir hosts. _Leishmania_ does not show antigenic variation like its relatives that inhabit the bloodstream; instead, it presents an extensive and complex array of species and strains and serotypes that evidently possess subtle differences in their surface proteins. In a recent study, Jaffe and McMahon-Pratt (7) used monoclonal antibodies to help sort out the complexities in the serotypes and isozyme types of _Leishmania_.

Unlike the African trypanosomes, _Leishmania_ can be cultured _in vitro_ without loss of infectivity. Culture forms, in fact, have formed the basis of a crude vaccine that has been in use in the Mediterranean area and parts of the Soviet Union for nearly 40 years. Recently, Anderson from John David's laboratory at Harvard University has obtained a monoclonal antibody that protects against infection by promastigotes (8).

GIARDIASIS

In addition to the poultry coccidia described by Danforth (loc. cit.), one other protozoan system may be yielding its secrets to analysis with monoclonal antibodies. Myer et al. at George Washington University and the National Institutes of Health have developed monoclonal antibodies to _Giardia_, the water-borne intestinal parasite of humans (9). The antibody, produced to whole cells, is specific for a surface antigen or antigens, and produces a four-fold increase in antibody-dependent cytotoxicity by macrophages.

SCHISTOSOMIASIS

The human blood fluke, or schistosome, is a master of antigenic disguise and a baffling subject for the monoclonal antibody technologist. These parasites are able to survive in

the host for many years, and the host is resistant to
reinfection as long as the adult worms are present. This
phenomenon, which has been called concomitant immunity, appears
to depend on the adult worm's ability to evade the host immune
response in rather specific ways. For example, schistosomes
change the nature of their surface antigens when they invade the
host as cercariae and transform into a migratory stage, the
schistosomule, which moves to the lungs, the liver, and
eventually to the intestinal blood vessels. Monoclonal
antibodies, together with biochemical studies, have shown that
there are different antigens on the surface of cercariae and
adults, but also that all stages share some common antigens.
The significance of these antigens in protective immunity is
unknown.

The second defense mechanism of the schistosome is the
existence of an additional outer bilayer on its syncytial
surface membrane. There is currently a great deal of interest
in differentiating the antigenic structure of the two bilayers,
but very little success has been achieved in this area.

Schistosomes have also been shown to acquire host antigens,
including histocompatibility antigens (10), blood group antigens
(11), and surface receptors for non-specific host
immunoglobulins (12). There have been reports of anticoagulant
factors (13) associated with the worm surface, and in our
laboratory (Robertson and Cain, unpublished) we have found
heparin-like polysaccharides that may be associated with the
integument. Monoclonal antibodies could contribute more to our
understanding of these mechanisms of immune evasion and
antigenic disguise.

Immunity to schistosomes appears to operate at the level of
the early schistosomule. At this time the parasite is
susceptible to antibody-dependent cytotoxicity mechanisms that
are not well understood. This susceptibility lasts for about 3
days, after which the schistosomule becomes refractory to immune
mechanisms. The best results in attempting to induce immunity
to schistosomes have come through experiments using attenuated
or irradiated schistosomules, as done by Hsu and Hsu (personal
communication) with _S. japonicum_ in cattle. For this reason,
most of the attempts to develop protective monoclonal antibodies
to schistosomes have used irradiated schistosomules as the
antigen.

Smith and coworkers in England, among the first to use
monoclonal antibodies against schistosomules, assessed
protection by complement-dependent cytotoxicity (14). Although
modest protection was achieved, there was no additional
information about the antigens involved. Zodda, working in
Phillips' laboratory at the University of Pennsylvania, produced
monoclonal antibodies that, when passively transferred to mice,
achieved 40 to 55% protection. The active agent appeared to be
a 135,000-Da protein, as shown by immunoprecipitation (15).
Gryzch (16) and Dissous (17), from Capron's laboratory in

France, have assayed the protective effect of their monoclonal
antibodies by eosinophil-dependent cytotoxicity and passive
transfer. Passive transfer experiments achieved up to 60%
protection; the active agent was a 38,000-Da protein that was
also detected by immunoprecipitation studies with immune serum
and was purified by immunoaffinity chromatography.

Another approach to the production of monoclonal antibodies
in schistosomes utilizes as the immunizing antigen some
preparation other than the schistosomule. Strand has used
glycoproteins purified by lectin-affinity chromatography of
total proteins from the cercarial stage. Three different
antigens from this mixture were subsequently precipitated with
immune human serum. One 35,000-Da protein was common to both
sexes, but the other two differed between males and females
(18). Two other proteins (120,000 and 175,000 Da) that Strand
and her colleagues have precipitated with the monoclonal
antibodies were shown by immunofluorescence to occur in spines
on the surface of schistosomes (19). Neither, however, was
actin, which is probably the major component of the schistosome
spine. Finally, this approach has yielded a 180,000-Da protein
that is common to all stages of the life cycle and is
immunoprecipitated with human serum (20).

Brink (personal communication) has isolated surface
material from the adult schistosome using the detergent NP-40.
Injection of this material into mice produced protection at a
level of about 60%. When immune serum from these mice was used
in a Western blot assay, at least five proteins were
detectable. Using a preparation isolated with a different
detergent, Triton X-100, we obtained close to the same amount of
protection (Cain and Oaks, unpublished data). By Western blot
analysis with immune mouse serum, we identified four proteins
(92 - 94,000, 76,000, 47,000 and 35 - 36,000 Da) that
corresponded to proteins found by Brink using similar
procedures. Moreover, three monoclonal antibodies to this
antigenic preparation were specific for 132,000-, 110,000-,
98,000-, 58,000-, 41,000- and 36,000-Da molecules on Western
blot analysis.

So far, the nature of protective antigens of schistosomes
has been sufficiently complex to make any conclusions tenuous at
best. In an effort to bring some order to this confusion, the
Edna McConnell Clark Foundation recently held a "wet" workshop
in which some 20 different labs sent or brought 93 monoclonal
antibodies, 54 semi-defined antigen preparations, and 23
characterized antisera for analysis. The attendees then used
various techniques for defining the antigens, antisera, and
monoclonals. The results of this heroic effort are too detailed
to account here, but they are summarized in the Journal of
Parasitology (21). One observation of interest from this
workshop was that partially protective monoclonal antibodies
from many laboratories all seemed to react to epitopes of
proteins of varying molecular weight, i.e., when Western blots

were done with a given monoclonal antibody, a large number
(usually > 6) of different antigens were localized. This
could reflect proteolytic degradation, or perhaps some other
type of antigen processing. Whatever the meaning of this
phenomenon, it certainly adds to the complexity of this research
area.

OTHER PARASITE INFESTATIONS

    Work with monoclonal antibodies to other helminth parasites
is generally limited but recent studies on _Fasciola_ _hepatica_
(22) produced six monoclonal antibodies that appeared to bind to
the same epitope or perhaps different epitopes on the same
50,000-Da polypeptide. This polypeptide correlated with one
immunoprecipitated by serum from hyperinfected hosts.
Morphological techniques were used to localize this antigen in
the outer tegument of the worm.
    Although higher in phylogenetic position, nematodes appear
to be last in line for monoclonal antibody studies. With the
exception of the studies of Gamble (loc. cit.), investigations
have been limited to the work of Despommier and Silberstein
(23), who have used monoclonal antibodies--also to _Trichinella_--
and immunoaffinity columns to purify three antigens of 48,000,
50 - 55,000, and 37,000 Da. The first two protected mice at
dose levels down to 0.1 µg. These same antibodies, used in
immunofluorescence studies, indicated that the antigens involved
were located in the stichosome, a collection of gland cells that
surround the anterior end of the digestive tract.
    Other nematodes, although not yet subjected to monoclonal
antibody studies, appear to produce immunoprotective proteins.
Jenkins and Wakelin (24) have used traditional techniques to
isolate a 30,000-Da protein which reduces the worm burdens of
Trichuris by 80 to 90%. Similar work by Stromberg on _Ascaris_
(25) and Rothwell and Merritt (26) on _Trichostrongylus_
colubriformis also suggest that protective antigens exist in
these species.
    Finally, dog heartworm, an important concern of veterinary
medicine over the years, may finally become amenable to some
kind of immunological attack. It has been recently reported by
Levy and coworkers (27) that sera from dogs immunized to
Dirofilaria will recognize four low-molecular-weight proteins by
Western blot analysis.
    The postscript to this rapid tour through the world of
protozoa and worms should be that the _ideal_ way to use
monoclonal antibodies to help identify immunoprotective antigens
is to first understand the mechanism of immunity and then
proceed in a rational way. But given human nature, and the
excitement of this new technology, the temptation to bypass the
rational approach has proven irresistible to many
investigators. Their results tend to substantiate the "law of

fortuitous findings"--that even a blind pig will pick up an acorn now and then! We should be encouraged, therefore, to use monoclonal antibodies even though we may not always have the best rational basis for doing so.

REFERENCES

1. Yoshida, N., Nussenzweig, R.S., Potocnjak, P., Nussenzweig, V., and Aikawa, M. (1980) Science 207:71-73.
2. Nardin, E.H., Zussenzweig, V., Nussenzweig, R.S., Collins, W.E., Tranakchit, K., Harinasuta, K., Tapchaisri, P., and Chomcharn, Y. (1982) J. Exp. Med. 156:20-30.
3. Godson, G.N., Ellis, J., Svec, P., Schlesinger, D.H., Nussenzweig, V. (1983) Nature 305:29-33.
4. Freeman, R.R., Trejdosewicz, A.J., and Cross, G.A.M. (1980) Nature 284:366-368.
5. Holder, A.A., and Freeman, R.R. (1981) Nature 294:361-364.
6. Rifkin, M.R. (1978) Proc. Natl. Acad. Sci. USA 75:3450-3454.
7. Jaffe, C.L., and McMahon-Pratt, D. (1983) J. Immunol. 131:1987-1993.
8. Anderson, S., David, J.R., and McMahon-Pratt, D. (1983) J. Immunol. 131:1616-1618.
9. Meier, T.J., Taylor, D.W., Keister, D.B., and Smith, P.D. (1983) Program, Am. Soc. Trop. Med. Hyg. p. 6.
10. Sher, A., Hall, B.F., and Vadas, M.A. (1978) J. Exp. Med. 148:46-57.
11. Goldring, O.L., Clegg, J.A., Smithers, S.R., and Terry, R.J. (1976) Clin. Exp. Immunol. 26:181-187.
12. Kemp, W.M., Brown, P.R., Merritt, S.C., and Miller, R.E. (1980) J. Immunol. 124:806-811.
13. Tsang, V.C.W., and Damian, R.T. (1977) Blood 49:619-633.
14. Smith, M.A., Clegg, J.A., Snary, D., and Trejdosiewicz, A.J. (1982) Parasitology 84:83-91.
15. Zodda, D.M., and Phillips, S.M. (1982) J. Immunol. 129:2326-2328.
16. Gryzch, J.-M., Capron, M., Bazin, H., and Capron, A. (1982) J. Immunol. 129:2739-2743.
17. Dissous, C., Gryzch, J.-M., and Capron, A. (1982) J. Immunol. 129:2232-2234.
18. Strand, M., McMillan, A., and Pan, X. (1982) Exp. Parasitol. 54:145-156.
19. Norden, A.P., Aronstein, W.S., and Strand, M. (1982) Exp. Parasitol. 54:432-442.
20. Aronstein, W.S., Norden, A.P., and Strand, M. (1983) Am. J. Trop. Med. Hyg. 32:334-342.
21. Colley, D.G. (1983) J. Parasitol. 69:45-48.
22. Hanna, R.E.B., and Trudgett, A.G. (1983) Parasite Immunol. 5:409-425.
23. Despommier, D., and Silberstein, D., 1983 meeting, Am. Soc. Parasitol.

24.   Jenkins, S.N., and Wakelin, B.E. (1983) Parasitology
      86:73-82.
25.   Stromberg, B.E. (1979) Int. J. Parasitol. 9:307-312.
26.   Rothwell, T.L. and Merritt, G.C. (1975) Int. J. Parasitol.
      5:453-460.
27.   Boto, W.O., Powers, K.G., Levy, D.A. 1983 meeting, Am. Soc.
      Trop. Med. Hyg.

UTILIZATION OF HYBRIDOMA ANTIBODIES IN THE STUDY OF AVIAN
COCCIDIAL ANTIGEN

HARRY D. DANFORTH
Protozoan Diseases Laboratory
Animal Parasitology Institute, Agricultural Research Service
U.S. Department of Agriculture, Beltsville, MD   20705

Hybridoma cell lines that secrete antibodies directed
against the sporozoite and merozoite stages of various avian
coccidia, protozoan parasites of the intestinal epithelium, were
produced by use of a modified fusion technique described by
Kennett et al. (1). These antibodies demonstrated different
binding patterns on air-dried sporozoites and merozoites, as
determined by the indirect immunofluorescent antibody (IFA)
test. These patterns varied from a general fluorescence of the
internal surface (similar to that seen when either developmental
stage was exposed to hyperimmune chicken serum) to fluorescence
seen on the tip, pellicle, and, in the case of the sporozoites,
the refractile body of the parasites (2). The antibodies that
produce these different patterns had different subclasses (IgM,
IgG1, IgG2a, IgG2b, and IgG3), indicating that the mouse had
responded to different antigenic determinants either in or on
the parasites.
     At present, more than 50 cloned antibody-secreting cell
lines have been produced against the various coccidial species,
and some of these have demonstrated either species specificity
or varying degrees of crossreactivity (3). The species-specific
monoclonal antibodies can be used in both the IFA test and
enzyme-linked immunosorbent assay (ELISA) to diagnose the
prevalent coccidial species in litter samples taken from broiler
chicken houses in large poultry producing areas. The
crossreactive monoclonal antibodies have shown IFA patterns that
are similar to the species-specific antibodies, so that no
correlation has been seen between a specific type of IFA pattern
and specificity or crossreactivity. A few of the crossreactive
hybridoma antibodies produced one type of IFA pattern with the
sporozoites that they were originally raised against and
different patterns with sporozoites of other coccidial species.
The antigenic determinants recognized by these antibodies are
evidently located in different areas of the parasites. Whether

this variation in IFA pattern with crossreactivity indicates the presence of homologous molecules in or on the different sporozoite species has not been determined yet.

In _in vitro_ studies, certain monoclonal antibodies raised against the sporozoite stage of _Eimeria tenella_ have been shown to inhibit the penetration and intracellular development of the parasite (4). Other antibodies were found to have no effect on the sporozoites. None of the antibodies studied completely inhibited cellular penetration by the sporozoites, but some were found to almost completely block development provided that the antibodies were continuously present in the cell culture medium.

One antibody, which was the most efficient in inhibiting penetration and development of the sporozoites, was further studied by transmission electron microscopy and labeling with a ferritin conjugate to determine the site of antibody-antigen interaction. Antibody-treated sporozoites showed a uniform layer of ferritin on the surface pellicle of the parasite. It is possible that such a surface antigen-antibody interaction may play a major factor in blocking invasion of host cells. Correlation of surface antigen-antibody interaction seen with several monoclonal antibodies with inhibition of penetration and development is now being attempted.

Initial studies have demonstrated that it is possible to isolate or identify sporozoite antigen present in gel filtration fractions. Efforts are now currently under way to utilize the monoclonal antibodies for antigen isolation and characterization. Because little is known about the antigen or antigens that may be involved in the immune response of birds against coccidial infection, the monoclonal antibodies have proven to be a powerful tool in understanding which antigens may be important. It is now possible to reduce the number of monoclonals needed for antigen characterization and allow for a concentrated effort in identifying the candidate antigens necessary producing an immune response in poultry.

REFERENCES

1.   Kennett, R.H., Denis, K.A., Tung, A.S., and Klinman,
     N.R. (1978) In: Current Topics in Microbiology and
     Immunology. Melchers, R., Potter, M., and Warner, N.L.
     (eds.) Springer-Verlag, New York.
2.   Danforth, H.D. (1982) J. Parasitol. 68:392-397.
3.   Danforth, H.D., and Augustine, P.C. (1983) Poult. Sci.
     62:2145-2151.
4.   Danforth, H.D. (1983) Am. J. Vet. Res. 44(9):1722-1727.

APPLICATIONS OF RECOMBINANT DNA TECHNOLOGY IN MODERN VACCINE
DEVELOPMENT

PETER T. LOMEDICO
Department of Molecular Genetics
Hoffmann-La Roche, Inc., Nutley, NJ   07110

As a first step in a subunit vaccine research program, it
is often necessary to clone the genes coding for the "protective
antigens" in order to study the detailed structural and
immunological properties of these proteins. I will discuss
three basic strategies used to clone genes from higher
organisms: a) If the desired protein can be isolated (e.g., by
antibody affinity methods) and sequenced, synthetic
oligonucleotide hybridization probes can be used to identify the
corresponding gene (1). b) Immunological screening of clone
banks (2). c) If the desired protein can be identified (e.g.,
immunologically) after mRNA translation, "positive selection"
methods can be used to screen for the cloned genes (3).

REFERENCES

1. Suggs, S.V., Wallace, R.B., Hirose, T., Kawashima, E.H., and
    Itakura, K. (1981) Proc. Natl. Acad. Sci. USA 78:6613-6617.
2. Helfman, D.M., Feramisco, J.R., Fiddes, J.C., Thomas, G.P.,
    and Hughes, S.H. (1983) Proc. Natl. Acad. Sci. USA 80:31-35.
3. Parnes, J.R., Velan, B., Felsenfeld, A., Ramanathan, L.,
    Ferrini, U., Appella, E., and Seidman, J.G. (1981) Proc.
    Natl. Acad. Sci. USA 78:2253-2257.

CHAPTER 15

CHARACTERIZATION OF IMMUNOLOGICAL RESPONSES

MONOCLONAL ANTIBODIES IN THE ANALYSIS OF IMMUNE RESPONSES IN SWINE

J.K. LUNNEY

Animal Parasitology Institute, Helminthic Diseases Laboratory
Agricultural Research Service, U.S. Department of Agriculture
Beltsville, Maryland 20705

The examination of cellular interactions involved in immune responses has been greatly aided by the availability of monoclonal antibodies that recognize unique surface antigens of cell subpopulations. In addition, monoclonal antibodies that distinguish immunologically and/or enzymatically active epitopes of proteins facilitate the dissection of the molecular components involved in these events. Panels of such monoclonal antibodies are readily available for work in humans, mice, and rats, but are only beginning to be developed for other species. Davis and his colleagues have prepared a panel of monoclonal antibodies that recognize interspecies cross-reactive determinants of cell surface molecules. This was accomplished by immunizing mice with cells from multiple species and selecting for binding activity on cells from still another species (1). We, on the other hand, have taken a more traditional approach, immunizing and testing for reactivity only on swine lymphocyte subpopulations. The results of our research are presented in an abstract in this volume (2) and in more detail elsewhere (3).

In this paper, I will present data on the identification of immune response genes in inbred miniature swine and discuss the potential usefulness of the monoclonal antibodies to swine lymphocyte subpopulations in the analysis of the cells involved in the _in vitro_ generation of these genetically controlled immune responses to antigens. I will also present data on the utilization of monoclonal antibodies that recognize interspecies cross-reacting determinants of cell surface molecules for the purification and characterization of porcine immune response associated (Ia) antigens.

EXPRESSION OF IMMUNE RESPONSE GENES IN INBRED SWINE

It has been known for many years that the ability of an individual to mount an antibody response to certain protein antigens is under genetic control that can be mapped to the immune response genes of the major histocompatibility complex (MHC) (4). In outbred swine, the antibody response to the protein hen egg lysozyme was shown to be under immune response gene control but the response to seven other antigens showed no such genetic control (5). In family studies these genes were shown to map to the swine major histocompatibility complex (MHC), termed the swine lymphocyte antigen (SLA) complex (5). Recently, we extended these studies in MHC inbred miniature swine of three independent SLA haplotypes, designated a, c, and d. Animals of the recombinant MHC haplotype g, which combines the SLA class I (SLA-A,B) genes of the c haplotype with the SLA class II (SLA-D) genes, were immunized with different antigens and analyzed for specific antibody response in an ELISA assay. The animals all made an equivalently high titered antibody response to sperm whale myoglobin but exhibited differential antibody responses to the two other antigens, lysozyme and (T,G)-A--L (7). dd and gg animals were high responders to lysozyme whereas cc animals were low responders, mapping this high antibody response to SLA-D^d region genes. For (T,G)-A--L the reverse was found. cc and cd animals were high responders whereas dd and gd animals were low responders, showing that the SLA-D^c genes were necessary for an animal to mount a high titered antibody response to (T,G)-A--L. Therefore, the ability of miniature swine to mount an antibody response to lysozyme or to (T,G)-A--L is under immune response gene control and, in agreement with results in other species (4), maps the genes controlling these responses to the D region of the MHC.

Analysis of the ability of peripheral blood lymphocytes (PBL) to be stimulated by these same antigens in secondary in vitro cellular responses has demonstrated the same immune response gene control, i.e., dd and gg PBL were stimulated in culture by lysozyme but not by (T,G)-A--L, whereas cc PBL were stimulated by (T,G)-A--L but not by lysozyme. Both cell types responded equivalently to third party antigens such as the purified protein derivative from the tuberculin bacilli in complete Freund's adjuvant (7). These results now set the stage for immunologic dissection of the cellular interactions mediating antigenic responses in swine. Our panel of monoclonal antibodies, should be useful in the analysis of the effects of T cells on these responses and particularly of the importance of helper T cells in these interactions. The macrophage-specific monoclonal antibodies should indicate whether these cells, or certain restricted subsets of macrophages, are necessary for in vitro stimulation of immune responses. Moreover, because the class II MHC, or Ia, antigens are known to be involved in these antigen-specific responses, various anti-Ia monoclonal

antibodies should be useful in indicating which epitope of the
Ia, or SLA-D, antigen is involved in antigen presentation.

USE OF MONOCLONAL ANTIBODY AFFINITY COLUMNS FOR Ia ANTIGEN
PURIFICATION

Because of the importance of the Ia, or SLA-D, antigens in
determining the ability of individuals to mount an immune
response to specific proteins, we became interested in purifying
these molecules for more detailed biochemical and structural
analyses. For this purification, monoclonal mouse anti-I-E
antibody affinity columns have been used. Earlier studies had
shown that antisera to a subset of mouse Ia antigens, the I-E
antigens, exhibited cross-reactions with Ia antigens of other
species (8). Using a panel of anti-I-E monoclonal antibodies,
we found that monoclonal antibodies to two of three epitopes of
the I-E molecule were cross-reactive with a subset of swine Ia
antigens indicating the evolutionary conservation of parts of
these molecules (9). One of these cross-reactive monoclonal
antibodies, 40D, was used to prepare an affinity column on which
detergent extracts of swine lymphoid plasma membranes were
applied. After washing the column, the Ia antigens were eluted
with sodium acetate buffer, pH 4.5, containing detergent. The
dialyzed column eluate contained up to 70% of the original Ia
activity, as measured by a cell ELISA assay (10). Sodium
dodecyl sulfate gel electrophoretic analysis revealed that the
eluted antigens contained few protein contaminants. This method
resulted in the purification of milligram quantities of swine Ia
antigens, which should be very useful for future biochemical and
x-ray crystallographic studies. A potential biological use of
these proteins is the analysis of the molecular basis of immune
response genes. With purified Ia antigens, we can examine
whether low responder macrophages can be transformed to the high
responder phenotype by the introduction into the macrophage cell
surface of Ia antigens purified from high responder membranes.

SUMMARY

Monoclonal antibodies are excellent tools for dissection of
the cellular interactions involved in immune responses. In
addition, they provide a reproducible source of antibody with
which to purify large amounts of specific cell surface antigens
for molecular analyses of their biological and chemical
properties.

REFERENCES

1. Davis, W.C., Perryman, L.E., McGuire, T.C. (1984) In:
    Hybridoma Technology in Agriculture and Veterinary
    Research. Stern, N.J., and Gamble, H.R. (eds.) Rowman and
    Allenheld, Totowa, New Jersey.
2. Lunney, J.K., Pescovitz, M.D., and Sachs, D.H. (1984) In:
    Hybridoma Technology in Agriculture and Veterinary
    Research. Stern, N.J., and Gamble, H.R. (eds.) Rowman and
    Allenheld, Totowa, New Jersey.
3. Pescovitz, M.D., Lunney, J.K., and Sachs, D.H. (1984)
    Unpublished data.
4. Moller, G. (ed.) (1976) Transplant. Rev. 30.
5. Vaiman, M., Metzger, J.-J., Renard, C., and Vila, J.-P.
    (1978) Immunogenetics 7:231.
6. Pennington, L.R., Lunney, J.K., and Sachs, D.H. (1981)
    Transplantation 31:66.
7. VanderPutten, D., Pescovitz, M.D., and Lunney, J.K. (1984)
    Unpublished data.
8. Shinohara, N., and Sachs, D.H. (1982) In: Ia Antigens,
    Vol. I, Mice. Ferrone, S., and David, C.S. (eds.) CRC
    Press, Boca Raton, Florida, p. 219.
9. Lunney, J.K., Osborne, B.A., Sharrow, S.O., Devaux, C.,
    Pierres, M., and Sachs, D.H. (1983). J. Immunol. 130:2786.
10. Lunney, J.K., and Osborne, B.A. (1984) Unpublished data.

EFFECTS OF SYSTEMIC IMMUNIZATION OF DAMS AND NEWBORN CALVES ON
PATHOGEN-SPECIFIC IMMUNOGLOBULIN PRODUCTION

WILLIAM A. FLEENOR
Department of Animal Sciences, The University of Arizona
Tucson, AZ 85721

Systemic immunization of dams at 6 weeks and again at 2
weeks prepartum increased pathogen-specific IgG levels in each
dam's colostrum and her calf's serum. Pathogen-specific IgG and
IgM concentrations in dams' sera and colostra were related to
subsequent pathogen-specific IgG and colostra was correlated to
neonatal sera concentrations. Conversely, no relationships were
evident between pathogen-specific IgA concentrations in dams'
sera or colostra and postcolostral pathogen-specific IgA
neonatal salivary concentrations.
Calves initially vaccinated at 3 days produced both a
primary and a secondary systemic pathogen-specific antibody
response, whereas calves initially vaccinated at 12 and 21 days
produced only secondary responses. Maternally derived
antibodies were found to suppress neonatal systemic antibody
production following primary immunization. They were also found
to influence secondary humoral immune responses, although in a
diminished capacity. Calves initially vaccinated at 3 and 12
days postpartum produced primary salivary pathogen-specific IgA
responses, although somewhat delayed and smaller compared to the
serum pathogen-specific IgA responses. All treatment groups
except controls exhibited secondary salivary pathogen-specific
IgA responses within 1 day of immunization.

IMMUNE RESPONSES TO AFRICAN TRYPANOSOMES

TERRY PEARSON
Department of Biochemistry and Microbiology
University of Victoria
Victoria, British Columbia, Canada V8W 2Y2

The immune response to trypanosomes has been primarily
studied in the mouse model. Comparison of the immune mechanisms
clearly defined in the mouse to those in cattle shows many
discrepancies. It is therefore important that greater emphasis
is placed on characterization of the immune response of bovidae
to trypanosomes to detect the relevant host-parasite
relationship. The insights gained will permit the development
of effective strategies for vaccination.

THE IMMUNE RESPONSE TO TRYPANOSOMA CRUZI

ALAN F. SHER
Immunology and Cell Biology Section
Laboratory of Parasitic Diseases
National Institute of Allergy and Infectious Diseases
Bethesda, MD 20205

Successful vaccination against parasitic infection depends on the induction in a genetically heterogeneous host population of immune responses that will protect against all variants of the parasite species in question. Whereas considerable information has been obtained in both experimental animals and man concerning the influence of host genotype of immune responsiveness, relatively little attention has been paid to parasite genetic factors affecting the susceptibility of the organism to immunologic attack. This subject is of great interest, not only with respect to the design of successful vaccination protocols, but also as an approach to elucidating mechanisms of immunity against parasites. Thus, parasite variants unsusceptible to host immunity can be utilized as "negative mutants" for identifying both target antigens and effector mechanisms directly relevant to the process of immune parasite destruction. In this article, I will discuss genetic variations occurring in Trypanosoma cruzi and Leishmania that affect either the antigenicity of the parasites or their susceptibility to cell-mediated killing by activated macrophages.
In the case of T. cruzi, a parasite that displays enormous genetic diversity (1), I have studied strain- and clone-related differences in the expression of a single prominent antigenic molecule present on the surface of the epimastigote and metacyclic trypomastigote states of the parasite. This component, first identified on the Y strain of T. cruzi by Snary and colleagues, is a 72,000-Da glycoprotein. The molecule can be readily identified and purified using a monoclonal antibody (WIC 29.26) that recognizes an epitope present on its carbohydrate portion (2). Recent studies have suggested that the 72,000-Da glycoprotein may be important both as a receptor involved in the regulation of parasite differentiation (3) and as a target antigen for vaccination (4).

To examine the possibility of intraspecific differences in the expression of the 72,000-Da surface protein, I have screened a total of 60 T. cruzi strains and clones for their reactivity with ^{125}I-labeled WIC 29.26 antibody. Significant binding of the radiolabeled antibody occurred with any two of the four strains and 23 of the 50 clones tested. A group of 10 clones derived from an isolate from a single patient (Miranda) was found to contain both reactive and non-reactive clones (Kirchhoff et al., unpublished data).

For confirmation of the results of the radioimmunoassay, reactive and non-reactive isolates were surface labeled by the iodogen procedure, solubilized, and assayed for the presence of the 72,000-Da glycoprotein by immunoprecipitation with WIC 29.26 antibody or a polyvalent antiserum produced in rabbits against this glycoprotein. Six of seven isolates previously shown to be non-reactive in the radioimmunoassay were found to contain surface components immunoprecipitable by both WIC 29.26 antibody and the polyvalent anti-glycoprotein serum. Of these six, one showed two antigenic components with electrophoretic mobilities clearly different from the 72,000-Da glycoprotein present on reactive isolates used as controls. Another of the non-reactive isolates tested was also found to lack a component immunoprecipitable with WIC 29.26 antibody but displayed molecules weakly reactive with polyvalent antiserum. Thus, although all of the isolates tested contain membrane proteins reactive with either WIC 29.26 or anti-glycoprotein antibodies, in only some membrane protein is the 29.26 epitope expressed in a form accessible on the surface of intact parasites.

The above findings provide evidence for heterogeneity in the expression of an important T. cruzi surface antigen. This epitopic variation, which may extend to other T. cruzi antigens, should be considered in the design of serodiagnostic as well as vaccination procedures. Finally, the results underscore an important drawback involved in the use of monoclonal antibodies in assays of major parasite antigens. Because of their exquisite specificity, these antibody reagents may be of limited use for establishing the presence or absence of important surface components in parasites in which genetic heterogeneity occurs.

REFERENCES

1. Dvorak, J.A. (1983) J. Cell. Biochem. (In press.)
2. Snary, D., Ferguson, A.J., Scott, M.T., and Allen, A.K. (1981) Mol. Biochem. Parasitol. 3:343.
3. Sher, A., and Snary, D. (1982) Nature 300:639.
4. Snary, D. (1983) Trans. R. Soc. Trop. Med. Hyg. (In Press.)

IMMUNE RESPONSE IN EQUINE PREGNANCY

DOUGLAS F. ANTCZAK
James A. Baker Institute for Animal Health
New York State College of Veterinary Medicine, Ithaca, NY

Pregnant mares make an unusually strong and reproducible immune response to antigens expressed by the fetal-placental unit. This immune response has many of the characteristics of classic allograft reactions. The response can be observed systemically in the form of cytotoxic antibody against paternally inherited fetal MHC antigens and locally as an accumulation of mononuclear cells around the endometrial cups. Endometrial cups are specialized structures derived from chorionic girdle cells of the equine trophoblast. The girdle cells invade the maternal endometrium between 36 and 38 days of gestation, differentiate to endometrial cup cells, and secrete pregnant mare's serum gonadotropin. The cytotoxic antibody response of pregnant mares carrying MHC-incompatible fetuses is first detectable about 2 weeks after the formation of the endometrial cups. However, the lymphocyte reaction around the endometrial cup cells occurs, even in MHC-compatible pregnancies, in the absence of detectable cytotoxic antibody to MHC antigens. This observation suggests that the cellular immune response to endometrial cup cells may be directed against non-MHC antigens, such as minor histocompatibility antigens or placental specific molecules. In an attempt to characterize the antigens against which this response is mounted, I prepared monoclonal antibodies to equine placental tissues. These antibodies were produced by immunizing rats with isolated endometrial cup tissue. After fusion of spleen cells from immunized rats with the mouse myeloma SP-2/0, the resulting hybrids were screened selectively for 1) non-reactivity with equine lymphocytes using an enzyme-linked immunosorbent assay (ELISA); 2) production of rat Ig, also using an ELISA assay; and 3) specific staining of frozen sections of endometrial cup tissue using immunoperoxidase labeling. Monoclonal antibodies giving five distinct reaction patterns to endometrial cup tissue sections have been cloned. One of these antibodies appears to identify a placental antigen that is expressed on endometrial cup cells but not on allantochorion. This monoclonal antibody may be useful for the purification of a molecule expressed only by endometrial cup cells. The purified molecule could be used to search for maternal immune responses to placental antigens in the horse.

CHAPTER 16

ABSTRACTS

Analysis of <u>Brucella</u> <u>abortus</u> Antigens by Use of Monoclonal
Antibodies, Sodium Dodecyl Sulfate-Polyacrylamide Gel
Electrophoresis (SDS-PAGE), and Two-Dimensional Gel
Electrophoresis

L. Garry Adams, Blair A. Sowa, Doris M. Hunter,
Jillaine M. Stiller, and Frederick C. Heck
Departments of Veterinary Pathology and of Veterinary
Microbiology and Parasitology, College of Veterinary Medicine,
Texas A&M University, College Station, TX 77843

   <u>Brucella</u> <u>abortus</u> strain 2308 grown on bacteriologic media
and inactivated with 1.25 mrads of ^{60}Co gamma radiation was
used to stimulate either <u>in</u> <u>vitro</u> BALB/c spleen cells grown in
thymocyte-conditioned medium for 5 days or the spleen cells of
BALB/c mice inoculated IP on days 0, 14, and IV on day 42. On
day 5 or day 46, respectively, 10^8 splenocytes were fused with
10^7 SP2/0 myeloma cells by use of 50% polyethylene glycol in
Dulbecco's minimal essential medium. The fused cells were
cultured in hypoxanthine-aminopterin-thymidine (HAT) medium for
10 to 21 days, and supernatants were assayed for anti-brucella
antibodies by an enzyme-linked immunosorbent assay (ELISA) that
included irradiated S2308 whole cells (WC), alkali-treated
lipopolysaccharide (LPS), or outer membrane proteins (MP) with
horseradish peroxidase (HRPO)-conjugated rabbit anti-mouse IgG.
Hybridomas that produced antibodies reacting with <u>Brucella</u>
antigens were recloned three times by limiting dilution and then
stored in liquid nitrogen. Ascites were produced in
pristane-primed BALB/c mice. Anti-WC, LPS, or MP monoclonal
antibodies from ascites that were purified by column
chromatography and isotyped by use of anti-mouse isotype ELISA
procedures were applied to nitrocellulose electroblots of
unidimensional SDS-polyacrylamide gel or two-dimensional gel
electrophoretic patterns of <u>B.</u> <u>abortus</u> S2308 and stained with
affinity-purified HRPO-conjugated goat anti-mouse IgG. Bands
were detected in the SDS-PAGE transblots corresponding with
membrane proteins and lipopolysaccharides of <u>B.</u> <u>abortus</u>, and
outer membrane proteins and proteins of undetermined origin were
detected in the transblot of the two-dimensional gel preparation
of <u>B.</u> <u>abortus</u> S2308.

Multiple Bacterial Etiologies in a Citrus Epidemic

Anne Alvarez, Albert Benedict*, and Carla Mizumoto*
Department of Plant Pathology, 3190 Maile Way,
and *Department of Microbiology, 207 Snyder Hall,
University of Hawaii, Honolulu, HI 96822

Serological relationships were examined among bacteria associated with a recent disease outbreak in lime groves of Mexico. The epidemic was thought to be citrus canker, a severe defoliating disease caused by Xanthomonas campestris pathovar (pv) citri. Known citrus canker strains A, B, and C from Japan, Argentina, and Brazil, respectively, reacted with anti-Xanthomonas monoclonal antibodies (MCA) XCc-1 and XCc-11, prepared against X. campestris pv. campestris. Among lime disease bacteria, only 1 of 41 isolates, selected on the basis of positive host responses in pathogenicity tests, reacted with anti-Xanthomonas MCA XCc-1 and XCc-11. To delineate the remaining bacteria expeditiously, mice were immunized with selected isolates, hybrids were initially screened with six isolates, and expanded cultures of all primary (non-cloned) antibody-forming hybridomas were tested with all isolates. The ratio of the myeloma cells to mouse spleen cells and plating densities were adjusted to yield one to three hybrids per microculture. Screening produced a large number of repetitive patterns representing hybridomas of the same specificities and permitted closely related bacteria to be grouped according to their surface antigens. The validity of patterns was ascertained by cloning representative cultures, and patterns formed by single hybridoma cultures were distinguished from those formed by mixtures. Non-overlapping repetitive patterns separated the bacteria into five tentative groups, suggesting that the epidemic involved a mixed etiology rather than a single agent.

Functional Epitopes of the Foot-and-Mouth Disease Virus
Outer Capsid Protein VPI

B. Baxt, D. O. Morgan, and B. H. Robertson
Plum Island Animal Disease Center, Agricultural Research Service,
U.S. Department of Agriculture, P.O. Box 848, Greenport, NY 11944

Foot-and-mouth disease virus (FMDV) structural protein VP1
elicits neutralizing and protective antibodies and is probably
the viral attachment protein that interacts with the cellular
receptor sites (CRS) on cultured cells. To study the
relationships between epitopes on the molecule related to
neutralization and reaction with the CRS, we prepared a series
of seven monoclonal antibodies against intact virions (type
$A_{12}119ab$), isolated VP1, or a CNBr fragment of VP1 and tested
their effects on viral adsorption to bovine kidney (BK) cells.
All of the antibodies selected neutralized virus, at various
efficiencies. Three of the antibodies caused a high degree of
viral aggregation as measured by velocity sucrose gradient
sedimentation, whereas the remaining four caused little or no
aggregation. Viral adsorption was determined by the attachment
of radiolabeled virus to BK cells. Two of the antibodies that
did not aggregate virus inhibited the adsorption of FMDV to
cells by 80 to 90% and were mapped between amino acids 169-179
on the VP1 molecule. The third antibody, which only reacted
with intact virus, did not inhibit adsorption to cells. Of the
three antibodies that caused viral aggregation, two inhibited
adsorption by 50 to 60%; the other by 60 to 70%. This residual
adsorption, however, was shown to be non-specific. These
results indicate that viral neutralization probably occurs by at
least three mechanisms: a) viral aggregation resulting in
non-specific adsorption to the cell surface; b) inhibition of
viral attachment to cells; and c) an unknown mechanism occurring
subsequent to adsorption. These may involve at least three
different epitopes on the VP1 molecule. Furthermore, the
results suggest that the functional domain on VP1 that interacts
with the CRS may be located between amino acids 169-179.

Rat Monoclonal Antibodies. I. Purification from _In Vitro_
Supernatant

Hervé Bazin, Louis-Marie Xhurdebise, Guy Burtonboy,
Anne-Marie Lebaco, Lieve De Clerco, and Francoise Cormont
Experimental Immunology Unit, Faculty of Medicine, University of
Louvain, Clos Chapelle aux Champs 30, Brussels 1200, Belgium

We have developed a technique to purify quickly and
efficiently rat monoclonal antibodies from _in vitro_ culture
supernatants based on the fact that more than 95% of rat
immunoglobulins carry κ light chains. A mouse monoclonal
antibody (MARK-1) with a suitable binding affinity for rat κ
light chains is immobilized on a solid support and used to
purify rat immunoglobulins. Milligrams of rat monoclonal
antibodies can be concentrated from culture supernatants in a
short time, thus yielding a high recovery. Fusion 1RG83F rat
non-secreting cell line and MARK-1 hybridoma (mouse monoAC
anti-rat κ chain) are both available upon requests to H. Bazin.

Rat Monoclonal Antibodies. II. Purification from <u>In Vivo</u> Serum or Ascitic Fluid

Hervé Bazin, Francoise Cormont, and Lieve De Clerco
Experimental Immunology Unit, Faculty of Medicine,
University of Louvain, Clos Chapelle aux Champs 30,
Brussels 1200, Belgium

We have developed a technique for purifying rat monoclonal antibodies from ascitic fluid or serum. The technique is based on two observations: First, about 95% of the rat immunoglobulin light chains are of the κ type and second, an allotype exists in the rat species, being located on the constant part of the κ light chain. By using a mouse monoclonal antibody having a suitable binding affinity to the IgK-1a allotype carried by the κ light chains of the LOU inbred strain of rat, we were able, by immunoaffinity chromatography, to purify the LOU.IgK-1a-bearing immunoglobulins from serum proteins, including the immunoglobulins of rats bearing the IgK-1b allotype. In practice, LOU-histocompatible hybridomas synthesizing the IgK-1a allotype are transplanted in congenic rats to the LOU inbred strain carrying the IgK-1b allotype, for LOU rats (which have the IgK-1a allotype of the κ light chains) and congenic LOU.IgK-1b rats (which have the Igk-1b allotype of the κ light chains) are fully histocompatible. Then, their serum or ascitic fluid is poured on an immunoabsorbent column in which a mouse monoclonal antibody (MARK-3) against the IgK-1a allotype is coupled. The serum proteins, including the immunoglobulins of the host, go directly through the column. An appropriate buffer can elute the monoclonal antibodies in a second step. The same technique could be employed for nearly all monoclonal antibodies, as well as for the LOU myeloma proteins. The method is rapid, efficient, and inexpensive. Its limitation is that the λ type carrying monoclonal antibodies are relatively rare.

Hybridoma Data Bank: A Data Bank on Cloned Cell Lines
and their Immunoreactive Products

Lois D. Blaine
American Type Culture Collection,
12301 Parklawn Drive, Rockville, MD 20852

The International Council of Scientific Unions Committee on
Data for Science and Technology (CODATA) has established a data
bank containing information on the worldwide availability of
cloned cell lines, their immunoreactive products, and related
services.

Policy and development of the data bank are the
responsibility of a Task Group of CODATA (Task Group on a
Hybridoma Data Bank). The Task Group is composed of
immunologists from a variety of countries. The funding of the
data bank is through grants from CODATA, the World Health
Organization, and the International Union of Immunological
Societies. The data bank is centered at the American Type
Culture Collection in Rockville, Maryland.

Storage and processing of the information in the Hybridoma
Data Bank are though the NIH Computer Center's IBM 370
computer. The newly created system closely follows the
prototype Microbial Information System (MICRO-IS), a joint
project for the National Institute for Dental Research and the
Food and Drug Administration.

The major portion of the data bank will comprise data
submitted by investigators who develop and distribute cloned
cell lines and immunoreactive products. Requests for
information from the data bank will be processed by data bank
personnel initially. Plans are under way for a network system
with nodes located in Japan and Europe.

The current version of the Hybridoma Data Bank consists of
115 unique data fields. It is flexible and open-ended and will
be able to handle rapid changes in hybridoma technology.

Application of Monoclonal Antibodies in Animal Production:
Pregnancy Diagnosis in Cattle

P. Booman, M. Tieman, D. F. M. van de Wiel,
J. M. Schakenraad, and W. Koops
Research Institute of Animal Production "Schoonoord,"
P.O. Box 501, 3700 AM, Zeist, The Netherlands

Monoclonal antibodies are proving useful in the development of diagnostic test kits for monitoring reproduction in farm animals. The ability to produce infinite quantities of monospecific antibodies makes possible the standardization of an enzyme immunoassay of milk progesterone for early diagnosis of pregnancy in cattle. The milk-progesterone test is based on the difference in progesterone concentrations in pregnant and non-pregnant cows.

For the production of monoclonal antibodies against progesterone, BALB/c mice were immunized with either 11α-hydroxyprogesterone hemisuccinate conjugated to bovine serum albumin or progesterone-7α-carboxyethylthioether conjugated to bovine thyroid globulin. After the spleen cells from immunized animals were fused with X63-Ag8.653 myeloma cells and cloned twice by serial dilutions, several positive and stable hybridoma lines were isolated. The monoclonal antibodies secreted by these hybridomas generally differed in their binding affinity for progesterone. The association constants (K_A) varied from 1.1×10^7 1/mol to 1.7×10^{10} 1/mol. Moreover, the antibodies had distinct specificities for a variety of steroids. Although all antibodies typically exhibited high cross-reactivity with the homologous steroid hapten, some of them bound to other steroids such as 5β-pregnane-3,20-dione or 4-pregnene-3,11,20-trione with comparable or even higher affinity. However, the antibody having the highest association constant for progesterone (1.7×10^{10} 1/mol) also showed excellent specificity. In a direct double-antibody solid-phase enzyme immunoassay using this antibody, the detection limit of the assay varied between 1 and 2 pg and the amount of progesterone causing a 50% reduction of initial binding in the standard curve varied between 15 and 20 pg. The assay could be completed in about 5 hours, and there was no need for extraction, centrifugation, or colorimetric measurement.

Use of Monoclonal Antibodies to Identify and Characterize
Distinct Structural Regions of 124-kDa Phytochrome from <u>Avena</u>

Susan M. Daniels and Peter H. Quail
Department of Botany, University of Wisconsin, Madison, WI 53706

The chromoprotein phytochrome is a photoreceptor present in
all higher plants that, upon conversion to its active form by
absorption of light, initiates numerous photomorphogenic
responses. The biochemical mechanism of phytochrome action
remains unknown. We are using monoclonal antibodies directed
against the native 124-kDa protein from <u>Avena</u> to probe the
molecular properties of phytochrome.
From about fifty ascites fluids derived from different
subcloned hybridoma cell lines, we obtained three general groups
of antibodies directed against three distinct regions of the
molecule. The results of immunoblot analysis of proteolytically
produced peptides indicate that one antibody group corresponds
to a domain at the amino terminus, present on the native 124-kDa
protein but absent from the proteolytically degraded 118-kDa
peptide. One of these antibodies also recognized undegraded
phytochrome from a dicot plant, zucchini. The second group
corresponds to the region of the molecule that contains the
chromophore. The third group strongly recognizes the region
adjacent to the caboxyl terminus. These monoclonal antibodies
also strongly recognize the protein when assayed by
enzyme-linked immunosorbent assay (ELISA). Additionally, we
have several antibodies that appear to be strictly
conformational in that strong recognition for the protein is
indicated by ELISA, but not by immunoblot analysis after sodium
dodecyl sulfate-polyacrylamide gel electrophoresis. The use of
these conformational antibodies to probe a possible difference
between the inactive (Pr) and the active (Pfr) forms of the
molecule will be discussed.

Monoclonal Antibodies for Determination of the
Structure-Dependent Function of the C-Terminal Region
of Tobacco Mosaic Virus (TMV) Coat Protein (TMV-P)

R. G. Dietzgen and Evamarie Sander
Institut Biologie II, Universität Tübingen,
Auf der Morgenstelle 28, D-7400 Tübingen,
Federal Republic of Germany

The use of monoclonal antibodies with highly characterized
properties is imperative for increasing the efficiency of
immunological diagnosis of plant viruses. With a monoclonal
antibody (clone 95) of characterized reactivity (enzyme-linked
immunosorbent assay, or ELISA) that was obtained with highly
purified and complete TMV vulgare as antigen, it was possible to
neutralize the infectivity of the virus by binding the monoclonal antibody to the C-terminal region of TMV-P. The monoclonal antibody no. 95 is of the IgG2a subclass with κ light
chains. The reactivity with the C-terminal tetrapeptide of
TMV-P vulgare was assessed by binding this monoclonal antibody
to the identical, but chemically synthesized, amino acid
sequence. No. 95 exhibited reactivity of comparable strength
not only with the nucleocapsid and the coat protein monomers of
TMV vulgare, but also with the nucleocapsid and the coat protein
monomers of TMV strains dahlemense and Holmes' Ribgrass. From
these results it can be concluded that the C-terminal tetrapeptide is exposed in comparable conformation on the surface of
both the nucleocapsid and the coat protein monomers of the three
TMV strains, constituting, therefore, a common epitope. The
binding of no. 95 to all three strains may be attributed to
comparable polarity and size of the exchanged amino acids and
the proposed flexible conformation of the C-terminal region.
    Compared with monoclonal antibodies of other clones directed
against the same epitope, no. 95 fulfilled the condition for
application in solid phase binding tests (ELISA) by having a
more than tenfold stronger specific activity for the viral
epitope. The C-terminal amino acid sequence containing the
described epitope appears to have a biological function in the
process of virus infection. This can be concluded from the
inhibition of infectivity up to 93.6% by the binding of no. 95
to TMV vulgare. For the assay, known molecular proportions of
purified monoclonal antibody and virus particles and Xanthi
tobacco were employed. Unlike no. 95, nos. 11 and 62 did not
neutralize infectivity, even though they reacted with the described C-terminal epitope, possibly because of the more than
tenfold lesser specific activity of nos. 11 and 62. For the
selection of the monoclonal antibody best suited for improvement
of sensitivity of virus diagnosis, the importance of the relation between the properties of the monoclonal antibodies and
virus structure is emphasized.

The Use of Monoclonal Antibodies Against Bovine Immunoglobulins
as a Means of Quantifying Bovine Immunoglobulins

D. M. Donahoe, S. Srikumaran, A. J. Guidry*, and R. A. Goldsby
Amherst College Laboratory for Hybridoma Research,
Amherst, MA 01002 and *Laboratory of Milk Secretion and Mastitis,
Agricultural Research Service, U.S. Department of Agriculture,
Beltsville, MD 20705

Hybrids formed from the fusion of spleen cells of a BALB/cJ
mouse inoculated with bovine immunoglobulins and an established,
nonsecreting murine cell line SP2/0 secreted monoclonal
antibodies to various classes of bovine Ig. These cells were
injected into mice to form a tumor, and the monoclonal
antibodies obtained by affinity purification of the mouse serum
were designated the DAS series. DAS2 is specific for bovine
IgG2, whereas DAS9 and DAS10 recognize all classes of bovine
Igs. These α-bovine Ig antibodies were used in a sandwich
radioimmunoassay (RIA) as a means to quantify the amount of
bovine Ig present in bovine serum, culture fluid, and buffered
saline solutions of known (Ig) employed as standards.
The bovine immunoglobulins used as standards for comparison
to bovine serum include monoclonal IgG2 and IgG1, and polyclonal
IgG and IgG2. Monoclonal bovine IgG2 and IgG1, LHRB-2 and
LHRB-4, respectively, were secreted from hybridomas of bovine
lymphoid cells and SP2/0 cells. Polyclonal IgG and IgG2 were
purified by conventional methods by A. J. Guidry. By coating
polyvinylchloride plates with a DAS α-bovine Ig, adding test
samples of bovine Ig, and then adding ^{125}I-labeled DAS9,
specific for all bovine Ig, the measure of counts per minute can
be used to determine the amount of bovine Ig in a preparation.

Production and Characterization of Anti-Zeatin Monoclonal
Antibodies

Tomas D. Hillson, Richard A. Van Deusen*,
Thomas A. Permar, and Richard C. Schultz
Department of Forestry, Iowa State University,
Ames, IA 50011 and *National Veterinary Services Laboratory,
Animal and Plant Health Inspection Service,
U.S. Department of Agriculture, Ames, IA 50010

Zeatin was selected for study to establish criteria and
procedures for the production of monoclonal antibodies specific
for plant hormones. Zeatin-riboside conjugated to bovine serum
albumin was used as the antigen. Hybridomas grew in 295 of 480
culture plate wells; of these, 149 were positive for anti-zeatin
antibodies. Sixty positive hybridomas each derived from a
single colony of cells were selected for further
characterization. A competitive enzyme-linked immunosorbent
assay (ELISA) was used to determine specificity of the
monoclonal antibodies. The results separated the monoclonal
antibodies into three classes based on their specificity: Class
I--monoclonal antibodies specific for zeatin and zeatin
riboside. (These monoclonal antibodies have the greatest
potential for use in developing an improved ELISA and methods
for affinity separation for zeatin.) Class II--monoclonal
antibodies binding with zeatin, zeatin riboside, dihydrozeatin,
and isopentyl adenine. (These monoclonal antibodies are
specific for any of the naturally occurring cytokinins, and they
will be useful in broad-spectrum isolation and characterization
of cytokinins and cytokinin-like compounds.) Class
III--monoclonal antibodies binding with any adenine or
substituted adenine. (These monoclonal antibodies were of no
interest to us, for they are specific for the adenine molecule
portion of the cytokinin.) Sixteen hybridomas secreting
monoclonal antibodies in classes I and II were selected and
frozen for future cloning. The culture media from them were
collected and the antibodies purified for further
characterization. Six hybridomas with high-affinity anti-zeatin
specific-antibody production were selected for cloning. These
results have provided us with specific monoclonal antibodies
against zeatin and a system for production and characterization
of monoclonal antibodies against other plant hormones.

Monoclonal Antibodies to Rabbit Platelets

John S. Kenney, Gayle M. Nakano, and Anna Mirkovich
Department of Immunology, Institute of Biological Sciences,
Syntex Research, Palo Alto, CA 94304

We sought to develop monoclonal antibodies specific for
rabbit platelet antigens. These antibodies would be useful in
defining the biological properties of platelets, the role of
platelets in disease, and the pharmacological aspects of
antiplatelet drugs. Spleen cells from mice immunized with
washed rabbit platelets were hybridized with either
P3/NS1/1-Ag4-1 or P3-X63-Ag 8.653 mouse myeloma cells.
Hybridoma culture supernatants were assayed for antibody by
solid-phase enzyme-linked immunosorbent assay (ELISA). Multiple
ELISAs were performed with rabbit platelets, red blood cells,
polymorphonuclear leukocytes, mononuclear leukocytes, or
fibroblasts as the solid-phase antigen. Hybridoma cultures
producing antibody specific for platelets alone were then
cloned, and the antibody was characterized. We have identified
19 different monoclonal antibodies, based upon the antibody
isotype and the pattern of binding in the ELISA--3 of isotype
IgG1, 6 of IgG2A, 6 of IgG2B, and 4 of IgM. Antibody binding to
platelets from other species has been determined with human,
guinea pig, and rat platelets. Of the 19 antibodies tested,
1 bound both human and guinea pig platelets, 1 bound human
platelets alone, and 2 bound guinea pig platelets alone. Thus,
it appears there are a limited number of rabbit platelet
antigens that are shared with platelets from other species and
that between species some platelet antigens are highly
restricted. The physicochemical characteristics of various
platelet antigens are being investigated.

Monoclonal Antibodies to Bovine Somatotropin:
Preparation and Characterization

Gwen G. Krivi and Edwin Rowold, Jr.
Monsanto Company/U4G, 800 N. Lindbergh Blvd., St. Louis, MO 63167

Twenty-nine stable hybridoma cell lines secreting monoclonal antibodies to bovine somatotropin have been produced and characterized. Five of the monoclonal antibodies cross-react with porcine and human as well as bovine somatotropins. Seven cross-react with porcine and bovine alone, and the rest are specific for bovine.

One of the monoclonal antibodies was used in the preparation of an immunoadsorbent reagent suitable for partial purification of pituitary or recombinant bovine somatotropin. Both the purified recombinant and pituitary bovine somatotropins showed a major band of MW of approximately 22,000 on SDS-PAGE and Western blot analysis. Antibody-purified bovine somatotropin retained activity in a rabbit liver receptor assay and in a radioimmunoassay (RIA). The immunoadsorbent will bind human and porcine somatotropins as well as bovine.

The monoclonal antibodies have also been used to develop a simple, fast and sensitive RIA for the detection of bovine somatotropin in bacterial and pituitary extracts. One of the antibodies was used to design a solid-phase RIA capable of detecting bovine somatotropin at concentrations of 10 ng/ml or greater. Other bovine pituitary hormones, FSH, LH, and prolactin, do not compete with ^{125}I-bovine somatotropin in the assay. Ovine somatotropin competed effectively with ^{125}I-bovine somatotropin for the antibody-binding sites. The assay can also be used to detect human, porcine, rat, and avian somatotropins although at greatly reduced sensitivity.

Complement-Fixing Monoclonal Antibodies Used to Estimate
Percentage of T and B Lymphocytes in Bovine Peripheral Blood

H. A. Lewin, L. L. Lasslo, W. C. Davis*, and D. Bernoco
Department of Veterinary Reproduction,
University of California, Davis, CA 95616, and
*Department of Veterinary Microbiology-Pathology,
Washington State University, Pullman, WA 99164

The specificities recognized by three monoclonal antibodies
were studied using the microcytotoxicity test. The H4
monoclonal antibody was previously shown to cross react with a
class II-like determinant on bovine, equine, ovine, and porcine
B lymphocytes (Lewin and Bernoco, Fed. Proc. 42:1230). Using
multiple linear regression analyses, we analyzed two monoclonal
antibodies, TH21A and B26A (produced by WD), as predictors of
peripheral T and B lymphocyte percentages (n = 185). Three
significantly different groups of animals were found upon
examination of H4 and TH21A reactions (p < .0001). In 92% of
the animals, H4 and TH21A seem to recognize an identical
lymphocyte subset (r = 0.93 p < .0001). Within this group,
the correlation coefficient between H4 and B26A was -0.94 (p <
.0001), and the correlation coefficient between TH21A and B26A
was -0.92 (p < .0001). In 5% of the animals, TH21A recognized
all B lymphocytes and H4 recognized a subpopulation of B
lymphocytes; in 3% of the animals, H4 recognized all B
lymphocytes and the TH21A monoclonal antibody recognized a
subpopulation of B lymphocytes. We interpret this as
recognition of separate epitopes. Thus the H4 and TH21A
monoclonal antibodies recognize bovine B lymphocytes and B26A
appears to be a pan-T lymphocyte reagent. These monoclonal
antibodies were used to estimate the mean percentages of B and T
lymphocytes in 33 animals from a herd with a low incidence of
bovine leukemia virus infection (%B = 35.0 $\pm$ 2.0; %T = 70.0 $\pm$
2.3).

(Supported by USDA Formula Funds AH-20.)

Peripheral T/B Lymphocyte Percentages in a Herd With a High
Incidence of Bovine Leukemia Virus (BLV) Infection

H. A. Lewin, L. L. Lasslo*, R. Ruppanner*, and D. Bernoco
Departments of Reproduction and of *Epidemiology and Preventive
Medicine, School of Veterinary Medicine,
University of California, Davis, CA 95616

Relationships between B lymphoctye percentage (%), age
(range 19 to 175 months), antibody to BLV-gp51 (a-BLV), and
lymphocytosis were examined in a herd with a high incidence of
BLV infection (80% were seropositive by agar gel immunodiffusion;
n = 206). The percentage of T and B lymphocytes was estimated
by use of three monoclonal antibodies: H4, TH21A and B26A.
Chi-square analyses showed that animals greater than 42 months
of age were more likely to be seropositive (p < .001).
Animals 19 to 42 months of age were more likely to have > 40%
B lymphocytes than a-BLV (p > .05). These two tests did not
differ significantly for animals older than 42 months. These
results suggest that a-BLV may not necessarily be a definitive
test for BLV infection for some animals (< 42 months, n = 10)
are seronegative and have > 40% B lymphocytes, whereas others
(n = 17) are seropositive and have < 40% B lymphocytes. In
younger animals, the proliferation of B lymphocytes may precede
seroconversion in BLV-infected animals. Alternatively, animals
that are seropositive and have < 40% B lymphocytes may
represent exposure to the virus and not infection. The logistic
model obtained for predicting lymphocytosis was lymphocytosis =
u + B + BLV (goodness of fit p = .98). The percentage of B
lymphocytes entered the model first and was the most significant
predictor of lymphocytosis. Odds ratios for the risk of
lymphocytosis were 18.5 for B $\geq$ 70% to B < 40%, and 2.9 for
B or 40-60% to B < 40%. Although age was not significant in
this model, chi-square analysis between age and lymphocytosis
showed that lymphocytosis was more frequent in animals older
than 55 months (p < .005). These results suggest that older
animals and animals with $\geq$ 70% B lymphocytes are at greater
risk of having lymphocytosis.

(Supported by USDA Formula Funds AH-20.)

Characterization of Monoclonal Antibodies to Swine
Peripheral Blood Lymphocyte (PBL) Subpopulations

Joan K. Lunney, Mark D. Pescovitz, and David H. Sachs
Transplantation Biology Section, Immunology Branch,
National Cancer Institute, National Institutes of Health,
Bldg. 10, Rm. 4B17, Bethesda, MD 20205

A panel of monoclonal antibodies to subpopulations of swine
PBL has been obtained from a fusion of SP2/0 myeloma variant
cells to spleen cells from mice immunized with SLA^{dd}
thymocytes. The hybridoma culture supernatants were screened
for reactivity on whole PBL or separated cell populations by
antibody-mediated cytotoxicity and by flow microfluorometric
analysis. Molecular weights of antigens recognized were
determined by immunoprecipitation of ^{125}I-labeled antigens.
The most interesting antibodies are summarized below. 74-11-40
(γ2b) recognizes a polymorphic determinant of class I major
histocompatibility complex antigens as determined by its lysis
of and binding to all SLA^{dd} cells but lack of reactivity with
SLA^{cc} PBL. Moreover, it precipitates molecules of 44,000 and
12,000 Da. 74-9-3 (μ) reacts with a common leukocyte antigen
that has apparent subunit molecular weights of 180,000, 195,000,
and 210,000 Da. 76-6-7 (μ) binds to both T cells and
macrophages but is unreactive with B cells, whereas 76-7-4 binds
only to B cells in a periphery but also recognizes a determinant
on a subpopulation of thymocytes. 74-22-15 (γ_1) and 76-5-28
(μ) both react specifically with macrophages and
granulocytes. 74-12-4 (γ2b) binds to a 55,000-Da protein on
helper T cells and blocks mixed lymphocyte reactions, whereas
76-2-11 (γ2a) binds to a 35,000-Da protein on cytotoxic T
cells and blocks the effector phase of cell-mediated
lympholysis. These antibodies should be very useful for
analysis of cellular interactions of swine lymphocytes and for
molecular analysis of their surface proteins.

Monoclonal Antibodies Specific for <u>Corynebacterium</u> <u>sepedonicum</u>, the Causative Agent of Potato Ring Rot

Wayne E. Magee, Claudia F. Beck, and Sandra S. Ristow
Department of Bacteriology and Biochemistry,
University of Idaho, Moscow, Idaho 83843

Three separate fusions were carried out using spleen cells from mice immunized with <u>C</u>. <u>sepedonicum</u> and the mouse myeloma P3X63Ag8.6543. A sensitive enzyme-linked immunosorbent assay (ELISA) procedure was developed for detecting antibodies to the bacterium. A large proportion of hybridoma cultures produced antibodies to the organism, but most also reacted with other closely related <u>Corynebacterium</u> sp. Cells from the most promising cultures were cloned several times by limiting dilution, and one hybridoma cell line has been characterized. The antibody was found to be IgG1 with a κ light chain. It was partially purified from ascites fluid from BALB/c mice carrying the hybridoma and tested for specificity. The antibody reacted strongly with several laboratory strains and recent field isolate of <u>C</u>. <u>sepedonicum</u> but showed little more than background reactivity toward <u>C</u>. <u>michiganense</u>, <u>C</u>. <u>insidiosum</u>, or <u>M</u>. <u>lacticum</u>. The standard ELISA procedure readily detected 10^4 to 10^5 organisms. An indirect fluorescence assay is being developed as the basis for a sensitive assay for the organism in diseased plants. These results show that it is feasible to produce highly specific monoclonal antibodies to important bacterial pathogens of plants, an area of hybridoma research that has received very little attention until now.

New Immunodiagnostic Techniques for Equine
Infectious Anemia Virus

Tatsuo Matsushita and James P. Porter
Syngene Products and Research, TechAmerica Group, Inc.,
225 Commerce Drive, Fort Collins, Colorado 80524

Horse sera from 1272 horses were examined by use of
enzyme-linked immunosorbent assay (ELISA) and the agar
immunodiffusion (AGID) test for detecting antibody against
equine infectious anemia virus (EIAV). Immunoblots were
performed on selected serum samples to identify the antigens
involved in the ELISA signal. Excellent correlation was found
between ELISA and AGID, with 100% of the positives and 99.3% of
the negatives correlating in the two tests at a single cut-off
point (0.4 optical density). The samples that did not correlate
showed interference from anti-bovine serum albumin antibodies by
immunoblots. In addition, the immunoblots showed that other
EIAV antigens besides p26 are involved in the ELISA signal of
some positive horse sera. Because of the ease of handling large
numbers of samples and an excellent correlation with the AGID
test, the ELISA provides an excellent testing procedure for
equine infectious anemia.

Development of Monoclonal Antibodies Against the Nitrogen-fixing
Microsymbiont, _Frankia_

Thomas A. Permar, Richard A. Van Deusen*,
Thomas D. Hillson, and Richard C. Schultz
Department of Forestry, Iowa State University, Ames, Iowa 50011,
and *National Veterinary Services Laboratory, Animal and Plant
Health Inspection Service, U.S. Department of Agriculture, Ames,
Iowa 50010

Monoclonal antibodies were developed against cell surface
antigens of the filamentous bacteria _Frankia_. Spleen cells of
BALB/c mice, previously injected with _Frankia_ strain CpI-1, were
fused with SP2/0 myeloma cells. The resulting hybridomas were
screened for antibody production by an indirect enzyme-linked
immunosorbent assay (ELISA). Positive hybridomas for strain
CpI-1 were then tested for specificity against _Frankia_ strains
ArI-4, MpI-1, and EuI-1 with the ELISA and immunofluorescent
assays. Results of previous work with polyclonal antibodies
developed in rabbits showed that in the immunofluorescent assay
the polyclonal antibodies reacted strongly with CpI-1 and ArI-3,
slightly with MpI-1, and not at all with EuI-1. The monoclonal
antibodies showed a strong fluorescence for CpI-1 and did not
react with the others. In the ELISA, polyclonal antibodies
showed a positive reaction with all strains except EuI-1,
whereas results with the monoclonal antibodies showed only
positive reaction with CpI-1.
The monoclonal antibodies are being used to study soil
population dynamics of _Frankia_ strain CpI-1 in a greenhouse
study. Pots inoculated with pure cultures of CpI-1 have been
subjected to various levels of moisture, pH, and organic matter
to determine the effects of these soil parameters on survival
and infectivity of strain CpI-1. In addition the effects of
different plant types on the soil population of CpI-1 are also
being examined.

Hybridoma Secreting Monoclonal Antibodies to Bovine
T-Cell Subpopulations

Eric D. Rabinovsky and Tsu-Ju (Thomas) Yang
Department of Pathobiology, University of Connecticut,
Storrs, CT 06268

Murine monoclonal antibodies have been raised against
bovine leukocytes by fusing spleen cells from BALB/c mice
(immunized with bovine T-cells) with P3-NAl/Ag4-1 (NS-1) mouse
myeloma cells. The enzyme-linked immunosorbent assay (ELISA)
was used to identify hybridomas secreting antibodies to nylon-
wool-non-adherent cells, which consist of 95 $\pm$ 3% T-cells as
determined by immunofluorescent technique (IF), employing goat
anti-bovine thymocyte serum. Nylon wool adherent cells
consisting of 75 $\pm$ 4% B cells and erythrocyte-antibody-comple-
ment (EAC) separated adherent cells, consisting of 92 $\pm$ 2%
B-cells, were used as controls of ELISA and IF, respectively.
Of 1500 wells plated, 1000 yielded hybridoma growth, 660
secreted antibodies, 22 contained specific reactivity of
T-cells, and 1 specific to B-cells. Seven hybridomas proved
stable throughout the subsequent cloning procedures and three
were used for this study.
Cells reactive with monoclonal antibodies were enumerated
using IF. Reagents 10F-2 and 5G-2 identified 95 and 60%,
respectively, of nylon wool non-adherent T cells, and 8 and 3%,
respectively, of EAC-purified cells. In addition, 10F-2 and
5G-2 react with 61 and 40% of lymphocytes from prescapular lymph
node, in which T-cells comprise 61% as determined by goat
anti-thymocyte serum. Both reagents are unreactive with
monocytes, erythrocytes, and platelets. Thus, the overall data
suggest that these monoclonal antibodies identify populations
and subpopulations of T-cells from peripheral blood and lymph
node. Since both 10F-2 and 5G-2 react with subpopulations of
predominantly large thymocytes, we are presently identifying the
location of reactive cells within the thymus. Also, we are
studying the functions of bovine T-cell subsets identified by
monoclonal antibodies.
Reagent 1C-2 reacts with 89 and 30% of EAC-purified B cells
and lymph node cells, respectively, but not with purified T
cells. Our data suggest that 1C-2 identified most B cells of
peripheral blood and lymph node, and studies are under way to
determine the nature of the B-cell marker.

(Supported by USDA-SEA Grant No. 59-2091-1-2-037-0 and by
Project NE-112.)

Epitope Mapping of Neutralizing Monoclonal Antibodies
Against Foot-and-Mouth Disease Virus Outer Capsid Protein

Betty H. Robertson, Donald O. Morgan and Douglas M. Moore
Plum Island Animal Disease Center, Agricultural Research Service,
U.S. Department of Agriculture, P.O. Box 848, Greenport, NY 11944

Monoclonal antibodies generated against purified A12
foot-and-mouth disease virus outer capsid protein (FMDV VP1) or
its 13-kDa cyanogen bromide fragment were assayed by a passive
mouse protection assay for their capacity to neutralize the
parent virus. Five of these antibodies were determined to be
effective in neutralizing viral infectivity. These monoclonal
antibodies were used to characterize the primary structure
epitopes involved in neutralization. Hybridoma supernatants
were initially assayed for their binding capacity against a
spectrum of labeled antigens using a solid-phase assay. The
labeled antigens included viral particles, 12S subunits,
trypsinized virus and 12S subunits, VP1, proteolytic fragments
derived from VP1, and a biosynthetically generated 32-amino-acid
residue region of VP1. Subsequently, competitive binding assays
were performed to further define the binding region using 140S
intact virus and 12S subunits as the labeled antigen. These
investigations revealed that there are at least two identifiable
primary structure epitopes of FMDV whose antibodies are capable
of neutralizing viral infectivity. One region is located within
the 32-residue fragment, positions 145 and 168, and the other
appears to be located between positions 169 and 179. Evidence
that amino acids 145 to 168 of A12 VP1 are responsible for a
neutralizing site on FMDV has been substantiated by protection
of swine after immunization with the biosynthetic 32-amino-acid
antigen and by passive protection of swine by a monoclonal
antibody identified as reacting with the 32-residue region.

A Monoclonal Antibody Specific for the K99 Fimbrial Adhesin in
Escherichia coli, Useful in the Reduction of Mortality in
Neonatal Pigs

P. L. Sadowski[1], R. A. Wilson[2], and D. M. Sherman[3]
[1]Molecular Genetics, Inc., Minnetonka, MN 55343;
[2]Dept. of Veterinary Science,
Pennsylvania State University, University Park, PA 16802;
[3]Department of Large Animal Clinical Sciences
College of Veterinary Medicine, University of Minnesota,
St. Paul, Minnesota 55108

A monoclonal antibody with specificity for the K99 fimbrial
adhesin has been obtained. This antibody will agglutinate
K99(+) E. coli grown at 37°C, but not K99(+) E. coli grown at
18°C. Agglutination could be demonstrated with all K99(+) E.
coli tested including the prototype strains B41, B42, B44, and
B85. No reactivity was observed against E. coli expressing the
K88ac, 987P, or F41 adhesins. The monoclonal antibody
immunoprecipitated a polypeptide that co-migrated with purified
K99 pilus. When orally administered to newborn pigs, this
antibody significantly reduced mortality after lethal K99(+) E.
coli challenge.

Monoclonal Antibodies Directed to Phytochrome
From Etiolated _Avena_ Seedlings
Do Not Quantitate Accurately Phytochrome From Green Avena

Yukio Shimazaki, Marie-Michele Cordonnier*, and Lee H. Pratt
Botany Department, University of Georgia, Athens, GA 30502,
and *Laboratoire de Physiologie Végétale, Université de Genève,
1211 Genève, Switzerland

Quantitation of phytochrome in green plants is not possible
by the common spectrophotometric assays because of their
relative insensitivity and of interference by chlorophyll.
Consequently, we have taken advantage of an enzyme-linked
immunosorbent assay (ELISA) for phytochrome quantitation. The
assay uses both monoclonal mouse and polyclonal rabbit
antibodies directed to phytochrome purified from etiolated _Avena_
seedlings. It is sensitive to less than 100 pg (<1 fmol) of
phytochrome and can be completed within 10 h. Nonspecific
interference by crude plant extracts with the assay was detected
and a method was developed to correct for it. After correction,
addition of known quantities of immunopurified _Avena_ phytochrome
to every test sample gave the expected increase in ELISA
activity, indicating not only the absence of any specific
interference with the assay, but also its validity.
Nevertheless, although the ELISA quantitated phytochrome
accurately in extracts of etiolated _Avena_, it detected a
decreasing proportion of photoreversible phytochrome during
greening. Only 7 to 10% of the photoreversibly detectable
phytochrome isolated from fully light-grown, green _Avena_ was
detected by the immunochemical assay. Although alternative
explanations of the data are possible, a working hypothesis is
that there are two immunochemically distinguishable pools of
photoreversible phytochrome produced by two different genes, one
expressed predominantly in the dark, the other in the light.

Studies of Conserved Antigenic Determinants Using Cross-Reactive
Monoclonal Antibodies Selected by Conserved Epitope Screening

E. H. Vance, S. Srikumaran, A. J. Guidry*, and R. A. Goldsby
Amherst College Laboratory for Hybridoma Research,
Amherst, MA 01002 and
*Laboratory of Milk Secretion and Mastitis,
Agricultural Research Service,
U.S. Department of Agriculture, Beltsville, MD 20705

Monoclonal antibodies that react with antigenic
determinants that are conserved across species are of
theoretical and practical interest. Such probes are useful for
investigating phylogenetic relationships and permit the
immunological identification and assay of an epitope in many
different species with a single defined probe. We have
constructed a number of hybridomas that secrete monoclonal
antibodies specific for bovine immunoglobulin by polyethylene
glycol (PEG)-assisted fusion of spleen cells from bovine Ig
immunized BALB/c mice with the HAT-sensitive established cell
line, SP 2/0. This monoclonal library was then examined by
conserved epitope screening (CES) to determine which monoclonal
antibodies were recognizing uniquely bovine epitopes and which
were reacting with determinants conserved in the Igs of other
vertebrates. Conserved epitope screening is conducted by
assaying monoclonal antibodies derived by immunization with an
antigen of a particular species against the homologous antigen
from a different species. Others (Davis et al.) have shown that
CES was a powerful strategy for the derivation of monoclonal
antibodies that reacted with cell surface antigens of a variety
of species. Here we show that CES permitted the selection of
six monoclonal antibodies that recognize conserved epitopes on
IgGs of different species. Enzymatic digestion, SDS-PAGE, and
electroblotting were used to investigate the sites of the
conserved epitopes recognized by these monoclonal antibodies.
The deliberate derivation of cross-reactive monoclonal
antibodies is a practical approach to the production of
versatile reagents for immunochemistry and immunoassay that may
be used in a number of species. Such conserved epitope-specific
probes permit the examination of the microevolution of
particular regions of protein molecules.

(This research was supported by a grant from the USDA.)

INDEX